Combinatorial
Optimization

Combinatorial Optimization

William J. Cook
William H. Cunningham
William R. Pulleyblank
Alexander Schrijver

A Wiley-Interscience Publication
JOHN WILEY & SONS, INC.
New York • Chichester • Weinheim • Brisbane • Singapore • Toronto

Copyright © 1998 by John Wiley & Sons, Inc.

Library of Congress Cataloging in Publication Data:

Combinatorial optimization / William J. Cook . . . [et al.].
        p.   cm. — (Wiley-Interscience series in discrete mathematics
    and optimization)
      "A Wiley-Interscience publication."
      Includes index.
      ISBN 0-471-55894-X (cloth : alk. paper)
      1. Combinatorial optimization.   I. Cook, William J., 1957–
    II. Series.
    QA402.5.C54523   1998
    519.7'6—dc21   97-35774
                                                        CIP

Printed in the United States of America
10 9 8 7 6 5 4 3

Contents

Preface

Combinatorial optimization is a lively field of applied mathematics, combining techniques from combinatorics, linear programming, and the theory of algorithms, to solve optimization problems over discrete structures. There are a number of classic texts in this field, but we felt that there is a place for a new treatment of the subject, covering some of the advances that have been made in the past decade. We set out to describe the material in an elementary text, suitable for a one semester course. The urge to include advanced topics proved to be irresistible, however, and the manuscript, in time, grew beyond the bounds of what one could reasonably expect to cover in a single course. We hope that this is a plus for the book, allowing the instructor to pick and choose among the topics that are treated. In this way, the book may be suitable for both graduate and undergraduate courses, given in departments of mathematics, operations research, and computer science. An advanced theoretical course might spend a lecture or two on chapter 2 and sections 3.1 and 3.2, then concentrate on 3.3, 3.4, 4.1, most of chapters 5 and 6 and some of chapters 8 and 9. An introductory course might cover chapter 2, sections 3.1 to 3.3, section 4.1 and one of 4.2 or 4.3, and sections 5.1 through 5.3. A course oriented more towards integer linear programming and polyhedral methods could be based mainly on chapters 6 and 7 and would include section 3.6.

The most challenging exercises have been marked in boldface. These should probably only be used in advanced courses.

The only real prerequisite for reading our text is a certain mathematical maturity. We do make frequent use of linear programming duality, so a reader unfamiliar with this subject matter should be prepared to study the linear programming appendix before proceeding with the main part of the text.

We benefitted greatly from thoughtful comments given by many of our colleagues who read early drafts of the book. In particular, we would like to thank Hernan Abeledo, Dave Applegate, Bob Bixby, André Bouchet, Eddie Cheng, Joseph Cheriyan, Collette Coullard, Satoru Fujishige, Grigor Gasparian, Jim Geelen, Luis Goddyn, Michel Goemans, Mark Hartmann, Mike Jünger, Jon Lee, Tom McCormick, Kazuo Murota, Myriam Preissmann, Irwin

Pressman, Maurice Queyranne, André Rohe, András Sebő, Éva Tardos, and Don Wagner. Work on this book was carried out at Bellcore, the University of Bonn, Carleton University, CWI Amsterdam, IBM Watson Research, Rice University, and the University of Waterloo.

Information on this book, including a list of errors, can be found on the web at `http://math.uwaterloo.ca/~whcunnin/bookpage.html`.

CHAPTER 1

Problems and Algorithms

1.1 TWO PROBLEMS

The Traveling Salesman Problem

An oil company has a field consisting of 47 drilling platforms off the coast of
Nigeria. Each platform has a set of controls that makes it possible to regulate
the amount of crude oil flowing from the wells associated with the platform
back to the onshore holding tanks. Periodically, it is necessary to visit certain
of the platforms, in order to regulate the rates of flows. This traveling is done
by means of a helicopter which leaves an onshore helicopter base, flies out to
the required platforms, and then returns to the base.

Helicopters are expensive to operate! The oil company wants to have a
method for routing these helicopters in such a way that the required platforms are visited, and the total flying time is minimized. If we make the
assumption that the flying time is proportional to the distance traveled, then
this problem is an example of the *Euclidean traveling salesman problem*. We
are given a set V of points in the Euclidean plane. Each point has a pair of
(x, y) coordinates, and the distance between points with coordinates (x_1, y_1)
and (x_2, y_2) is just $\sqrt{(x_1 - x_2)^2 + (y_1 - y_2)^2}$. We wish to find a simple circuit
(or *tour*) passing through all the points in V, for which the length is minimized. We call such a tour *optimal*. In this case, V consists of the platforms
to be visited, plus the onshore base.

> ### Euclidean Traveling Salesman Problem
>
> *Input*: A set V of points in the Euclidean plane.
> *Objective*: Find a simple circuit passing through the points for which the sum of the lengths of the edges is minimized.

There are many methods that attempt to solve this problem. Most simple ones share the characteristic that they do not work very well, either from a point of view of solution quality or of running time. For example, suppose we wish simply to try all possible solutions, and then select the best. This will certainly find the shortest circuit. However, if $|V| = n$, then there are $(n-1)!/2$ different possible solutions (Exercise 1.1). Suppose we have at our disposal a computer capable of evaluating a single possibility in one nanosecond ($= 10^{-9}$ seconds). If we had only 23 platforms to visit, then it would take approximately 178 centuries to run through the possible tours!

Suppose, on the other hand, that we require a faster method, but which need not be guaranteed to produce the optimal solution. The "Nearest Neighbor Algorithm" proceeds as follows. Pick any starting point. Go to the nearest point not yet visited. Continue from there to the nearest unvisited point. Repeat this until all points have been visited, then return to the starting point. The result of applying this to a sample problem (from Gerd Reinelt's TSPLIB) is given in Figure 1.1. Notice that although each move is locally the

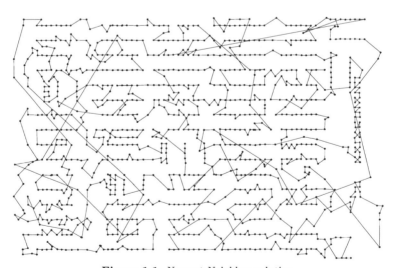

Figure 1.1. Nearest Neighbor solution

best possible, the overall result can be quite poor. First, it is easy to omit a point, which must be visited later at great expense. Second, at times you may "paint yourself into a corner" where you are forced to make a long move to reach the nearest point where you can continue the tour.

The Matching Problem

A designer of logic circuits will use a plotter to draw a proposed circuit, so that it can be visually checked. The plotter operates by moving a pen back and forth and, at the same time, rolling a sheet of paper forwards and backwards beneath the pen. Each color of line is drawn independently, with a pen change before each new color. The problem is to minimize the time required to draw the figure. This time consists of two parts: "pen-down" time, when actual drawing is taking place, and "pen-up" time, when the pen is not contacting the paper, but is simply moving from the end of one line to be drawn to the start of another. Surprisingly, often more than half of the time is spent on pen-up movement. We have very little control over the pen-down time, but we can reduce the pen-up time considerably.

For example, suppose we wish to draw the circuit illustrated in Figure 1.2. Note first that the figure we are drawing is connected. This simplifies things,

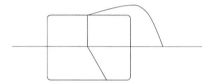

Figure 1.2. Circuit diagram

for reasons we discuss later. Can you convince yourself that some amount of pen-up motion is necessary? We define a *node* of the figure to be a point where two or more lines meet or cross, or where one line ends. In other words, it is a point of the figure from which a positive number of lines, other than two, emanates. We call a node *odd* if there is an odd number of lines coming out, and *even* otherwise. See Figure 1.3.

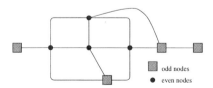

Figure 1.3. Odd and even nodes

One of the oldest theorems of graph theory implies that there will always be an even number of odd nodes. Another old theorem, due to Euler, states that the figure can be traced, returning to the starting point, with no pen-up motion if and only if it is connected and there are no odd nodes.

We minimize pen-up motion by finding a set of new lines that we can add to the figure turning every odd node into an even node, and such that the total traversal time of the new lines is as small as possible.

Let $t(p, q)$ be the time required to draw a line from point p to q (pen-up or pen-down). If we make the assumption that $t(p, q)$ is proportional to the Euclidean distance between p and q, then t satisfies the *triangle inequality*: for any points p, q, r, we have $t(p, r) \leq t(p, q) + t(q, r)$. This is also satisfied, for example, when $t(p, q)$ is proportional to whichever direction of motion—horizontal or vertical—is greater.

Whenever t satisfies the triangle inequality, the optimal set of new lines will pair up the odd nodes. In the Euclidean case, the problem of finding these lines is an example of the *Euclidean matching problem*.

Euclidean Matching Problem

Input: A set V of points in the Euclidean plane.

Objective: Find a set of lines, such that each point is an end of exactly one line, and such that the sum of the lengths of the lines is minimized.

If the original figure is not connected, then we may add the extra lines and obtain a figure having no odd nodes, but which itself is not connected. In this case, some amount of extra pen-up motion is necessary. Moreover, the problem of minimizing this motion includes the Euclidean traveling salesman problem as a special case. For suppose we have an instance of the Euclidean traveling salesman problem which we wish to solve. We draw a figure consisting of one tiny circle in the location of each point. If we take one of these circles to be the pen's home position, then the problem of minimizing pen-up time is just the traveling salesman problem, assuming pen travel time is proportional to Euclidean distance.

Some Similarities and Differences

The Euclidean traveling salesman problem and the Euclidean matching problem are two prominent models in combinatorial optimization. The two problems have several similarities. First, each involves selecting sets of lines connecting points in the plane. Second, in both cases, the number of feasible solutions is far too large to consider them all in a reasonable amount of time. Third, most simple heuristics for the problems do not perform very well.

There is a major difference between the problems lurking under the surface, however. On the one hand, there exists an efficient algorithm, due to Edmonds, that will find an optimal solution for any instance of the Euclidean matching problem. On the other hand, not only is no such algorithm known for the Euclidean traveling salesman problem, but most researchers believe that there simply does not exist such an algorithm!

The reason for this pessimistic view of the Euclidean traveling salesman problem lies in the theory of computational complexity, which we discuss in Chapter 9. Informally, the argument is that if there would exist an efficient algorithm for the Euclidean traveling salesman problem, then there would also exist such an efficient algorithm for *every* problem for which we could check the feasibility of at least one optimal solution efficiently. This last condition is not very strong, and just about every combinatorial optimization problem satisfies it.

Throughout the book, we will be walking the line between problems that are known to have efficient algorithms, and problems that are known to be just as difficult as the Euclidean traveling salesman problem. Most of our effort will be spent on describing models that lie on the "good side" of the dividing line, including an in-depth treatment of matching problems in Chapters 5 and 6. Besides being very important on their own, these "good" models form the building blocks for attacks on problems that lie on the "bad side." We will illustrate this with a discussion of the traveling salesman problem in Chapter 7, after we have assembled a toolkit of optimization techniques.

1.2   MEASURING RUNNING TIMES

Although the word "efficient," which we used above, is intuitive and would suffice for some purposes, it is important to have a means of quantifying this notion. We follow established conventions by estimating the efficiency of an algorithm by giving upper bounds on the number of steps it requires to solve a problem of a given size. Before we make this precise, a few words of warning are in order. Bounding the number of steps only provides an estimate of an algorithm's efficiency and it should not be taken as a hard and fast rule that having a better bound means better performance in practice. The reason for this is the bounds are taken over all possible instances of a given problem, whereas in practice you only want your algorithm to solve the instances you have in hand as quickly as possible. (It may not really concern you that some pathological examples could cause your algorithm to run and run and run.) A well-known example of this phenomenon is the simplex method for linear programming: it performs remarkably well on wide classes of problems, yet there are no good bounds on its behavior in general. It is fair to say, however, that this idea of the complexity of an algorithm does often point out the advantages of one method over another. Moreover, the widespread use of this notion has led to the discovery of many algorithms that turned out to be not only superior in a theoretical sense, but also much faster in practice. With this in mind, let us define more precisely what we mean by "giving upper bounds on the number of steps."

The concept of an algorithm can be expressed in terms of a Turing Machine or some other formal model of computation (see Chapter 9), but for now the intuitive notion of an algorithm as a list of instructions to solve a problem is sufficient. What we are concerned with is: How long does an algorithm take to solve a given problem? Rapid changes in computer architecture make it nearly pointless to measure all running times in terms of a particular machine. For this reason we measure running times on an abstract computer model where we count the number of "elementary" operations in the execution of the algorithm. Roughly speaking, an elementary operation is one for which the amount of work is bounded by a constant, that is, it is not dependent on the size of the problem instance. However, for the arithmetic operations of addition, multiplication, division, and comparison, we sometimes make an exception to this rule and count such operations as having unit cost, that is, the length of the numbers involved does not affect the cost of the operation. This is often appropriate, since the numbers occurring in many algorithms do not tend to grow as the algorithm proceeds. A second, more precise, model of computation counts the number of "bit operations": the numbers are represented in binary notation and the arithmetic operation is carried out bit by bit. This is more appropriate when the length of the numbers involved significantly affects the complexity of a problem (for example, testing whether a number is prime).

A combinatorial optimization problem usually consists of a discrete structure, such as a network or a family of sets, together with a set of numbers (which may represent costs or capacities, for example). We measure the size of such a problem by the length of an encoding of the structure (say in binary notation) plus the size of the set of numbers. (Either each number is counted as a single unit [when we are counting arithmetic operations] or we count the number of digits it takes to write each number in binary notation [when we are counting bit operations].) This measure, of course, depends on the particular encoding chosen, but if one is consistent with the types of encodings used, a robust measure can be obtained. Furthermore, in most cases the various choices of an encoding will differ in size by only a constant factor. So given an instance of a problem, we measure its size by an integer n, representing the number of bits in the encoding plus the size of the set of numbers. We can therefore make statements like, "the number of steps is bounded by $5n^2 + 3n$."

When analyzing an algorithm, we are mainly interested in its performance on instances of large size. This is due to the obvious reason that just about any method would solve a problem of small size. A superior algorithm will really start to shine when the problem sizes are such that a lesser method would not be able to handle the instances in any reasonable amount of time. Therefore, if an algorithm has a running-time bound of $5n^2 + 3n$, we would often ignore the $3n$ term, since it is negligible for large values of n. Furthermore, although a bound of $5n^2$ is clearly better than a bound of $17n^2$, it probably would not make the difference between being able to solve an instance of a problem and not being able to solve it. So we normally concentrate on the magnitude of the bound, describing $5n^2 + 3n$ as "order n^2." There is a formal notation for this: If $f(n)$ and $g(n)$ are positive real-valued functions on the set of nonnegative integers, we say $f(n)$ is $O(g(n))$ if there exists a constant $c > 0$ such that $f(n) \leq c \cdot g(n)$ for all large enough values of n. (The notation $O(g(n))$ is read "big oh of $g(n)$".) Thus $5n^2 + 3n$ is $O(n^2)$ and $35 \cdot 2^n + n^3$ is $O(2^n)$.

As an example of these ideas, consider once again the Nearest Neighbor Algorithm for the traveling salesman problem. We described this method as a fast (but sometimes poor) alternative to enumerating all $(n-1)!/2$ possible tours. We can quantify this easily with the big oh notation.

Let's first consider the arithmetic operation model. An instance of the Euclidean traveling salesman problem can be specified by giving the (x, y) coordinates of the n points to be visited. So the size of an instance is simply $2n$.

An easy (albeit somewhat inefficient) way to implement the Nearest Neighbor Algorithm is to set up an n-element array, where each object in the array has the three fields $\boxed{\; x \mid y \mid \text{mark} \;}$. We initialize the array by placing the (x, y) coordinates of each point v_i in the ith object and setting all the mark fields to 0. A general pass of the algorithm takes a point v_j (say $j = 1$ on the first pass), scans through all n objects, computing the distance from v_i to v_j for all points v_i having mark equal to 0, while keeping track of the point v_{i*} that has the least such distance. We then output v_{i*} as the next point in the tour, set v_{i*}'s mark field to 1 and continue the search from v_{i*}. The algorithm terminates when we have visited all n points.

The initialization pass takes $3n$ elementary operations (excluding an approximately equal number of loop and control operations). A general pass takes n steps to check the mark field, plus at most $n-1$ distance calculations, each of which takes 3 additions, 2 multiplications, and 1 comparison (to keep track of the minimum distance). (Notice that we do not need to calculate a square root in the distance calculation, since we need only compare the values $(x_j - x_i)^2 + (y_j - y_i)^2$ to find the point v_{i*} of minimum distance to v_j.) Since we execute the general step $n - 1$ times, we obtain an upper bound of $3n + (n-1)(n + 6(n-1))$ operations. That is, the Nearest Neighbor Algorithm takes $O(n^2)$ arithmetic operations.

To analyze the algorithm in the bit operation model, we need to measure the size of the input in a way that takes into acount the number of bits in the (x, y) coordinates. A standard estimate is $2nM$, where M is the maximum of $1 + \lceil \log(|x| + 1) \rceil$ and $1 + \lceil \log(|y| + 1) \rceil$ amongst the (x, y) coordinates. (We take logs with base 2. If t is a rational number, then $\lceil t \rceil$ is the smallest integer that is greater than or equal to t and $\lfloor t \rfloor$ is the greatest integer that is less than or equal to t.) The number of elementary operations in the algorithm only changes in the fact that we must now read M-bit-long numbers (so the initialization takes $2nM + n$ steps), and compute and compare the values

$(x_j - x_i)^2 + (y_j - y_i)^2$ bitwise (which takes $O(M^2)$ operations). So a quick estimate of the number of bit operations is $O(n^2 M^2)$.

Our main goal will be to present algorithms that, like the Nearest Neighbor Algorithm, have running-time bounds of $O(n^k)$ for small values of k, whenever possible. Such "polynomial-time algorithms" have the nice property that their running times do not increase too rapidly as the problem sizes increase. (Compare n^3 and 2^n for $n = 100$.)

The above analysis shows that the Nearest Neighbor Algorithm is, in fact, polynomial-time in both the arithmetic model and the bit model. This will occur very often in the book. Typically, we will work with the arithmetic model, but a simple computation will show that the sizes of the numbers appearing in the algorithm do not grow too fast (that is, if t is the number of bits in the problem, then all of the numbers appearing will be $O(t^k)$ for some fixed k), and so a polynomial-time bound in the arithmetic model will directly imply a polynomial-time bound in the bit model. Indeed, throughout the text, whenever we say that an algorithm runs in "polynomial time," we implicitly mean that it runs in polynomial time in the bit model. It should be noted, however, that there are important problems (such as the linear programming problem) for which polynomial-time algorithms in the bit model are known, but no algorithm is known that is polynomial-time in the arithmetic model.

We will discuss the issue of computational complexity further in Chapter 9.

Exercises

1.1. Show that there are $(n - 1)!/2$ distinct tours for a Euclidean traveling salesman problem on n points.

1.2. Suppose we have a computer capable of evaluating a feasible solution to a traveling salesman problem in one nanosecond $(= 10^{-9}$ seconds). How large a problem could we solve in 24 hours of computing time, if we tried all possible solutions? How would the size increase if we had a machine ten times faster? One hundred times faster?

CHAPTER 2

Optimal Trees and Paths

2.1 MINIMUM SPANNING TREES

A company has a number of offices and wants to design a communications network linking them. For certain pairs v, w of offices it is feasible to build a direct link joining v and w, and there is a known (positive) cost c_{vw} incurred if link vw is built. The company wants to construct enough direct links so that every pair of offices can communicate (perhaps indirectly). Subject to this condition, the company would like to minimize the total construction cost.

The above situation can be represented by a diagram (Figure 2.1) with a point for each office and a line segment joining v and w for each potential link. Notice that in this setting, unlike that of the Euclidean traveling salesman problem, we do not have the possibility of a direct connection between every pair of points. Moreover, the cost that we associate with the "feasible" pairs of points need not be just the distance between them. To describe such optimization problems more accurately we use the language of graph theory.

An (undirected) *graph* G consists of disjoint finite sets $V(G)$ of *nodes*, and $E(G)$ of *edges*, and a relation associating with each edge a pair of nodes, its *ends*. We say that an edge is *incident* to each of its ends, and that each end is *adjacent* to the other. We may write $G = (V, E)$ to mean that G has node-set V and edge-set E, although this does not define G. Two edges having the same ends are said to be *parallel*; an edge whose ends are the same is called a *loop*; graphs having neither loops nor parallel edges are called *simple*. We may write $e = vw$ to indicate that the ends of e are v and w. Strictly speaking, this should be done only if there are no parallel edges. In fact, in most applications, we can restrict attention to simple graphs. A *complete*

9

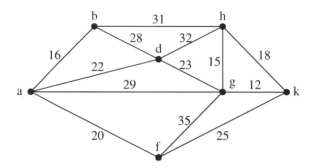

Figure 2.1. Network design problem

graph is a simple graph such that every pair of nodes is the set of ends of some edge.

A *subgraph* H of G is a graph such that $V(H) \subseteq V(G)$, $E(H) \subseteq E(G)$, and each $e \in E(H)$ has the same ends in H as in G. Although in general the sets of nodes and edges do not determine the graph, this is so when we know the graph is a subgraph of a given graph. So given a graph G and subsets P of nodes and Q of edges, we may refer unambiguously to "the subgraph (P, Q) of G." For $A \subseteq E$, $G \backslash A$ denotes the subgraph H obtained by *deleting* A, that is, $V(H) = V$ and $E(H) = E \backslash A$. Similarly, we can delete a subset B of V, if we also delete all edges incident with nodes in B. The resulting subgraph is denoted by $G \backslash B$ or by $G[V \backslash B]$; it may be referred to as the subgraph of G *induced* by $V \backslash B$. For $a \in V$ or E, we may abbreviate $G \backslash \{a\}$ to $G \backslash a$. A subgraph H of G is *spanning* if $V(H) = V(G)$.

Our standard name for a graph is G, and we often abbreviate $V(G)$ to V and $E(G)$ to E. We usually reserve the symbols n and m to denote $|V|$ and $|E|$, respectively. We extend this and other notation to subscripts and superscripts. For example, for graphs G' and G_1, we use n' to denote $|V(G')|$ and V_1 to denote $V(G_1)$.

A *path* P in a graph G is a sequence $v_0, e_1, v_1, \ldots, e_k, v_k$ where each v_i is a node, each e_i is an edge, and for $1 \leq i \leq k$, the ends of e_i are v_{i-1} and v_i. We say that P is *from v_0 to v_k*, or that it is a (v_0, v_k)-*path*. It is *closed* if $v_0 = v_k$; it is *edge-simple* if $e_1, \ldots, e_k$ are distinct; it is *simple* if $v_0, \ldots, v_k$ are distinct; it is a *circuit* if it is closed, $v_0, \ldots, v_{k-1}$ are distinct, and $k \geq 1$. We remark that if there is a path from u to v, then there is a simple one. The *length* of P is k, the number of edge-terms of P. The graph G is *connected* if every pair of nodes is joined by a path. A node v of a connected graph G is a *cut node* if $G \setminus v$ is not connected.

The requirement in the communications network design problem is that the subgraph consisting of all the centers and of the subset of links that we choose to build be connected. Suppose that each edge e of a graph G has a positive cost c_e, and the cost of a subgraph is the sum of the costs of its edges. Then the problem is:

Connector Problem

Given a connected graph G and a positive cost c_e for each $e \in E$, find a minimum-cost spanning connected subgraph of G.

Using the fact that the costs are positive, we can show that an optimal subgraph will be of a special type. First, we make the following observation.

Lemma 2.1 *An edge $e = uv$ of G is an edge of a circuit of G if and only if there is a path in $G \backslash e$ from u to v.* ∎

It follows that if we delete an edge of some circuit from a connected graph, the new graph is still connected. So an optimal solution to the connector problem will not have any circuits. A graph having no circuit is called a *forest*; a connected forest is called a *tree*. Hence we can solve the connector problem by solving the *minimum spanning tree* (MST) problem:

Minimum Spanning Tree Problem

Given a connected graph G and a real cost c_e for each $e \in E$, find a minimum cost spanning tree of G.

We remark that the connector problem and the MST problem are equivalent for positive edge costs. If we allow negative costs, this is no longer true. We shall solve the minimum spanning tree problem for arbitrary edge-costs. The possibility of negative costs in the connector problem is the subject of Exercise 2.7.

A second useful observation is the following.

Lemma 2.2 *A spanning connected subgraph of G is a spanning tree if and only if it has exactly $n - 1$ edges.* ∎

We leave its proof as Exercise 2.4.

Surprisingly simple algorithms will find a minimum spanning tree. We describe two such algorithms, both based on a "greedy" principle—that is, they make the cheapest choice at each step.

Kruskal's Algorithm for MST

Keep a spanning forest $H = (V, F)$ of G, with $F = \emptyset$ initially.
At each step add to F a least-cost edge $e \notin F$ such that H remains a forest.
Stop when H is a spanning tree.

If we apply Kruskal's Algorithm to the graph of Figure 2.1, edges are chosen in the order gk, gh, ab, af, ad, dg. This method was first described by Kruskal [1956]. The second algorithm is known as "Prim's Algorithm", and was described in Jarník [1930], Prim [1957], and Dijkstra [1959].

Prim's Algorithm for MST

Keep a tree $H = (V(H), T)$ with $V(H)$ initially $\{r\}$ for some $r \in V$, and T initially $\emptyset$.
At each step add to T a least-cost edge e not in T such that H remains a tree.
Stop when H is a spanning tree.

If Prim's Algorithm is applied to the graph of Figure 2.1 with $r = a$, edges are chosen in the order ab, af, ad, dg, gk, gh.

We show, first, that these algorithms do find a minimum spanning tree, and, second, that they have efficient implementations.

Validity of MST Algorithms

We begin with a fundamental characterization of connectivity of a graph. For a graph $G = (V, E)$ and $A \subseteq V$, we denote by $\delta(A)$ the set $\{e \in E: e \text{ has an end in } A \text{ and an end in } V \backslash A\}$ and by $\gamma(A)$ the set $\{e \in E : \text{both ends of } e \text{ are in } A\}$. A set of the form $\delta(A)$ for some A is called a *cut* of G.

Theorem 2.3 *A graph $G = (V, E)$ is connected if and only if there is no set $A \subseteq V$, $\emptyset \neq A \neq V$, with $\delta(A) = \emptyset$.*

Proof: It is easy to see that, if $\delta(A) = \emptyset$ and $u \in A$, $v \notin A$, then there can be no path from u to v, and hence, if $\emptyset \neq A \neq V$, G is not connected.

We must show that, if G is not connected, then there exists such a set A. Choose $u, v \in V$ such that there is no path from u to v. Define A to be $\{w \in V: \text{there exists a path from } u \text{ to } w\}$. Then $u \in A$ and $v \notin A$, so $\emptyset \neq A \neq V$. We claim $\delta(A) = \emptyset$. For, if not, suppose that $p \in A$, $q \notin A$, and $e = pq \in E$. Then adding e, q to any path from u to p gives a path from u to q, contradicting the fact that $q \notin A$. ∎

The following result allows us to show that both of the above minimum spanning tree algorithms (and, incidentally, a variety of others) work correctly. Let us call a subset A of edges of G *extendible* to an MST if A is contained in the edge-set of some MST of G.

Theorem 2.4 *Suppose that $B \subseteq E$, that B is extendible to an MST, and that e is a minimum-cost edge of some cut D satisfying $D \cap B = \emptyset$. Then $B \cup \{e\}$ is extendible to an MST.*

Before proving Theorem 2.4, we use it to prove that both algorithms are correct.

Theorem 2.5 *For any connected graph G with arbitrary edge costs c, Prim's Algorithm finds a minimum spanning tree.*

Proof: We begin by showing that at each step, $\delta(V(H))$ is the set of edges f such that adding f to T preserves the tree property of H. This follows from the fact that adding f creates a circuit if and only if both ends of f are in $V(H)$, by Lemma 2.1, and adding f makes H not connected if and only if neither end of f is in $V(H)$, by Theorem 2.3. Hence the algorithm chooses $e \in \delta(V(H))$ such that c_e is minimum. Now $\delta(V(H))$ cannot be empty until H is spanning, since G is connected. Therefore, the final H determined by the algorithm is a spanning tree of G. Moreover, since $\emptyset$ is extendible to an MST, at each step of the algorithm $B = T$, e, and $D = \delta(V(H))$ satisfy the hypotheses of Theorem 2.4. Therefore, the edge-set of the spanning tree H constructed by the algorithm is extendible to an MST, and hence H *is* an MST. ∎

For each node v of a graph G, let C_v be the set of nodes w such that there is a (v, w)-path in G. It is easy to see that $v \in C_w$ if and only if $w \in C_v$, so every node is in exactly one such set. The subgraphs of G of the form $G[C_v]$ are called the *components* of G. Obviously if G is connected, then it is its only component.

Theorem 2.6 *For any connected graph G with arbitrary edge costs c, Kruskal's Algorithm finds a minimum spanning tree.*

Proof: Let $S_1, \ldots, S_k$ be the node-sets of the components of $H = (V, F)$ at a given step of the algorithm. Then $f \in E$ can be added to F and preserve the forest property of H if and only if, by Lemma 2.1, the ends of f are in different S_i. In particular, any element of $\delta(S_i)$, for some i, has this property. It follows that the algorithm does construct a spanning tree, since if H is not connected and there is no such edge f, then $\delta(S_i) = \emptyset$ and $\emptyset \neq S_i \neq V$, which would imply that G is not connected. Moreover, if e is an edge chosen by the algorithm, B is the edge-set of the current spanning forest H when e is chosen, S_i is the node-set of a component of H containing an end of e, and $D = \delta(S_i)$, then $c_e = \min\{c_f : f \in D\}$. Hence, since $\emptyset$ is extendible to an MST, each $E(H)$ occurring in the algorithm is extendible to an MST by Theorem 2.4. It follows that the tree constructed by the algorithm is an MST. ∎

Finally, we need to provide a proof of Theorem 2.4. We use the following lemma, whose proof is left as an exercise.

Lemma 2.7 *Let $H = (V, T)$ be a spanning tree of G, let $e = vw$ be an edge of G but not H, and let f be an edge of a path in T from v to w. Then the subgraph $H' = (V, (T \cup \{e\}) \backslash \{f\})$ is a spanning tree of G.* ∎

Proof of Theorem 2.4: Let $H = (V, T)$ be an MST such that $B \subseteq T$. If $e \in T$, then we are done, so suppose not. Let P be a path in H from v to w, where $vw = e$. Since there is no path in $G \backslash D$ from v to w, there is an edge f of P such that $f \in D$. Then $c_f \geq c_e$, and so by Lemma 2.7, $(V, (T \cup \{e\}) \backslash \{f\})$ is also an MST. Since $D \cap B = \emptyset$, it follows that $f \notin B$, so $B \cup \{e\}$ is extendible to an MST, as required. ∎

Efficiency of Minimum Spanning Tree Algorithms

Let us begin by describing a standard way to store a graph $G = (V, E)$ in a computer. We keep for each $v \in V$ a list L_v of the edges incident with v, and the other end of each edge. (Often the latter is enough to specify the edge.) If there is a cost associated with each edge, this is also stored with the edge. Notice that this means that each edge and cost is stored twice, in two different lists. In all complexity estimations we assume that $n = O(m)$ and that $m = O(n^2)$. Situations in which these assumptions do not hold are usually trivial, from the point of view of the problems we consider. Prim's Algorithm can be restated, using an observation from the proof of Theorem 2.5, as follows.

Prim's Algorithm

Initialize $H = (V(H), T)$ as $(\{r\}, \emptyset)$;
While H is not a spanning tree
    Add to T a minimum-cost edge from $\delta(V(H))$.

Here is a straightforward implementation of this algorithm. We keep $V(H)$ as a *characteristic vector* x. (That is, $x_u = 1$ if $u \in V(H)$, and $x_u = 0$ if $u \notin V(H)$.) At each step we run through E, checking for each $f = uv$ whether $f \in \delta(V(H))$ by checking whether $x_u \neq x_v$, and if so comparing c_f to the current minimum encountered. So e can be chosen in $O(m)$ time. Then x is updated by putting $x_v = 1$ where v is the end of e for which x_v was 0. This will be done $n - 1$ times, so we have a running time of $O(nm)$.

Now we describe the improvement to this running time found by Prim and Dijkstra. We keep, for each $v \notin V(H)$, an edge $h(v)$ joining v to a node of H such that $c_{h(v)}$ is minimum. Then e can be chosen as the $h(v)$ that has smallest cost. The advantage of this is that only $O(n)$ elementary steps are required to choose e. The disadvantage is that the values $h(v)$ need to be changed after each step. Say that w was added to $V(H)$ and v remains in $V \backslash V(H)$. Then $h(v)$ may have to be changed, but only if there is an edge $f = wv$ with $c_f < c_{h(v)}$. We can do all of these changes by going through L_w once, which is again $O(n)$ work. So we do $O(n)$ elementary steps per step of the algorithm and get a running time of $O(n^2)$, an improvement on

$O(nm)$. Further improvements are presented, for example, in the monograph of Tarjan [1983].

Now we turn to the implementation of Kruskal's Algorithm. Notice that, once an edge $e = vw$ becomes unavailable to add to F, that is, H contains a path from v to w, it remains so. This means that finding the next edge to be added can be done simply by considering the edges in order of cost. That is, Kruskal's Algorithm can be restated, as follows.

Kruskal's Algorithm for MST

Order E as $\{e_1, \ldots, e_m\}$, where $c_{e_1} \leq c_{e_2} \leq \ldots \leq c_{e_m}$;
Initialize $H = (V, F)$ as $(V, \emptyset)$;
For $i = 1$ to m
    If the ends of e_i are in different components of H
        Add e_i to F.

Therefore, implementation of Kruskal's Algorithm requires first sorting m numbers. This can be accomplished in $O(m \log m)$ time by any one of a number of sorting algorithms.

To do the other step, we keep the partition of V into "blocks": the node-sets of components of H. The operations to be performed are $2m$ "finds": steps in which we find the block P containing a given v, and $n - 1$ "merges": steps in which two blocks P, Q are replaced by $P \cup Q$, because an edge uv with $u \in P$, $v \in Q$ has been added to F. We keep for each v the name $block(v)$ of the block containing v, so each merge can be done by changing $block(v)$ to $block(u)$ for every $v \in P$ and some $u \in Q$. It is important always to do this for the smaller of the two blocks being merged, that is, we take $|P| \leq |Q|$ (and so we need to keep the cardinalities of the blocks). To find the elements of blocks quickly, we keep each block also as a linked list. After a merge, the lists can also be updated in constant time. It is easy to see that the main work in this phase of the algorithm is the updating of $block(v)$, which could require as much as $n/2$ elementary steps for a single merge. However, it can be proved that for each v, $block(v)$ changes at most $\log n$ times. See Exercise 2.14. Therefore, the total work in the second phase of the algorithm is $O(m \log n) = O(m \log m)$, and we get a running time for Kruskal's Algorithm of $O(m \log m)$. Again, a discussion of further improvements can be found in Tarjan [1983].

Minimum Spanning Trees and Linear Programming

There is an interesting connection between minimum spanning trees and linear programming. Namely, there is a linear-programming problem for which every minimum spanning tree provides an optimal solution. This fact will be useful in Chapter 7, in connection with the traveling salesman problem.

Consider the following linear-programming problem. (For any set A and vector $p \in \mathbf{R}^A$ and any $B \subseteq A$, we use the abbreviation $p(B)$ to mean $\sum(p_j : j \in B)$. We denote the set of real numbers by $\mathbf{R}$, the set of integers by $\mathbf{Z}$, the set of nonnegative integers by $\mathbf{Z}_+$.

$$\text{Minimize } c^T x \tag{2.1}$$

subject to

$$x(\gamma(S)) \leq |S| - 1, \text{ for all } S, \ \emptyset \neq S \subset V \tag{2.2}$$

$$x(E) = |V| - 1 \tag{2.3}$$

$$x_e \geq 0, \text{ for all } e \in E. \tag{2.4}$$

(Do not be alarmed by the number of constraints.) Let S be a nonempty subset of nodes, let T be the edge-set of a spanning tree, and let x^0 be the characteristic vector of T. Notice that $x^0(\gamma(S))$ is just $|T \cap \gamma(S)|$, and since T contains no circuit, this will be at most $|S| - 1$. Also $x^0 \geq 0$ and $x^0(E) = |V| - 1$, so x^0 is a feasible solution of (2.1). Moreover, $c^T x^0 = c(T)$, that is, this feasible solution has objective value equal to the cost of the corresponding spanning tree. So, in particular, the optimal objective value of (2.1) is a lower bound on the cost of an MST. But in fact these two values are equal, a theorem of Edmonds [1971].

Theorem 2.8 *Let x^0 be the characteristic vector of an MST with respect to costs c_e. Then x^0 is an optimal solution of (2.1).*

Proof: We begin by writing an equivalent form of (2.1) that is easier to deal with. For a subset A of the edges, let $\kappa(A)$ denote the number of components of the subgraph (V, A) of G. Consider the problem

$$\text{Minimize } c^T x \tag{2.5}$$

subject to

$$x(A) \leq |V| - \kappa(A), \text{ for all } A \subset E \tag{2.6}$$

$$x(E) = |V| - 1 \tag{2.7}$$

$$x_e \geq 0, \text{ for all } e \in E. \tag{2.8}$$

We claim that the two problems have exactly the same feasible solutions, and thus the same optimal solutions. It is easy to see that every constraint of the form $x(\gamma(S)) \leq |S| - 1$ is implied by an inequality of the type (2.6); namely take $A = \gamma(S)$ and observe that $\kappa(\gamma(S)) \geq |V \setminus S| + 1$. On the other hand, we will show that every constraint of the form (2.6) is implied by a combination of constraints from (2.2) and (2.4). Let $A \subseteq E$, and let $S_1, \ldots, S_k$ be the node-sets of the components of the subgraph (V, A). Then $x(A) \leq \sum_{i=1}^k x(\gamma(S_i)) \leq \sum_{i=1}^k (|S_i| - 1) = |V| - k$.

Now it is enough to show that x^0 is optimal for problem (2.5), and further, it is enough to show that this is true where x^0 is the characteristic vector of a spanning tree T generated by Kruskal's Algorithm. We will show in this case that x^0 is optimal by showing that Kruskal's Algorithm can be used to compute a feasible solution of the dual linear-programming problem to (2.5) that satisfies complementary slackness with x^0. It is easier to write the dual of (2.5) if we first replace minimize $c^T x$ by the equivalent maximize $-c^T x$. Now the dual problem is

$$\text{Minimize } \sum_{A \subseteq E} (|V| - \kappa(A)) y_A \tag{2.9}$$

subject to

$$\sum (y_A : e \in A) \geq -c_e, \text{ for all } e \in E \tag{2.10}$$

$$y_A \geq 0, \text{ for all } A \subset E. \tag{2.11}$$

Notice that y_E is not required to be nonnegative. Let $e_1, \ldots, e_m$ be the order in which Kruskal's Algorithm considers the edges. (Here we are following the second version of the statement of the algorithm.) Let R_i denote $\{e_1, \ldots, e_i\}$ for $1 \leq i \leq m$. Here is the definition of our dual solution y^0. We let $y_A^0 = 0$ unless A is one of the R_i, we put $y_{R_i}^0 = c_{e_{i+1}} - c_{e_i}$, for $1 \leq i \leq m - 1$, and we put $y_{R_m}^0 = -c_{e_m}$. It follows from the ordering of the edges that $y_A^0 \geq 0$ for $A \neq E$. Now consider the constraints (2.10). Then, where $e = e_i$, we have

$$\sum (y_A^0 : e \in A) = \sum_{j=i}^{m} y_{R_j}^0 = \sum_{j=i}^{m-1} (c_{e_{i+1}} - c_{e_i}) - c_{e_m} = -c_{e_i} = -c_e.$$

In other words, all of these inequalities hold with equality. So we now know that y^0 is a feasible solution to (2.9), and also that the complementary slackness conditions of the form, $x_e^0 > 0$ implies equality in the corresponding dual constraint, are satisfied. There is only one more condition to check, that $y_A^0 > 0$ implies that x^0 satisfies (2.6) with equality. For this, we know that $A = R_i$ for some i. If (2.6) is not an equality for this R_i, then there is some edge of R_i whose addition to $T \cap R_i$ would decrease the number of components of $(V, T \cap R_i)$. But such an edge would have ends in two different components of $(V, R_i \cap T)$, and therefore would have been added to T by Kruskal's Algorithm. Therefore, x^0 and y^0 satisfy the complementary slackness conditions. It follows that x^0 is an optimal solution to (2.5), and hence to (2.1). ∎

Notice that, since any spanning tree that provides an optimal solution of the linear-programming problem must be an MST, and since the proof used only the fact that T was generated by Kruskal's Algorithm, we have actually given a second proof that Kruskal's Algorithm computes an MST.

Exercises

2.1. Find a minimum spanning tree in the graph of Figure 2.2 using: (a) Kruskal's Algorithm; (b) Prim's Algorithm with the indicated choice of r.

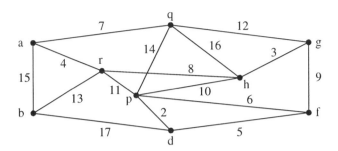

Figure 2.2. MST exercise

2.2. Find a dual solution to the linear-programming problem 2.1 for the graph of Figure 2.2.

2.3. Show that we may assume in the MST problem that the input graph is simple.

2.4. Prove Lemma 2.2.

2.5. Prove Lemma 2.7.

2.6. Give an algorithm to find a minimum-cost forest of a graph, where edge-costs are not assumed to be positive.

2.7. Give an algorithm to solve the connector problem where negative costs are allowed.

2.8. Show that any MST problem can be reduced to an MST problem with positive edge-costs.

2.9. Prove that if $H = (V, T)$ is an MST, and $e \in T$, then there is a cut D with $e \in D$ and $c_e = \min\{c_f : f \in D\}$.

2.10. Prove that a spanning tree $H = (V, T)$ of G is an MST if and only if for every $e = vw \in E\backslash T$ and every edge f of a (v, w) path in T, $c_e \geq c_f$.

2.11. Show that the following algorithm finds an MST of a connected graph G. Begin with $H = G$. At each step, find (if one exists) a maximum-cost edge e such that $H\backslash e$ is connected, and delete e from H. Try this algorithm on the example of Exercise 2.1.

2.12. Show that there is an O(m) algorithm to find *some* spanning tree of a connected graph.

2.13. In the implementation of Prim's Algorithm, suppose we keep for each $v \in V(H)$ an edge $h(v)$ joining v to a node of $V\backslash V(H)$ whose cost is minimum. Does this idea lead to an O(n^2) running time?

2.14. For the implementation of Kruskal's Algorithm described in the text, show that for each $v \in V$, $block(v)$ is changed at most $\log n$ times.

2.15. Here is another way to do finds and merges in Kruskal's Algorithm. Each block S has a distinguished node $name(S) \in S$. Each $v \in S$ different from $name(S)$ has a predecessor $p(v) \in S$ such that evaluating $p(v)$, then $p(p(v)), \ldots$, we eventually get to *name(S)*. With each *name(S)*, we also keep $|S|$. Show how this idea can be used to implement Kruskal's Algorithm, so that the running time is $O(m \log m)$.

2.16. Suppose that, instead of the *sum* of the costs of edges of a spanning tree, we wish to minimize the *maximum* cost of an edge of a spanning tree. That is, we want the most expensive edge of the tree to be as cheap as possible. This is called the *minmax spanning tree problem*. Prove that every MST actually solves this problem. Is the converse true?

2.17. Here is a different and more general way to solve the minmax spanning tree problem of Exercise 2.16. Show that the optimal value of the objective is the smallest cost c_e such that $\{f : f \in E, \ c_f \leq c_e\}$ contains the edge-set of a spanning tree of G. Use this observation and the result of Exercise 2.12 to design an $O(m^2)$ algorithm. Can you improve it to $O(m \log m)$?

2.2   SHORTEST PATHS

Suppose that we wish to make a table of the minimum driving distances from the corner of Bay Street and Baxter Road to every other street corner in the city of Bridgetown, Barbados. By this we mean that the routes must follow city streets, obeying the directions on one-way streets. We can associate a graph with the "network" of the city streets, but the notion of direction imposed by the one-way streets leads to the idea of a directed graph.

A *directed graph* or *digraph* G consists of disjoint finite sets $V = V(G)$ of *nodes* and $E = E(G)$ of *arcs*, and functions associating to each $e \in E$ a *tail* $t(e) \in V$ and a *head* $h(e) \in V$. In other words, each arc has two *end* nodes, to which it is said to be *incident*, and a direction from one to the other. The street map of Bridgetown defines a digraph whose nodes are the street corners. There is an arc for each section of street joining (directly) two corners, and for each direction in which it is legal to drive along it.

The terminology and notation of digraph theory is similar to that of graph theory. In fact, to every digraph there corresponds a graph, obtained by letting the arcs be the edges and ignoring the arc directions. Whenever we use a digraph term or notation without definition, it means what it does for the associated undirected graph. Hence we get immediately notions like *loop* and *path* in a digraph. In addition, notions of *subdigraph* and *deletion* of arcs and nodes are defined exactly in analogy with corresponding terms for graphs. But some differences also appear. Two arcs of a digraph are *parallel* if they have the same head and the same tail, and a digraph is *simple* if it has no loops or parallel arcs. Hence a digraph may be simple as a digraph, but not as a graph. When we write $e = vw$ for an arc of a digraph G, we mean that $v = t(e)$, $w = h(e)$. An arc e_i of a path $P : v_0, e_1, v_1, \ldots, e_k, v_k$ is *forward* if $t(e_i) = v_{i-1}$ and $h(e_i) = v_i$ and is *reverse* otherwise. A path in which every arc is forward is a *directed path* or *dipath*. A *dicircuit* is a dipath that is also a circuit. If each $e \in E$ has a real cost c_e, the *cost* $c(P)$ (with respect to c) of the dipath P is defined to be $\sum_{i=1}^{k} c_{e_i}$. In Figure 2.3 a digraph with arc-costs is represented pictorially.

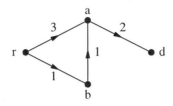

Figure 2.3. A Digraph with arc-costs

Shortest Path Problem

Input: A digraph G, a node $r \in V$, and a real cost vector $(c_e : e \in E)$.
Objective: To find, for each $v \in V$, a dipath from r to v of least cost (if one exists).

There are many direct applications of shortest path problems. We shall see that there are also many more difficult problems in combinatorial optimization for which solution algorithms use shortest path algorithms as subroutines.

One reason that a least-cost dipath to some $v \in V$ may not exist, is that G has no dipath at all from r to v. We could modify the algorithms we shall describe to detect this, but it is more convenient to be able to assume that it never happens. One way to do this is to check for this condition in advance by a graph-theoretic method. Instead, however, we can modify the given G so that there is an arc from r to v for every $v \in V$. Where this requires adding a new arc, we assign it a sufficiently large cost (how large?) so that a least-cost dipath will include it only if there was no dipath from r to v in the original digraph G. So we assume that dipaths exist from r to all the nodes. Notice that we may also assume that G is simple. (Why?)

Here is the basic idea behind all the methods for solving the shortest path problem. Suppose we know that there exists a dipath from r to v of cost y_v for each $v \in V$, and we find an arc $vw \in E$ satisfying $y_v + c_{vw} < y_w$. Since appending vw to the dipath to v gives a dipath to w, we know that there is a cheaper dipath to w, of cost $y_v + c_{vw}$. In particular, it follows that if y_v, $v \in V$, is the *least cost* of a dipath to v, then y satisfies

$$y_v + c_{vw} \geq y_w, \text{ for all } vw \in E. \tag{2.12}$$

We call $y = (y_v : v \in V)$ a *feasible potential* if it satisfies (2.12) and $y_r = 0$. Notice that (2.12) is the essential requirement, since subtracting y_r from each y_v preserves (2.12) and makes $y_r = 0$. Feasible potentials provide lower bounds for shortest path costs, as the following result shows.

Proposition 2.9 *Let y be a feasible potential and let P be a dipath from r to v. Then $c(P) \geq y_v$.*

Proof: Suppose that P is $v_0, e_1, v_1, \ldots, e_k, v_k$, where $v_0 = r$ and $v_k = v$. Then

$$c(P) = \sum_{i=1}^{k} c_{e_i} \geq \sum_{i=1}^{k} (y_{v_i} - y_{v_{i-1}}) = y_{v_k} - y_{v_0} = y_v.$$

∎

Here is another simple but useful observation. Since we want dipaths from r to many other nodes, it may seem that the paths might use many arcs altogether. In fact, however, all the shortest paths can be assumed to use just one arc having head v for each node $v \neq r$. The reason is that *subpaths of shortest paths are shortest paths*, that is, if v is on the least-cost dipath P from r to w, then P splits into a dipath P_1 from r to v and a dipath P_2 from v to w. Obviously, if P_1 is not a least-cost dipath from r to v, then replacing it by a better one would also lead to a better dipath to w. Hence, the only arc having head v that we really need is the last arc of *one* least-cost dipath to v. Moreover, because there will be exactly $n - 1$ such arcs, and the corresponding subgraph contains a path from r to every other node, it is the arc-set of a spanning tree of G. So, just as in the connector problem of Section 2.1, the solution takes the form of a spanning tree of G. However, there are two crucial differences. First, not every spanning tree provides a feasible solution to the shortest path problem: We need a *directed spanning tree rooted at r*, meaning that it contains a dipath from r to v for every $v \in V$. Second, our objective here, in terms of the spanning tree, is not to minimize the sum of the costs of its arcs; see Exercise 2.18.

Ford's Algorithm

Proposition 2.9 provides a stopping condition for a shortest path algorithm. Namely, if we have a feasible potential y and, for each $v \in V$, a dipath from r to v of cost y_v, we know that each dipath is of least cost. Moreover, we have already described the essence of a "descent" algorithm — if y describes dipath costs and we find an arc vw violating (2.12), we replace y_w by $y_v + c_{vw}$. We can initialize such an algorithm with $y_r = 0$ and $y_v = \infty$ for $v \neq r$. Here $y_v = \infty$ simply means that we do not yet know a dipath to v, and ∞ satisfies $a + \infty = \infty$ and $a < \infty$ for all real numbers a. Since we wish, at termination of the algorithm with an optimal y, to obtain also the optimal dipaths, we add one more refinement to the algorithm. The arcs vw of a least-cost dipath will satisfy $y_v + c_{vw} = y_w$, so the last arc of the optimal path to w will be the arc vw such that y_w was most recently lowered to $y_v + c_{vw}$. Moreover, the least-cost dipath to w must consist of a least-cost dipath to v with the arc vw appended, so knowing this "last-arc" information at each node allows us to trace (in reverse) the optimal dipath from r (because G is simple). For this reason, we keep a "predecessor" $p(w)$ for each $w \in V$ and set $p(w)$ to v whenever y_w is set to $y_v + c_{vw}$. Let us call an arc vw violating (2.12) *incorrect*. To *correct* vw means to set $y_w = y_v + c_{vw}$ and to set $p(w) = v$. To *initialize* y and p means to set $y_r = 0$, $p(r) = 0$, $y_v = \infty$ and $p(v) = -1$ for $v \in V \backslash \{r\}$. (We are using $p(v) = -1$ to mean that the predecessor of v is not (yet) defined, but we want to distinguish r from such nodes. We are assuming that $0, -1 \notin V$.) The resulting algorithm is due to Ford [1956].

Ford's Algorithm

Initialize y, p;
While y is not a feasible potential
    Find an incorrect arc vw and correct it.

On the digraph of Figure 2.3, Ford's Algorithm might execute as indicated in Table 2.1. At termination, we do have a feasible potential y and paths of

	Start		$vw = ra$		$vw = rb$		$vw = ad$		$vw = ba$		$vw = ad$	
	y	p	y	p	y	p	y	p	y	p	y	p
r	0	0	0	0	0	0	0	0	0	0	0	0
a	∞	-1	3	r	3	r	3	r	2	b	2	b
b	∞	-1	∞	-1	1	r	1	r	1	r	1	r
d	∞	-1	∞	-1	∞	-1	5	a	5	a	4	a

Table 2.1. Ford's Algorithm applied to the first example

cost y_v given (in reverse) by tracing the values of p back to r, and so we have solved this (trivial) example. Notice that we must have $y_v = y_{p(v)} + c_{p(v)v}$ at termination, but that this need not be true at all times — consider $v = d$ after the fourth iteration. In fact, $y_v \geq y_{p(v)} + c_{p(v)v}$ holds throughout. (Proof: It held with equality when y_v and $p(v)$ were assigned their current values and after that $y_{p(v)}$ can only decrease.)

Figure 2.4 shows a second example and Table 2.2 represents the first few iterations of Ford's Algorithm on that instance. It shows a situation in which the algorithm goes very badly wrong.

It is not hard to see that vw can be chosen as

$$ab, \ bd, \ da, \ ab, \ bd, \ da, \ ab, \ldots$$

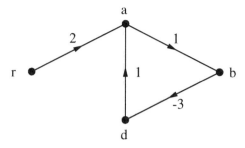

Figure 2.4. A second example

	Start		$vw = ra$		$vw = ab$		$vw = bd$		$vw = da$		$vw = ab$	
	y	p	y	p	y	p	y	p	y	p	y	p
r	0	0	0	0	0	0	0	0	0	0	0	0
a	∞	-1	2	r	2	r	2	r	1	d	1	d
b	∞	-1	∞	-1	3	a	3	a	3	a	2	a
d	∞	-1	∞	-1	∞	-1	0	b	0	b	0	b

Table 2.2. Ford's Algorithm applied to second example

indefinitely, that the algorithm will not terminate, and that certain y_v will become arbitrarily small (that is, will go to $-\infty$). This should not be such a surprise, since if we are asked to find a least-cost dipath from r to a we can repeat the sequence a, b, d as many times as we like before stopping at a. That is, there are arbitrarily cheap dipaths to a, so *there is no* least-cost one. It is apparent that if (G, c) has a negative-cost closed dipath (as in this example), then there exist nodes to which no least-cost dipath exists. In addition to wanting an algorithm that always terminates (quickly, we hope!), we want the algorithm to recognize when a negative-cost dicircuit exists. (Exercise: There is a negative-cost closed dipath if and only if there is a negative-cost dicircuit.)

In fact, not only are there many important applications in which negative costs really arise, but there are several in which a negative-cost dicircuit is actually the object of interest. As an example, consider the common situation of currency exchange rates where, for each pair v, w of currencies, we are quoted a rate r_{vw}, the number of units of currency w that can be purchased for one unit of currency v. Notice that if we convert a unit of currency 1 into currency 2, and then convert all of that into currency 3, we shall have $r_{12}r_{23}$ units of currency 3. This suggests looking for a sequence $v_0, v_1, v_2, \ldots, v_k$ of currencies with $v_0 = v_k$ such that $\prod_{i=1}^{k} r_{v_{i-1}v_i} > 1$, for on the associated sequence of exchanges we would make money. We form a digraph G whose nodes are the currencies, with an arc vw of cost $c_{vw} = -\log r_{vw}$ for each pair v, w. Then such a sequence is money-making if and only if the associated closed dipath of G has cost $-\sum_{i=1}^{k} \log r_{v_{i-1} v_i} = -\log \left(\prod_{i=1}^{k} r_{v_{i-1} v_i} \right) < 0.$ So we can check for the existence of a money-making sequence by checking for the existence of a negative-cost dicircuit in G.

Validity of Ford's Algorithm

We shall show that, provided no negative-cost dicircuit exists, Ford's Algorithm does terminate and that it terminates with least-cost dipaths. This algorithm is itself too crude to be directly useful. But all of the algorithms we treat later are refinements of it, so it is worthwhile (but a bit tedious) to establish its properties. The main step is the following.

Proposition 2.10 *If (G, c) has no negative-cost dicircuit, then at any stage of the execution of Ford's Algorithm we have:*

(i) If $y_v \neq \infty$, then it is the cost of a simple dipath from r to v;

(ii) If $p(v) \neq -1$, then p defines a simple dipath from r to v of cost at most y_v.

Proof: Let y_v^j be the value of y_v after j iterations. We know that y_v^j is the cost of a dipath (if $y_v^j \neq \infty$), so suppose that the dipath is not simple. Then there is a sequence $v_0, v_1, \ldots, v_k$ of nodes with $v_0 = v_k$ and iteration numbers $q_0 < q_1 < \ldots < q_k$ such that

$$y_{v_{i-1}}^{q_{i-1}} + c_{v_{i-1} v_i} = y_{v_i}^{q_i}, \quad 1 \leq i \leq k.$$

The cost of the resulting closed dipath is

$$\sum c_{v_{i-1} v_i} = \sum \left(y_{v_i}^{q_i} - y_{v_{i-1}}^{q_{i-1}} \right) = y_{v_k}^{q_k} - y_{v_0}^{q_0}.$$

But y_{v_k} was lowered at iteration q_k, so this dipath has negative cost, a contradiction, and (i) is proved. Notice that it follows from (i) that $y_r = 0$.

The proof of (ii) is similar. If p does not define a simple dipath from r to v, then there is a sequence $v_0, v_1, \ldots, v_k$ with $v_0 = v_k$ and $p(v_i) = v_{i-1}$ for $1 \leq i \leq k$. The cost of the resulting closed dipath is ≤ 0 since $c_{p(v)v} \leq y_v - y_{p(v)}$ always holds. But consider the most recent predecessor assignment on the dipath; say $y_{p(v)}$ was lowered. Then the above inequality is strict, so we have a negative-cost closed dipath, a contradiction.

Finally, we need to show that the simple dipath to v has cost at most y_v. Let the dipath be $v_0, e_1, v_1, \ldots, e_k, v_k$ where $v_0 = r$, $v_k = v$ and $p(v_i) = v_{i-1}$ for $1 \leq i \leq k$. Then its cost is $\leq \sum (y_{v_i} - y_{v_{i-1}}) = y_v - y_r = y_v$, as required. ∎

Theorem 2.11 *If (G, c) has no negative-cost dicircuit, then Ford's Algorithm terminates after a finite number of iterations. At termination, for each $v \in V$, p defines a least-cost dipath from r to v of cost y_v.*

Proof: There are finitely many simple dipaths in G. Therefore, by Proposition 2.10, there are a finite number of possible values for the y_v. Since at each step one of them decreases (and none increases), the algorithm terminates. At termination, for each $v \in V$, p defines a simple dipath from r to v of cost $\leq y_v$. But no dipath to v can have smaller cost than y_v by Proposition 2.9. ∎

A consequence of the correctness of Ford's Algorithm is the following fundamental fact. Notice that it applies even without any assumption about the existence of dipaths.

Theorem 2.12 *(G, c) has a feasible potential if and only if it has no negative-cost dicircuit.*

Proof: We have already observed that if G has a feasible potential, then it can have no negative-cost dicircuit. Now suppose that G has no negative-cost dicircuit. Add a new node r to G with arcs from r to v of cost zero for every $v \in V$. Where G' is the new digraph and c' is the new cost vector, (G', c') has no negative-cost dicircuit, because no dicircuit of G' goes through r. Now we can apply Ford's Algorithm to (G', c'), and since there is a dipath from r to all other nodes, it will terminate with a feasible potential, which clearly gives a feasible potential for (G, c). ∎

If there is no least-cost dipath to some node v, it is because there are arbitrarily cheap *nonsimple* dipaths to v. So it is natural to ask why we do not try to find a least-cost simple one. (One exists, because the number of simple dipaths is finite.) However, this problem is difficult (unless there is no negative-cost dicircuit) in the same sense that the traveling salesman problem is difficult. In fact, a solution to it could be used quite directly to solve the Euclidean traveling salesman problem (Exercise 2.26).

We shall see that Ford's Algorithm, although it can be modified to recognize the existence of negative-cost dicircuits and hence made finite in all cases (Exercise 2.20), does not have acceptable efficiency. (See Exercise 2.25.) We shall discuss a number of refinements that have better efficiency, although several of them work only in special cases. All of them specify more narrowly the order in which the arcs are considered in the basic step of the algorithm. However, there is one simple observation that can be made for the case in which the arc-costs are integers. Then each step of Ford's Algorithm decreases some y_v by at least 1, since all of these values are integer or ∞. We let C denote $2 \max(|c_e| : e \in E) + 1$. Then we can prove the following.

Proposition 2.13 *If c is integer-valued, C is as defined above, and G has no negative-cost dicircuit, then Ford's Algorithm terminates after at most Cn^2 arc-correction steps.* ∎

The proof of Proposition 2.13 is left to Exercise 2.27. Several of the other exercises investigate better bounds that can be obtained via arguments that assume integral arc-costs and work with the size of the numbers. We shall see in the text that there are good bounds that do not depend on the size of the costs.

Feasible Potentials and Linear Programming

We have seen that feasible potentials provide lower bounds for dipath costs. But in fact at termination of Ford's Algorithm we have a feasible potential and dipaths for which equality holds. One possible statement of this fact is the following.

Theorem 2.14 *Let G be a digraph, $r, s \in V$ and $c \in \mathbf{R}^E$. If there exists a least-cost dipath from r to v for every $v \in V$, then*

$$\min\{c(P) : P \text{ a dipath from } r \text{ to } s\} = \max\{y_s : y \text{ a feasible potential}\}.$$

∎

We wish to point out the connection between this statement and linear-programming duality. The maximization in the theorem statement is obviously a linear-programming problem. It is convenient to drop the requirement that $y_r = 0$ and write that linear-programming problem as:

$$\text{Maximize } y_s - y_r \tag{2.13}$$
$$\text{subject to}$$
$$y_w - y_v \leq c_{vw}, \text{ for all } vw \in E.$$

Where b_v is defined to be 1 if $v = s$, -1 if $v = r$ and 0 otherwise, the dual linear-programming problem of (2.13) is

$$\text{Minimize } \sum(c_e x_e : e \in E) \tag{2.14}$$
$$\text{subject to}$$
$$\sum(x_{wv} : w \in V, \ wv \in E) - \sum(x_{vw} : w \in V, \ vw \in E) = b_v, \text{ for all } v \in V$$
$$x_{vw} \geq 0, \text{ for all } vw \in E.$$

The Duality Theorem says that if one of the optimal values in (2.13), (2.14) exists, then they both do, and they are equal. Notice that any dipath P from r to s provides a feasible solution to (2.14), as follows. Define $(x_e^P : e \in E)$ by: x_e^P is the number of times that arc e is used in P. (In particular, if P is simple, then x^P is $\{0, 1\}$-valued, and is the characteristic vector of P.) Then the objective function of (2.14) for $x = x^P$ is just the cost of P. Therefore, Theorem 2.14 implies that, when shortest paths exist, (2.14) has an optimal solution that is the characteristic vector of a simple dipath. As we shall see (Chapter 7), this result is equivalent to the statement that the vertices of the polyhedron of feasible solutions to (2.14) are characteristic vectors of simple dipaths.

Since we have solved the linear-programming problem (2.14) with Ford's Algorithm, one might wonder whether there is any connection between that algorithm and the simplex algorithm. The simplex algorithm keeps a set T of "basic" arcs (corresponding to the variables in (2.14) that are basic), a feasible solution x of (2.14), and a vector $y \in \mathbf{R}^V$ satisfying

$$x_e = 0, \text{ for all } e \notin T \tag{2.15}$$

$$y_w - y_v = c_{vw}, \text{ for all } vw \in T. \tag{2.16}$$

In each iteration it proceeds to a new such set by replacing one of the arcs in the set by one outside the set. The set T of basic arcs must correspond to a maximal linearly independent set of columns (that is, a column basis) $\{a_e : e \in T\}$ of the constraint matrix $A = \{a_e : e \in E\}$ of the equality constraints of (2.14). This matrix is called the *incidence matrix* of G. Its column bases can be characterized in a very nice way. We state the result here and leave the proof to Exercise 2.28.

Proposition 2.15 *Let G be a connected digraph and $A = \{a_e : e \in E\}$ be its incidence matrix. A set $\{a_e : e \in T\}$ is a column basis of A if and only if T is the arc-set of a spanning tree of G.* ∎

Ford's Algorithm, once it has found paths to all nodes, does have such a set T, namely, $\{p(v)v : v \in V \setminus \{r\}\}$. (It is possible to require that the simplex method for (2.14) keep such a directed spanning tree.) Moreover, the dipath from r to s determined by p uses only arcs from T, so its characteristic vector x^P satisfies (2.15). However, for this T, (2.16) becomes $y_{p(v)} + c_{p(v)v} = y_v$, a relation that is generally *not* enforced by Ford's Algorithm. Notice that enforcing this (and $y_r = 0$) would mean that the dipath to v determined by p would have cost exactly y_v. A spanning tree encountered by the simplex method need not have the property that every node other than r is the head of exactly one arc, but if it does encounter such a tree, then there is always a choice of the arc to delete (namely, the arc of the tree having the same head as the incoming arc) so that the property is kept. So (a version of) the simplex method moves from spanning tree to spanning tree, as does Ford's Algorithm, but the former method keeps the path costs determined by the current tree. In fact Ford's Algorithm may do a correction step on an arc of the form $p(v)v$, so that the tree does not change, but y does. In this sense, each step of the simplex algorithm could be regarded as a sequence of steps of Ford's Algorithm, one ordinary step followed by several steps that do not change the tree, until y "catches up" to the tree. We will learn more about such "network" versions of the simplex method in Chapter 4.

Refinements of Ford's Algorithm

The basic step of Ford's Algorithm could be written as

> Choose an arc e;
> If e is incorrect, then correct it;

Notice that, assuming that we store the values of y and p appropriately, we can perform each basic step in constant time. But the number of basic steps depends crucially on the order in which arcs e are chosen. Suppose that arcs are chosen in the sequence $f_1, f_2, f_3, \ldots, f_\ell$, which we denote by $\mathcal{S}$. (In general, there will be repetitions.) There are choices for $\mathcal{S}$ that result in very bad performance of the algorithm. (For example, see Exercise 2.25.) The basic idea for looking for good choices for $\mathcal{S}$ is simple. Denote by P the dipath $v_0, e_1, v_1, \ldots, e_k, v_k$ from $r = v_0$ to $v = v_k$. After the first time that Ford's Algorithm considers the arc e_1 we will have $y_{v_1} \leq y_r + c_{e_1} \leq c_{e_1}$. After the first subsequent time that the algorithm considers e_2, we will have $y_{v_2} \leq y_{v_1} + c_{e_2} \leq c_{e_1} + c_{e_2}$. Continuing, once $e_1, e_2, \ldots, e_k$ have been considered *in that order*, we will have $y_v \leq c(P)$. We say that P is *embedded* in $\mathcal{S}$ if its arcs occur (in the right order, but not necessarily consecutively) as a subsequence of $\mathcal{S}$. Our discussion can be summarized as follows.

Proposition 2.16 *If Ford's Algorithm uses the sequence $\mathcal{S}$, then for every $v \in V$ and for every path P from r to v embedded in $\mathcal{S}$, we have $y_v \leq c(P)$.* ∎

It follows that, if $\mathcal{S}$ has the property that for every node v there is a least-cost dipath to v embedded in $\mathcal{S}$, then $\mathcal{S}$ will bring the algorithm to termination. We want $\mathcal{S}$ to have this property, and we also want $\mathcal{S}$ to be short, since its length will be the running time of the algorithm.

The Ford-Bellman Algorithm

A simple way to use Proposition 2.16 is to observe that *every* simple dipath in G is embedded in $\mathcal{S}_1, \mathcal{S}_2, \ldots, \mathcal{S}_{n-1}$, where for each i, $\mathcal{S}_i$ is an ordering of E. When we use such an ordering in Ford's Algorithm we speak of a sequence of "passes" through E. Since each arc is handled in constant time per pass, we get a shortest path algorithm that runs in time $O(mn)$. We call it the Ford-Bellman Algorithm, because Bellman [1958] seems to have been the first to prove a polynomial bound for such an algorithm. We want the algorithm also to recognize whether there exists a negative-cost dicircuit. We know that, if there is no negative-cost dicircuit, then $n - 1$ passes are sufficient to determine a feasible potential. Therefore, if y is not a feasible potential after $n - 1$ passes, then there exists a negative-cost dicircuit.

Ford-Bellman Algorithm

Initialize y, p;
Set $i = 0$;
While $i < n$ and y is not a feasible potential
    Replace i by $i + 1$;
    For $e \in E$
        If e is incorrect
            Correct e.

Theorem 2.17 *The Ford-Bellman Algorithm correctly computes a least-cost dipath from r to v for all $v \in V$ (if $i < n$ at termination), or correctly detects that there is a negative-cost dicircuit (if $i = n$ at termination). In either case it runs in time $O(mn)$.* ∎

We shall see that the running time of $O(mn)$ can be improved if special assumptions are made about G or c. However, in the general case, no better bound is currently known. Here we mention some further refinements that speed up the algorithm in practice. Most of them are based on the natural idea of scanning nodes, that is, considering consecutively all the arcs having the same tail. (We point out that the simplex method is not based on scanning. For another such example, see Exercise 2.42.)

A usual representation of a digraph is to store all the arcs having tail v in a list L_v. To *scan* v means to do the following:

    For $vw \in L_v$
        If vw is incorrect
            Correct vw;

A natural further refinement of the Ford-Bellman Algorithm is to replace the last three lines in its statement by

    For $v \in V$
        Scan v;

It is obvious that, if y_v has not decreased since the last time v was scanned, then v need not be scanned. Taking advantage of this observation saves time. One way to do that is to keep a set Q of nodes to be scanned, adding a node v to Q when y_v is decreased (if $v \notin Q$) and choosing the next node to be scanned from Q (and deleting it from Q). Initially $Q = \{r\}$, and the algorithm terminates when Q becomes empty. We keep Q both as a list and

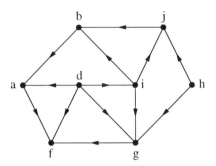

Figure 2.5. Digraph having a topological sort

a characteristic vector, so that we can add, delete, and check membership in Q in constant time. In order to do the first two operations in constant time, we add and delete at the ends of Q. If we add at one end, the "back," and delete at the other, the "front," we are keeping Q as a "first in, first out" list or a *queue*. In this case it can be checked (Exercise 2.32) that the algorithm is equivalent to a refinement of the Ford-Bellman Algorithm and so has a running time of $O(mn)$. This refinement also works well in practice, but there are some variants that are even faster. The paper of Gallo and Pallottino [1986] contains more information.

Acyclic Digraphs

Suppose that the nodes of G can be ordered from left to right so that all arcs go from left to right. More precisely, suppose that there is an ordering $v_1, v_2, \ldots, v_n$ of V so that $v_i v_j \in E$ implies $i < j$. We call such an ordering a *topological sort*. In the digraph of Figure 2.5, d, h, i, g, j, b, a, f is a topological sort.

If we order E in the sequence $\mathcal{S}$ so that $v_i v_j$ precedes $v_k v_\ell$ if $i < k$, then *every* dipath of G is embedded in $\mathcal{S}$. It follows that Ford's Algorithm will solve the shortest path problem in just one pass through E. There is a simple description of the class of digraphs for which this observation works. It is obvious that if G has a topological sort, then it has no dicircuit at all (and hence no negative-cost dicircuit); in other words, G is *acyclic*. Conversely, we claim that every acyclic digraph has a topological sort. To see this, first observe that each acyclic digraph has a candidate for v_1, that is, a node v such that $uv \in E$ for *no* $u \in V$. (Why?) Moreover, since $G \backslash v$ is acyclic, this can be repeated. This idea can be turned into an $O(m)$ algorithm to find a topological sort (Exercise 2.33). Notice that, if $r = v_i$ with $i > 1$, then there can be no dipath from r to $v_1, \ldots, v_{i-1}$, so these can be deleted. Hence we may assume that $v_1 = r$.

Shortest Paths in an Acyclic Digraph

Find a topological sort $v_1, \ldots, v_n$ of G with $r = v_1$;
Initialize y, p;
For $i = 1$ to n
    Scan v_i.

Theorem 2.18 *The shortest path problem on an acyclic digraph can be solved in time $O(m)$.* ∎

Nonnegative Costs

In many applications of the shortest path problem we know that $c \geq 0$. In fact, probably this is the situation more often than not, so this is an extremely important special case. Again it is possible to design a correct "one-pass" algorithm. Moreover, the ordering is determined from an ordering of the nodes as in the acyclic case. However, this ordering is computed during the course of execution. Namely, if $v_1, v_2, \ldots, v_i$ have been determined and scanned, then v_{i+1} is chosen to be the unscanned node v for which y_v is minimum. In this situation we have the following result.

Proposition 2.19 *For each $w \in V$, let y'_w be the value of y_w when w is chosen to be scanned. If u is scanned before v, then $y'_u \leq y'_v$.*

Proof: Suppose $y'_v < y'_u$ and let v be the earliest node scanned for which this is true. When u was chosen to be scanned, we had $y'_u = y_u \leq y_v$, so y_v was lowered to a value less than y'_u after u was chosen to be scanned but before v was chosen. So y_v was lowered when some node w was scanned, and it was set to $y'_w + c_{wv}$. By choice of v, $y'_w \geq y'_u$ and since $c_{wv} \geq 0$, we have $y'_v \geq y'_u$, a contradiction. ∎

We claim that after all nodes are scanned, we have $y_v + c_{vw} \geq y_w$ for all $vw \in E$. Suppose not. Since this was true when v was scanned, it must be that y_v was lowered after v was scanned, say while q was being scanned. But then $y_v = y'_q + c_{qv} \geq y'_v$ since q was scanned later than v and $c_{qv} \geq 0$, a contradiction. So the following algorithm, due to Dijkstra [1959], is valid.

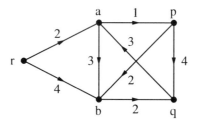

Figure 2.6. Example for Dijkstra's Algorithm

Dijkstra's Algorithm

Initialize y, p;
Set $S = V$;
While $S \neq \emptyset$
    Choose $v \in S$ with y_v minimum;
    Delete v from S;
    Scan v.

For example, in the digraph of Figure 2.6, the nodes will be scanned in the order r, a, p, b, q.

Actually, one can slightly improve the algorithm by observing that, for $w \notin S$, $y_v + c_{vw} \geq y_w$ follows from $y_v \geq y_w$. So the test that $y_v + c_{vw} < y_w$ could be done only for $w \in S$. The running time of the algorithm is $O(m)$ plus the time to find v. But this simple step requires considerable time: $k - 1$ comparisons when $|S| = k$ and $n - 1 + n - 2 + \ldots + 1 = O(n^2)$ comparisons in all. So the running time of a straightforward implementation is $O(n^2)$.

Theorem 2.20 *If $c \geq 0$, then the shortest path problem can be solved in time $O(n^2)$.* ∎

A number of improvements are discussed in Tarjan [1983].

Shortest path problems with nonnegative costs arise frequently in applications, so it is convenient to have a notation for the time required to solve them. We use $S(n, m)$ for the time needed to solve a nonnegative-cost shortest path problem on a digraph having n nodes and m arcs.

Feasible Potentials and Nonnegative Costs

If we happen to know a feasible potential y, we can use it to transform the cost vector c to a nonnegative one c'. Namely, put $c'_{vw} = c_{vw} + y_v - y_w$. This does not affect the least-cost dipaths, since any (r, s)-dipath P satisfies $c'(P) = c(P) + y_r - y_s$. Hence Dijkstra's Algorithm can be used.

There are several applications of this simple idea. We shall see an important one in Chapter 4. Meanwhile, here is another useful one. The "all pairs" shortest path problem is to find a least-cost dipath from r to v for every choice of r and v. There are direct algorithms for this problem but, from the point of view of running time, it seems to be better just to use a standard algorithm n times. Hence we get a running time of $O(nS(n,m))$ in the case of nonnegative costs, and $O(n^2m)$ in general. But the latter time can be improved. We find a feasible potential in time $O(nm)$ with Ford-Bellman, then transform to nonnegative costs, and then use Dijkstra n (or $n-1$) times, resulting in an overall running time of $O(nS(n,m))$.

Unit Costs and Breadth-First Search

The problem of finding a dipath from r to v having as few arcs as possible is, of course, a kind of shortest path problem, namely, it is the case where all arc-costs are 1. It is interesting to see how Dijkstra's Algorithm can be improved in this situation.

Proposition 2.21 *If each $c_e = 1$, then in Dijkstra's Algorithm the final value of y_v is the first finite value assigned to it. Moreover, if v is assigned its first finite y_v before w is, then $y_v \leq y_w$.*

Proof: Notice that these statements are obviously true for $v = r$. If $v \neq r$, the first finite value assigned to y_v is $y'_w + 1$, where y'_w is the final value of y_w. Moreover, any node j scanned later than w has $y'_j \geq y'_w$ by Proposition 2.19, so y_v will not be further decreased. Similarly, any node q assigned its first finite y_q after v, will have $y_q = y'_j + 1 \geq y'_w + 1 = y_v$.   ∎

When picking $v \in S$ such that y_v is minimum, we choose among the set Q of those unscanned nodes v having y_v finite. Proposition 2.21 tells us that we can simply choose the element of Q that was added to Q first, that is, we can keep Q as a queue, and v can be found in constant time. So Dijkstra's Algorithm has a running time of $O(m)$ in this case. Notice that we no longer need to maintain the y_v (although we may want to). This algorithm is often called *breadth-first search*.

Breadth-first Search

Initialize p;
Set $Q = \{r\}$;
While $Q \neq \emptyset$
    Delete v from the front of Q;
    For $vw \in L_v$
        If $p(w) = -1$
            Add w to the back of Q;
            Set $p(w) = v$.

Exercises

2.18. Show by an example that a spanning directed tree rooted at r can be of minimum cost but not contain least-cost dipaths to all nodes. Also show the converse, that it may contain least-cost dipaths but not be of minimum cost.

2.19. Show by an example that a subpath of a shortest simple dipath need not be a shortest simple dipath, if a negative-cost dicircuit exists.

2.20. Modify Ford's Algorithm (in a simple way) so that it always terminates, recognizing the existence of a negative-cost dicircuit if there is one.

2.21. Solve the shortest path problem for the digraph described by the following lists, using (a) Ford-Bellman using node-scanning and a queue; (b) the acyclic algorithm; (c) Dijkstra. $V = \{r, a, b, d, f, g, h, j, k\}$, and for each $v \in V$ the elements of the list L_v are the pairs (w, c_{vw}) for which $vw \in E$. $L_r : (a, 2), (k, 7), (b, 5)$. $L_a : (d, 8), (f, 4)$. $L_b : (k, 3), (f, 2)$. $L_d : (h, 5)$. $L_f : (g, 3), (j, 7)$. $L_g : (h, 4), (j, 3)$. $L_j : (k, 4), (h, 3)$. $L_k : (d, 2), (h, 9), (g, 6), (f, 1)$. ($L_h$ is empty.)

2.22. We are given a digraph $G = (V, E)$, $c \in \mathbf{R}^E$, and disjoint sets R, $S \subseteq V$. The problem is to find a least-cost dipath joining a node in R to a node in S. Show that this problem can be reduced to an ordinary shortest path problem.

2.23. Suppose that we are given a shortest path problem on a digraph G such that a node w is incident with exactly two arcs. Explain how the solution of a shortest path problem on a smaller digraph yields the solution to the given problem.

2.24. There are certain street corners in Bridgetown such that the street on which a car leaves the intersection may depend on the street on which it entered (for example, "no left turn"). How can a digraph, and arc costs, be defined so that the dipaths correspond to legal routes?

2.25. (Edmonds) Consider the digraph G_k of Figure 2.7. Show that Ford's Algorithm (in fact, the simplex method) can take more than 2^k steps to solve the shortest path problem on G_k. Hint: Use induction. Try to make the algorithm solve the problem on G_{k-1} twice.

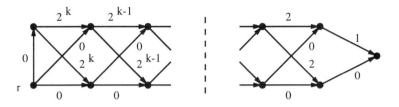

Figure 2.7. A bad example for the Ford and simplex algorithms

2.26. Prove that the problem of finding a least-cost simple dipath joining two fixed nodes in a digraph is hard, assuming only that the traveling salesman problem is hard.

2.27. Prove Proposition 2.13.

2.28. Prove Proposition 2.15. Hint: To prove that if T does not contain the arc-set of a circuit, then the corresponding columns are linearly independent, use the fact that if a forest has at least one arc, then there is a node incident to just one of its arcs.

2.29. Generalize Proposition 2.9 in the following way. Suppose that we have dipath costs y_v, $v \in V$ such that for every arc vw we have $y_w - c_{vw} - y_v \leq K$. Prove that for each node v, y_v is within Kn of being optimal.

2.30. A "scaling" approach to improving the complexity of Ford's Algorithm for integral arc-costs might work as follows. For some integer p suppose that in "stage p," we do correction steps only for arcs vw that satisfy $y_w - y_v - c_{vw} \geq 2^p$. If there are no more arcs of this sort, then we decrease p by 1. When we get to $p = 0$ we are just doing Ford's Algorithm. Use Exercise 2.29 to prove a bound on the number of steps in each stage after the first one. Then choose the first value of p so that the bound also applies for the first stage. What is the overall running time?

2.31. Here is a variant on the approach of doing only large correction steps in Ford's Algorithm. Suppose that at each step we choose to correct the arc that maximizes $y_w - c_{vw} - y_v$. (Of course, this requires some extra work to identify the arc.) Let $gap(k)$ denote the difference between the value after k iterations of $\sum(y_v : v \in V)$ and its minimum value. What will happen in the first $n - 1$ iterations? To analyze the number of subsequent iterations, use the result of Exercise 2.29 to prove that $gap(k+1) \leq gap(k)(1 - 1/n^2)$, and hence prove a bound on the number of steps. (You may need the inequality $1 - x \leq e^{-x}$.)

2.32. Prove that the version of Ford's Algorithm that uses node-scanning and a queue to store the nodes to be scanned, has a running time of $O(nm)$.

2.33. Give an $O(m)$ algorithm to find in a digraph either a dicircuit or a topological sort.

2.34. We are given numbers $a_1, \ldots, a_n$. We want to find i and j, $1 \le i \le j \le n + 1$, so that $\sum_{k=i}^{j-1} a_k$ is minimized. Give an $O(n)$ algorithm.

2.35. Suppose that we are given tasks $t_1, t_2, \ldots, t_k$. Each task t_i has a processing time p_i. For certain pairs (i, j), t_i *must precede* t_j, that is, the processing of t_j cannot begin until the processing of t_i is completed. We wish to schedule the processing of the tasks so that all of the tasks are completed as soon as possible. Solve this problem as a maximum feasible potential problem on an acyclic digraph.

2.36. Give an example to show that Dijkstra's Algorithm can give incorrect results if negative costs are allowed.

2.37. Consider the least cost path problem for undirected graphs. Show that if the costs can be assumed to be nonnegative, then this problem can be solved by reducing it to a digraph problem. When costs are allowed to be negative, what difficulty arises?

2.38. Consider the problem of finding a minimum cost dipath with an odd (even) number of arcs from r to s in a digraph G having nonnegative arc costs. Notice that the dipath may not be simple. Show how to solve this problem by solving a shortest path problem in a digraph having two nodes for each node different from r and s.

2.39. Consider the *minmax path problem*: Given a digraph G with arc-costs and nodes r and s, find an (r, s) dipath P whose maximum arc-cost is as small as possible. Show how the idea of Exercise 2.17 can be applied to solve this problem. What is the running time?

2.40. Try to adapt Dijkstra's Algorithm to solve the minmax path problem. Prove that your algorithm works and give the running time.

2.41. Describe a direct all-pairs shortest-path algorithm based on the following idea. Given a set $S \subseteq V$, let y_{vw}, for v, $w \in V$, denote the least cost of a (v, w)-dipath whose internal nodes are from S. Compute this information beginning with $S = \emptyset$, adding one node at a time to S until $S = V$.

2.42. Consider the following refinement of Ford's Algorithm. Let $v_1, \ldots, v_n$ be an ordering of V, with $r = v_1$. Split E into E_1 and E_2, where $E_1 = \{v_i v_j : i < j\}$. Now order E_1 into a sequence S_1, so that $v_i v_j$ precedes $v_k v_\ell$ if $i < k$ and order E_2 into a sequence S_2 so that $v_i v_j$ precedes $v_k v_\ell$ if $i > k$. Now use the sequence $S_1, S_2, S_1, S_2, \ldots$ in Ford's Algorithm. How does the running time compare to that of Ford-Bellman?

CHAPTER 3

Maximum Flow Problems

3.1 NETWORK FLOW PROBLEMS

A company produces tires in a number of large factories and sells them in hundreds of retail outlets. Each retail outlet j has a known monthly requirement b_j of tires, and each factory i has a known production capacity a_i of tires per month. The tires must be shipped from factories to outlets. For those pairs (i, j) for which this is possible, a unit shipping cost c_{ij} is incurred. The problem is to arrange a minimum-cost shipping pattern, satisfying the requirements from the available production. Such a "transportation problem" is a classical instance of a (minimum-cost) network flow problem. Network flow theory is probably the single area of combinatorial optimization that has received the most attention. Many of its fundamental results were developed in the 1950s, but it continues to be a subject of basic research, and to find new applications.

We treat minimum-cost network flow problems in Chapter 4. The subject of this chapter is an important subclass, maximum flow problems. Maximum flow problems admit especially fast and attractive solution algorithms, as well as having many nice applications. They are related to the minimum-cost flow problem in the following way: The problem of deciding whether a minimum-cost flow problem has a feasible solution is equivalent to a maximum flow problem. Another important special class of minimum-cost flow problems are the familiar shortest path problems of the previous chapter. Their relation to the more general class is this: We can test whether a given feasible solution is optimal by solving a shortest path problem.

The minimum cut problem is a fundamental optimization problem intimately related to the maximum flow problem. It has many applications in

37

its own right, several of which we describe. We also present, for certain kinds
of minimum cut problems, algorithms that do not depend on solving flow
problems.

3.2 MAXIMUM FLOW PROBLEMS

Suppose that we want to send as many trucks as possible from one point
r in a street network to another point s. The restriction is that, for each
street segment e, there is an upper bound u_e on the number of trucks that are
allowed to use e. If we formulate this problem on the corresponding digraph
G, it is one of finding a family $(P_1, \ldots, P_k)$ of (not-necessarily-distinct) (r, s)-
dipaths in G, such that each arc e is an arc of at most u_e of the dipaths and
such that k is maximized. Notice that there is no harm in assuming that each
P_i is simple. In addition, we may assume that G is simple. (Why?) Let x_e
denote $|\{i : P_i \text{ uses } e\}|$. Notice that, for each node $v \neq r, s$, any P_i must
enter and leave v the same number of times. Therefore, x satisfies

$$\Sigma(x_{wv} : w \in V, wv \in E) - \Sigma(x_{vw} : w \in V, vw \in E) = 0,$$
$$\text{for all } v \in V \backslash \{r, s\} \qquad (3.1)$$
$$0 \leq x_{vw} \leq u_{vw}, \text{ for all } vw \in E \qquad (3.2)$$
$$x_{vw} \text{ integer, for all } vw \in E. \qquad (3.3)$$

Moreover, the number k of dipaths satisfies $k = \Sigma(x_{ws} : w \in V, ws \in E) - \Sigma(x_{sw} : w \in V, sw \in E)$. For example, in Figure 3.1 the dipaths having node
sequences r, p, a, s; r, q, b, s and r, q, a, p, b, s satisfy the limitations given by
the numbers u_e and generate the numbers x_e, where u_e, x_e are indicated on
arc e. Let us call a vector x an (r, s)-*flow*, or just a *flow* if it satisfies (3.1),
and a *feasible flow* if it also satisfies (3.2). The left-hand side of (3.1) is the
net flow into v, or the *excess of x at* v, and we abbreviate it to $f_x(v)$. The
condition $f_x(v) = 0$ requires "conservation of flow" at v. The special nodes
r and s where conservation of flow is not required are called the *source* and
the *sink*, respectively. We call $f_x(s)$ the *value* or *amount* of x. It turns out
that flows are easier to work with than families of paths, so it is nice to know
that one can recover paths from integral flows. This is the content of our first
result.

Proposition 3.1 *There exists a family* $(P_1, \ldots, P_k)$ *of* (r, s)-*dipaths such
that* $|\{i : P_i \text{ uses } e\}| \leq u_e$ *for all* $e \in E$ *if and only if there exists an integral
feasible* (r, s)-*flow of value* k.

Proof: We have already seen that a family of dipaths determines a corresponding flow, so let us assume we have the flow x. It is convenient to assume
that x is "acyclic," that is, that there is no dicircuit C, each of whose arcs
e has $x_e > 0$. If such a dicircuit exists, we can decrease x_e by 1 on all arcs

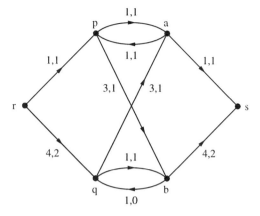

Figure 3.1. A feasible flow of value 3

C. The new x clearly remains a feasible integral flow of value k, and we can
repeat this procedure until no such dicircuit exists. If $k = 0$, there is nothing
to do. If $k \geq 1$, we can find an arc vs with $x_{vs} \geq 1$. Then, provided that
$v \neq r$, it follows from (3.1) that there is an arc wv with $x_{wv} \geq 1$. If $w \neq r$,
the argument can be repeated producing an arc pw with $x_{pw} \geq 1$. This process produces distinct nodes, since x is acyclic, so eventually we get a simple
(r, s)-dipath P_k on each of whose arcs e, we have $x_e \geq 1$. Now we decrease
x_e by 1 for each arc e of P_k. The new x is an integral feasible flow of value
$k - 1$, and the process can be repeated. ∎

It follows that we can solve the "path packing" problem if we can solve the
following problem.

Maximum Integral Flow Problem

Maximize $f_x(s)$
subject to
$f_x(v) = 0$, for all $v \in V \setminus \{r, s\}$
$0 \leq x_e \leq u_e$, for all $e \in E$
x_e integral, for all $e \in E$.

The *maximum flow problem* is the same problem without the restriction of
integrality. We shall see that the integrality restriction causes no additional
difficulties. The numbers u_e are called *capacities*. We allow them to be nonnegative real numbers or ∞. The latter just means that there is no (explicit)
upper bound on x_e. One remark: We use "u is integral" to mean that the
finite components of u are integers.

We remark that the decomposition of integral flows into paths described above can be extended to nonintegral flows. We define a *path flow* to be a vector $x \in \mathbf{R}^E$ such that, for some (r, s)-dipath P and some nonnegative real number α, $x_e = \alpha$ for each arc e of P, and $x_e = 0$ for every other arc of G. A *circuit flow* is defined in the same way with "dicircuit" replacing "(r, s)-dipath." The following result can be proved by similar methods to those above.

Proposition 3.2 *Every nonnegative (r, s)-flow is the sum of at most m flows, each of which is a path flow or a circuit flow.* ∎

Maximum Flows and Minimum Cuts

There is a natural way to get upper bounds for the maximum value of a flow. We call a set $\delta(R) = \{vw : vw \in E, v \in R, w \notin R\}$ for some $R \subseteq V$, a *cut*. (Notice that this defines both δ and "cut" differently than for the associated undirected graph.) An (r, s)-*cut* is a cut for which $r \in R$, $s \notin R$. For any $A \subseteq V$ we use $\overline{A}$ to denote $V \backslash A$. Finally, for $v \in V$, we use $\delta(v)$ as an abbreviation for $\delta(\{v\})$, and $\delta(\overline{v})$ as an abbreviation for $\delta(\overline{\{v\}})$.

Proposition 3.3 *For any (r, s)-cut $\delta(R)$ and any (r, s)-flow x, we have*

$$x(\delta(R)) - x(\delta(\overline{R})) = f_x(s).$$

(For example, for the flow x of Figure 3.1 and $R = \{r, p, q\}$ we have $x(\delta(R)) - x(\delta(\overline{R})) = 4 - 1 = 3$.)

Proof: We add the equations $f_x(v) = 0$ for $v \in \overline{R} \backslash \{s\}$ as well as the identity $f_x(s) = f_x(s)$. Obviously the right-hand side of the sum is just $f_x(s)$, the value of x. For any arc vw with $v, w \in R$, x_{vw} occurs in none of the equations added, so it does not occur in the sum. If $v, w \in \overline{R}$, then x_{vw} occurs in the equation for v with a coefficient of -1, and in the equation for w with a coefficient of $+1$, so it has a coefficient of 0 in the sum. If $v \in R$, $w \notin R$, then x_{vw} occurs in the equation for w with a coefficient of 1 and so occurs with a coefficient of 1 in the sum. Similarly, if $v \notin R$, $w \in R$, then x_{vw} occurs in the sum with a coefficient of -1. So the left-hand side of the sum is $x(\delta(R)) - x(\delta(\overline{R}))$, as required. ∎

Corollary 3.4 *For any feasible (r, s)-flow x and any (r, s)-cut $\delta(R)$, we have*

$$f_x(s) \le u(\delta(R)).$$

Proof: This follows from Proposition 3.3, since $x(\delta(R)) \le u(\delta(R))$ and $x(\delta(\overline{R})) \ge 0$. ∎

(For example, choosing $R = \{r, p, q\}$ in Figure 3.1, we have $f_x(s) \leq 8$.) Therefore, the maximum flow value is bounded by the minimum cut capacity. So if we can find a flow and a cut such that the value of the flow is equal to the capacity of the cut then we know that the flow is maximum. The famous *Max-Flow Min-Cut Theorem* states that this can always be done. It was discovered by Ford and Fulkerson [1956] and by Kotzig [1956].

Theorem 3.5 *(Max-Flow Min-Cut Theorem) If there is a maximum (r, s)-flow, then*

$$\max\{f_x(s) : x \ a \ feasible \ (r, s)\text{-}flow\} = \min\{u(\delta(R)) : \delta(R) \ an \ (r, s)\text{-}cut\}.$$

We can prove Theorem 3.5 and at the same time present the basic idea for an algorithm to solve the two optimization problems. The proof of Proposition 3.1 suggests one way to proceed. Given a feasible flow x, try to find one of larger value by finding an (r, s)-dipath P for which $x_e < u_e$ for each arc e of P. Then increase x by the same value on all arcs of P. However, the flow of Figure 3.1 is not maximum (as we shall see), but there is no such dipath P, so this idea is not enough to solve the problem. A slight generalization is better: We call a path x-*incrementing* if every forward arc e has $x_e < u_e$ and every reverse arc e has $x_e > 0$. An x-*augmenting* path is an (r, s)-path that is x-incrementing. Given an x-augmenting path we can raise x_e by some positive ε on each forward arc and lower x_e by ε on each reverse arc; this yields a flow of larger value. For example, in Figure 3.1, r, q, a, p, b, s is the node-sequence of an x-augmenting path. Augmenting the flow by $\varepsilon = 1$ results in a flow of value 4. Because there is a cut, $\delta(\{r, q, a\})$ of capacity 4, we know by Corollary 3.4 that the maximum flow value (and the minimum cut value) is 4.

Proof of Max-Flow Min-Cut Theorem 3.5: By Corollary 3.4, we need only show that there exists a feasible flow x and a cut $\delta(R)$ such that $f_x(s) = u(\delta(R))$. Let x be a flow of maximum value. Let R denote $\{v \in V : $ there exists an x-incrementing path from r to $v\}$. Clearly $r \in R$, and moreover $s \notin R$, since there can be no x-augmenting path. For every arc $vw \in \delta(R)$ we must have $x_{vw} = u_{vw}$, since otherwise adding vw to the x-incrementing path from r to v would yield such a path to w, but $w \notin R$. Similarly, for every arc $vw \in \delta(\overline{R})$ we have $x_{vw} = 0$. But then by Proposition 3.3, $f_x(s) = x(\delta(R)) - x(\delta(\overline{R})) = u(\delta(R))$, and we are done. ∎

Essentially the same argument proves the next two results as well.

Theorem 3.6 *A feasible flow x is maximum if and only if there is no x-augmenting path.*

Proof: Obviously if x is maximum there is no x-augmenting path. If there is no x-augmenting path, then the construction of the proof of Theorem 3.5 yields a cut $\delta(R)$ with $f_x(s) = u(\delta(R))$, so that x is maximum, by Corollary 3.4. ∎

Theorem 3.7 *If u is integral and there exists a maximum flow, then there exists a maximum flow that is integral.*

Proof: Choose an integral flow x of maximum value. If there is an x-augmenting path, then since x and u are integral, the new flow can be chosen integral, contradicting the choice of x. Hence there is no x-augmenting path, and so x is a maximum flow, by Theorem 3.6.   ∎

Finally, we can combine the max flow-min cut theorem with the proof of Corollary 3.4 to prove the following.

Corollary 3.8 *If x is a feasible (r, s)-flow and $\delta(R)$ is an (r, s)-cut, then x is maximum and $\delta(R)$ is minimum if and only if*

$$x_e = u_e, \; \text{for all } e \in \delta(R) \; \text{and } x_e = 0, \; \text{for all } e \in \delta(\overline{R}).$$

∎

The Augmenting Path Algorithm

An algorithm for finding a maximum flow and a minimum cut is now immediate. It is the classical maximum flow algorithm of Ford and Fulkerson. Beginning with any feasible flow x ($x = 0$ will do), repeatedly find an x-augmenting path P and augment x by the maximum value permitted. Of course, this value is $\min(\varepsilon_1, \varepsilon_2)$, where $\varepsilon_1 = \min(u_e - x_e : e$ forward in $P)$ and $\varepsilon_2 = \min(x_e : e$ reverse in $P)$. We call ε the x-*width* of P. If such a path with x-width ∞ is found, then there is no maximum flow, and the algorithm terminates. If there is no x-augmenting path, then x is a maximum flow and the set R of nodes reachable by an x-incrementing path from r determines a minimum cut; again, the algorithm terminates.

We need a method for searching for augmenting paths, but we can use the same ideas as we used earlier for finding dipaths. In fact, we can make the connection explicit. Define an *auxiliary digraph* $G(x)$, depending on G, u and the current flow x, as follows. Put $V(G(x)) = V$ and put $vw \in E(G(x))$ if and only if $vw \in E$ and $x_{vw} < u_{vw}$ or $wv \in E$ and $x_{wv} > 0$. (It is possible for both of these to happen, and then it is convenient to put two parallel arcs vw into $E(G(x))$.) Figure 3.2 shows $G(x)$ for the G, x of Figure 3.1. Obviously (r, s)-dipaths in $G(x)$ correspond to x-augmenting paths. It follows that each iteration of the maximum flow algorithm can be performed in $O(m)$ time. (In particular, one could use breadth-first search.)

So the question of the running time of the algorithm depends directly on the number of augmentations required. The most obvious bound on the number of augmentations applies only to the case of integral capacities, but is often useful. It is immediate from the observation that when u is integral, x remains integral, and each augmentation is of value at least 1.

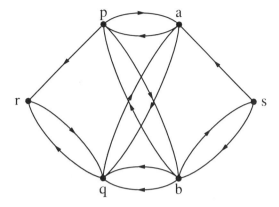

Figure 3.2. An auxiliary digraph

Theorem 3.9 *If u is integral and the maximum flow value is $K < \infty$, then the maximum flow algorithm terminates after at most K augmentations.* ▮

A consequence of Theorem 3.9 is that if each capacity is a rational number or ∞, and there exists a maximum flow, then the algorithm terminates. (Proof: There exists an integer D such that each component of Du is an integer or ∞, and the algorithm behaves in the same way for this scaled problem.) Surprisingly, if we allow irrational numbers as capacities, the algorithm can fail to terminate. In fact, Ford and Fulkerson [1962] showed that it can converge to a flow value different from the maximum one. (Of course, it is hard to imagine how irrational capacities would arise in practical problems.)

For a number of applications, the bound provided by Theorem 3.9 is perfectly satisfactory. However, in general the augmenting path algorithm cannot be considered acceptable. Figure 3.3 shows one difficulty that it can encounter. Here there is no maximum flow, but if each augmenting path uses arc ab, then each augmentation will be of value 1, and the algorithm will never terminate. It is quite easy to check in advance whether there exists no maximum flow (Exercise 3.3), so one could argue that this difficulty could be avoided. However, if we replace each infinite capacity in the example by a large integer M, then the algorithm terminates, but only after $2M$ augmentations. This example shows that the bound of Theorem 3.9 can actually be attained. It is natural to try to avoid this sort of problem by looking only for augmenting paths of large x-width. In fact, this kind of approach does decrease the dependence of the running time on the value of the maximum flow (in case of integral capacities). It is the subject of Exercises 3.11–3.14. Here we present something even better: a bound on the number of augmentations that does not depend on the capacities at all.

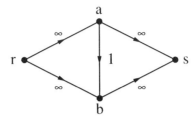

Figure 3.3. A bad example for the augmenting path algorithm

Shortest Augmenting Paths

We call an x-augmenting path *shortest* if it has the minimum possible number of arcs. By restricting attention to shortest augmenting paths we can improve the running time of the maximum flow algorithm dramatically. This result is due to Dinits [1970] and to Edmonds and Karp [1972].

Theorem 3.10 *If each augmentation of the augmenting path algorithm is on a shortest augmenting path, then there are at most nm augmentations.*

Notice that the theorem makes no hypothesis about either the existence of a maximum flow or the rationality of the components of u. So it handles all of the difficulties mentioned in the previous subsection. In addition, it is no harder to find a shortest augmenting path than to find *an* augmenting path—breadth-first search will accomplish it in time $O(m)$. (For this reason Edmonds and Karp called their modification "so simple that it is likely to be incorporated innocently into a computer implementation.") It follows that we have a polynomial-time algorithm for the maximum flow problem.

Corollary 3.11 *The augmenting path algorithm with breadth-first search solves the maximum flow problem in time $O(nm^2)$.* ∎

To prove Theorem 3.10, we consider a typical augmentation from flow x to flow x' determined by augmenting path P having node-sequence $v_0, v_1, \ldots, v_k$. Let $d_x(v, w)$ be the least *length* (least number of arcs) of a (v, w)-dipath in $G(x)$. ($d_x(v, w) = \infty$ if none exists.) Clearly, $d_x(r, v_i) = i$ and $d_x(v_i, s) = k - i$ for each i. Also, if an arc vw of $G(x')$ is *not* an arc of $G(x)$, then $v = v_i$, $w = v_{i-1}$ for some i.

Lemma 3.12 *For each $v \in V$, $d_{x'}(r, v) \geq d_x(r, v)$ and $d_{x'}(v, s) \geq d_x(v, s)$.*

Proof: Suppose that there exists a node v such that $d_{x'}(r, v) < d_x(r, v)$, and choose such v so that $d_{x'}(r, v)$ is as small as possible. Clearly $d_{x'}(r, v) > 0$. Let P' be an (r, v) dipath in $G(x')$ of length $d_{x'}(r, v)$ and let w be the second-last node of P'. Then

$$d_x(r, v) > d_{x'}(r, v) = d_{x'}(r, w) + 1 \geq d_x(r, w) + 1. \qquad (3.4)$$

It follows that wv is an arc of $G(x')$ but not of $G(x)$, for otherwise $d_x(r, v) \leq d_x(r, w) + 1$, so $w = v_i$, $v = v_{i-1}$ for some i. But then (3.4) implies that $i - 1 > i + 1$, a contradiction. So the first statement is proved. The proof that $d_{x'}(v, s) \geq d_x(v, s)$ is similar. ∎

It follows from Lemma 3.12 that the refined maximum flow algorithm proceeds through a sequence of at most $n - 1$ "stages," during each of which all augmentations are on paths of a fixed length k. So we need to obtain information on the number of augmentations during a stage. Let $\tilde{E}(x)$ denote $\{e \in E : e$ is an arc of a shortest x-augmenting path$\}$.

Lemma 3.13 *If $d_{x'}(r, s) = d_x(r, s)$, then $\tilde{E}(x')$ is a proper subset of $\tilde{E}(x)$.*

Proof: Let $k = d_x(r, s)$ and suppose that $e \in \tilde{E}(x')$. Then e induces an arc vw of $G(x')$ and $d_{x'}(r, v) = i - 1$, $d_{x'}(w, s) = k - i$ for some i. Therefore, $d_x(r, v) + d_x(w, s) \leq k - 1$, by Lemma 3.12. Now suppose that $e \notin \tilde{E}(x)$. Then $x_e \neq x'_e$, so e is an arc of P, a contradiction. Therefore, $\tilde{E}(x') \subseteq \tilde{E}(x)$. Now there is an arc e of P such that e is forward and $x'_e = u_e$ or e is reverse and $x'_e = 0$. Therefore, any x'-augmenting path using e must use it in the opposite direction from P, so its length, for some i, will be at least $i + k - i + 1 + 1 = k + 2$, so $e \notin \tilde{E}(x')$. The result follows. ∎

Proof of Theorem 3.10: It follows from Lemma 3.13 that there can be at most m augmentations per stage. Since there are at most $n - 1$ stages, there are at most nm augmentations in all. ∎

In Section 3.4 we shall see other maximum flow algorithms that improve on the $O(nm^2)$ running time of the shortest augmenting path algorithm. In the next section we describe a number of applications of the maximum flow problem and the minimum cut problem.

Exercises

3.1. Prove Proposition 3.2.

3.2. Find a maximum flow and a minimum cut for the network given in Figure 3.4.

3.3. Prove directly that there is no maximum flow if and only if there is an (r, s)-dipath, each of whose arcs e has $u_e = \infty$.

3.4. Construct networks with integral capacities having: (a) many integral maximum flows and many minimum cuts; (b) many integral maximum flows and a unique minimum cut; (c) many minimum cuts and a unique maximum flow.

3.5. Use Corollary 3.8 and knowledge of one maximum flow in Figure 3.1 to show that there is a unique minimum cut.

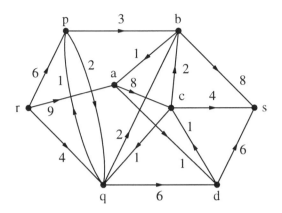

Figure 3.4. Maximum flow exercise

3.6. Use Corollary 3.8 and knowledge of one minimum cut in Figure 3.1 to show that there is a unique maximum flow.

3.7. Let $\delta(R_1)$ and $\delta(R_2)$ be minimum cuts. Prove that $\delta(R_1 \cap R_2)$ and $\delta(R_1 \cup R_2)$ are also minimum cuts.

3.8. Let x^1 and x^2 be maximum flows, and suppose that there is an x^1-incrementing (r, v)-path. Prove that there is also an x^2-incrementing (r, v)-path.

3.9. Given digraph G, $u \in \mathbf{R}^E$, and $r, s \in V$, find R, $r \in R \subseteq V \backslash \{s\}$ such that $u(\delta(R)) - u(\delta(\overline{R}))$ is minimized. (Hint: This is much easier than the minimum cut problem.)

3.10. Suppose that we perform an augmentation on an augmenting path of maximum x-width. Can the maximum x-width with respect to the new flow be larger?

3.11. Suppose that we have a feasible flow x such that, for some nonnegative number K, there is no x-augmenting path of x-width greater than K. Prove that $f_x(s)$ is within Km of the maximum value of a feasible flow. (Hint: How would you show this for $K = 0$?)

3.12. Suppose that we apply the scaling approach of Exercise 2.30 to the integral-capacity maximum flow problem by searching in stage p for augmenting paths of x-width at least 2^p. Derive the best bound on the number of augmentations you can for such an approach. You will need the result of Exercise 3.11.

3.13. Another approach to improving the augmenting path algorithm is to choose at each step an augmenting path of maximum x-width. Give a good bound on the number of augmentations, using the method of Exercise 2.31, and the result of Exercise 3.11.

3.14. How efficiently can an augmenting path of maximum x-width be found?

3.15. Suppose that the augmenting path algorithm always chooses an augmenting path having as few reverse arcs as possible. Prove that the number of augmentations will be $O(mn)$. Give an $O(m)$ algorithm for finding each augmentation.

3.3   APPLICATIONS OF MAXIMUM FLOW AND MINIMUM CUT

In this section we discuss a number of applications of maximum flow and minimum cut. Several more applications appear in later chapters.

Bipartite Matchings and Covers

We are given disjoint sets P of men and Q of women, and the pairs (p, q) that like each other. The *marriage problem* is to arrange as many (monogamous) marriages as possible with the restriction that married people should (at least initially!) like each other. We can associate with the input a graph $G = (V, E)$ such that $V = P \cup Q$ and $E \subseteq \{pq : p \in P, q \in Q\}$. Such graphs, that is, ones in which there is a partition of the nodes into two parts such that every edge has its ends in different parts, are called *bipartite*. (Sometimes $\{P, Q\}$ is called a *bipartition* of G.) The marriage problem asks for a *matching* of G of maximum size, that is, a subset M of E such that no two edges in M share an end. A bipartite graph appears in Figure 3.5, and the thick edges constitute a matching of size 6; we shall see that there is no larger one. Although the problem of finding a maximum matching in a general graph is more difficult (and is treated in Chapter 5), that for bipartite graphs is an easy application of maximum flows.

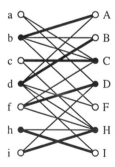

Figure 3.5. A bipartite graph and a matching

In fact, the bipartite matching problem was solved by Kőnig [1931] long before the development of network flow theory. He discovered a characterization of the maximum size of a matching that can be thought of as a prototype of the Max-Flow Min-Cut Theorem. (He also introduced, in the restricted setting of bipartite matching, the notion of flow-augmenting path.) Again, the basic idea is a general method to provide bounds, in this case, an upper bound for the maximum size of a matching. A *cover* of a graph G is a set C of nodes such that every edge of G has at least one end in C. For any matching M and any cover C, each edge $vw \in M$ has an end in C, but because matching edges cannot have an end in common, the corresponding nodes of C are all distinct. Therefore, $|M| \leq |C|$. It follows that, if we can find a matching M and a cover C with $|M| = |C|$, then we know that M is maximum. For the graph of Figure 3.5, the black nodes form a cover of size 6 and so the displayed matching is indeed of maximum size. Kőnig proved that, for bipartite graphs, it is always possible to make this kind of argument.

Theorem 3.14 *(Kőnig's Theorem) For a bipartite graph G,*

$$\max\{|M| : M \ a \ matching\} = \min\{|C| : C \ a \ cover\}.$$

We shall show how the Max-Flow Min-Cut Theorem implies Kőnig's Theorem, and how a maximum flow algorithm provides an efficient algorithm for constructing a maximum matching and a minimum cover. Given G with bipartition $\{P, Q\}$, we form a digraph G' with capacity vector u as follows. $V' = V \cup \{r, s\}$, where r, s are new nodes. For each edge pq of G with $p \in P$, $q \in Q$ there is a (directed) arc pq of G' with capacity ∞. For each $p \in P$ there is an arc rp of capacity 1. For each $q \in Q$ there is an arc qs of capacity 1. For the graph of Figure 3.5 we show the corresponding flow network in Figure 3.6.

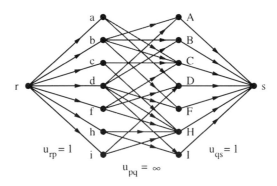

Figure 3.6. Flow network for bipartite matching

Let x be an integral feasible flow in G' of value k. In fact, this implies that x is $\{0, 1\}$-valued. (Why?) Define $M \subseteq E$ by: $pq \in M$ if $x_{pq} = 1$ and $pq \notin M$ if $x_{pq} = 0$. Then M is a matching of G, and $|M| = k$. Now suppose that we are given a matching M of G. Define $(x_{vw} : vw \in E')$ by: If $v \in P$, $w \in Q$ then $x_{vw} = 1$ if $vw \in M$ and 0 if $vw \notin M$; if $v = r$, $w \in P$, then $x_{vw} = 1$ if there is an edge of M incident with w and $x_{vw} = 0$ otherwise; if $v \in P$, $w = s$, then $x_{vw} = 1$ if there is an edge of M incident with v and $x_{vw} = 0$ otherwise. Then x is an integral feasible flow of value $|M|$ in G'. Hence we can find a maximum cardinality matching in G by solving the maximum flow problem on G'. There will be at most $|P| \leq n$ augmentations, by Theorem 3.9, since the maximum matching size is at most $|P|$. So we get an algorithm for maximum bipartite matching having running time $O(mn)$.

Now consider a minimum cut $\delta'(\{r\} \cup A)$ where $A \subseteq V$. Since it has finite capacity, there can be no edge of G from $A \cap P$ to $Q \backslash A$. Therefore, every edge of G is incident with an element of $C = (P \backslash A) \cup (Q \cap A)$. That is, C is a cover. Moreover, the capacity of the cut is $|P \backslash A| + |Q \cap A| = |C|$, so C is a cover of cardinality equal to the maximum size of a matching. This proves König's Theorem, and shows that the algorithm also finds a minimum-cardinality cover.

There are many other related applications. Some of them are investigated in the exercises.

Optimal Closure in a Digraph

There are many applications in which we want to choose an optimal subset of "projects," where each project has a benefit. This benefit may be positive, negative, or zero. There is no restriction on the number of projects to be chosen, but there are restrictions of the form: If project v is to be chosen, then project w must be chosen also. If we model the projects as the nodes of a digraph G and the restrictions (v, w) as its arcs, then we must choose a maximum benefit set $A \subseteq V$ such that $\delta(A) = \emptyset$. We call such a subset A a *closure* of G. We remark that the problem is trivial if either all the benefits are nonnegative (V will be optimal) or all are nonpositive ($\emptyset$ will be optimal).

A classical application of this form is in the design of an open-pit mine. Here the region under consideration is divided into 3-dimensional blocks. For each block v there is a known estimated net profit b_v associated with excavating block v. The constraints come from the fact that it is not possible to excavate a block without also excavating those above it. The definition of "above" will depend on restrictions on the steepness of the sides of the pit.

It turns out that the optimal closure problem can be reduced to a minimum cut problem, an observation due to Picard [1976]; in earlier work, Rhys [1970] solved an important special case. Given G and b, define a digraph G' and capacity vector u as follows. Put $V' = V \cup \{r, s\}$ for new nodes r, s. For each $v \in V$ with $b_v > 0$, G' has an arc rv with $u_{rv} = b_v$. For each $v \in V$ with $b_v < 0$, G' has an arc vs with $u_{vs} = -b_v$. The remaining arcs of G' are just the arcs of G, each with capacity ∞. Figure 3.7 summarizes this construction. It is easy to see that any finite-capacity (r, s)-cut $\delta'(R)$ in G' will be such that $R = \{r\} \cup A$, where A is a closure of G. (In particular, a minimum-capacity cut will have this property.) Any closure $A \subseteq V$ determines an (r, s)-cut $\delta'(A \cup \{r\})$ having capacity $\Sigma(b_v : v \notin A, \ b_v > 0) - \Sigma(b_v : v \in A, \ b_v < 0)$. By adding $\Sigma(b_v : v \in A, \ b_v \geq 0)$ to both terms, this can be rewritten as

$$\Sigma(b_v : v \in V, \ b_v \geq 0) - b(A).$$

Since the first term of the latter expression is constant, that is, does not depend on A, this expression is minimized when $b(A)$ is maximized. In summary, we simply find a minimum cut $\delta'(A \cup \{r\})$ of G', and A is a maximum-weight closure.

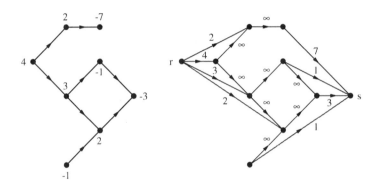

Figure 3.7. Flow network for the optimal closure problem

Elimination of Sports Teams

Sports writers are fond of using the term "mathematically eliminated" to refer to a team that cannot possibly finish the season in first place. More formally, let us say that the Buzzards are *eliminated* if, no matter what the outcome of the remaining games, they cannot finish with the most wins (even in a tie). For convenience, we assume that there are no tie games.

The simplest situation in which the Buzzards are eliminated (and the only one of which sportswriters seem to be aware!) is illustrated in Table 3.1. In this case even if the Buzzards win all their remaining games, they will have fewer wins than the Anteaters already have. Notice that in this case we can see that the Buzzards are eliminated, no matter what pairs of teams are involved in the remaining games. A more interesting situation involves the data of Table 3.2. (The additional columns of the table indicate the number of remaining games against various opponents. Notice that there may be other teams that we have not included in the table.) Here it is possible for the Buzzards to finish with as many wins as the Anteaters, if the Anteaters lose all of their remaining games and the Buzzards win all of theirs. However, in this case the Banana Slugs must finish with more wins than the Buzzards, since they win their remaining games against the Anteaters. Thus, although we cannot be sure which team will finish first, we can see that the Buzzards are eliminated.

Team	Wins	To Play
Anteaters	33	8
Buzzards	28	4

Table 3.1. A simple example of elimination

Team	Wins	To Play	A.	B.	B.S.	R.
Anteaters	33	8	–	1	6	1
Buzzards	29	4	1	–	0	3
Banana Slugs	28	8	6	0	–	1
Fighting Ducks	27	5	1	3	1	–

Table 3.2. A second example of elimination

We will show that in any situation in which a team is eliminated, there is a simple reason, as in the previous examples. Let T denote the set of teams other than the Buzzards. For each $i \in T$, let w_i denote the number of wins for team i, and for $i, j \in T$ with $i \neq j$, let r_{ij} denote the number of remaining games between teams i and j. We will need also a notation for the set $\{\{i, j\} : \{i, j\} \subseteq T, i \neq j, r_{ij} > 0\}$; we denote it by P. Finally, let M denote the number of wins for the Buzzards at the end of the season if they win all their remaining games.

Let A be a subset of T. Since every game between two teams in A is won by one of them, the total number of wins for teams in A at the end of the season is at least $w(A) + \sum(r_{ij} : \{i, j\} \subseteq A, \{i, j\} \in P)$. If this number is bigger than $M|A|$, then the average number of wins of teams in A at the end of the season is more than M. But M is the most wins that the Buzzards can hope to have, so at least one team in A will finish with more wins than the Buzzards. In summary, the Buzzards are eliminated if there exists $A \subseteq T$ such that

$$w(A) + \sum(r_{ij} : \{i, j\} \subseteq A, \{i, j\} \in P) > M|A|. \qquad (3.5)$$

This criterion for elimination is general enough to include the arguments used in the examples above. In the first example, A consists of the Anteaters alone, in the second, of the Anteaters and the Banana Slugs.

We will show that if the Buzzards have been eliminated there is a set A with property (3.5). To prove this we make use of the fact that, if the Buzzards are *not* eliminated, there is a set of possible outcomes of the remaining games so that the Buzzards finish with the most wins. Let y_{ij} denote the (unknown) number of wins for team i over team j in the remaining games between them. Then if the Buzzards are not eliminated, there must exist values for the y_{ij} satisfying

$$y_{ij} + y_{ji} = r_{ij}, \text{ for all } \{i,j\} \in P \tag{3.6}$$

$$w_i + \sum(y_{ij} : j \in T, \ j \neq i) \leq M, \text{ for all } i \in T$$

$$y_{ij} \geq 0, \text{ for all } \{i,j\} \in P$$

$$y_{ij} \text{ integral, for all } \{i,j\} \in P.$$

We create a flow network $G = (V, E)$ as follows. $V = T \cup P \cup \{r, s\}$. For each $i \in T$, there is an arc (r, i) having capacity $M - w_i$. For each $i \in T$ and $j \in T$ with $\{i,j\} \in P$, there are arcs $(i, \{i,j\})$ and $(j, \{i,j\})$ with capacity ∞, and there is an arc $(\{i,j\}, s)$ with capacity r_{ij}. The network arising from the data of Table 3.2 is illustrated in Figure 3.8. Now suppose that in this network there is an integral feasible (r, s)-flow of value $\sum(r_{ij} : \{i,j\} \in P)$. Then if we put y_{ij} equal to the flow on arc $(i, \{i,j\})$, we get a solution to (3.6). Conversely, a solution to (3.6) yields such an integral feasible flow, by assigning flow y_{ij} to arc $(i, \{i,j\})$ and then defining the flows on arcs incident to the source and sink to satisfy conservation of flow.

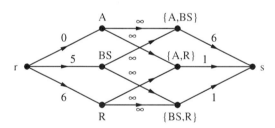

Figure 3.8. Flow network for the elimination problem

It follows that we can determine whether the Buzzards are eliminated by solving a maximum (integral) flow problem. If they are not eliminated, a maximum flow will determine a set of outcomes for the remaining games in which the Buzzards finish first. Now we show that if the Buzzards are eliminated, then a minimum cut will determine a set A satisfying (3.5). Let $\delta(S)$ be a minimum (r, s)-cut. By the Max-Flow Min-Cut Theorem its capacity is less than $\sum(r_{ij} : \{i,j\} \in P)$. Let $A = T \backslash S$. We claim that $S = \{r\} \cup (T \backslash A) \cup \{\{i,j\} \in P : i \text{ or } j \notin A\}$. First, if i or j is not in A but $\{i,j\} \notin S$, then $\delta(S)$ has capacity ∞. Second, if $\{i,j\} \in S$ and $i, j \in A$, then deleting $\{i,j\}$ from S decreases the capacity of $\delta(S)$ by r_{ij}. In either case, $\delta(S)$ is not a minimum cut, a contradiction. Now it is easy to compute the capacity of $\delta(S)$. It is

$$M|A| - w(A) + \sum(r_{ij} : \{i,j\} \in P, \{i,j\} \not\subseteq A).$$

This is less than $\sum(r_{ij} : \{i,j\} \in P)$ if and only if A satisfies (3.5), and we are done.

Flow Feasibility Problems

An example of what might be called a flow feasibility problem is that of deciding, given (G, u, r, s, k), whether there exists a feasible flow from r to s of value at least k. Of course, this is essentially a restatement of the maximum flow problem. We have solved this problem in two senses. First, we have given a good algorithm that will construct such a flow if one exists. Second, we have given a good characterization for its existence. (Such a flow exists if and only if every (r, s)-cut has capacity at least k.) A number of useful and interesting flow feasibility problems will be solved here. Although some of them appear to be significantly more general, all of them can be reduced to the above problem, and solved (in the two senses) by the maximum flow algorithm and the Max-Flow Min-Cut Theorem.

As a first example, consider the problem of deciding whether the transportation model mentioned at the beginning of this chapter has a feasible solution. This problem can be restated as: Given a bipartite graph $G = (V, E)$ with bipartition $\{P, Q\}$ and vectors $a \in \mathbf{Z}_+^P$, $b \in \mathbf{Z}_+^Q$, to find $x \in \mathbf{R}^E$ satisfying

$$\sum(x_{pq} : q \in Q, \ pq \in E) \le a_p, \text{ for all } p \in P \qquad (3.7)$$
$$\sum(x_{pq} : p \in P, \ pq \in E) = b_q, \text{ for all } q \in Q$$
$$x_{pq} \ge 0, \text{ for all } pq \in E$$
$$x_{pq} \text{ integral, for all } pq \in E.$$

The method for converting this to a flow problem is similar to the one used for bipartite matching. Form digraph G' with $V' = V \cup \{r, s\}$. Each $pq \in E$ gives rise to an arc pq of G' with $u_{pq} = \infty$. For each $p \in P$ there is an arc rp with $u_{rp} = a_p$. For each $q \in Q$, there is an arc qs with $u_{qs} = b_q$. This construction is illustrated in Figure 3.9. It is easy to see that there exists $x \in \mathbf{Z}^E$ satisfying (3.7) if and only if there is an integral feasible flow in G' from r to s of value $\Sigma(b_q : q \in Q)$. Thus we can find such a solution, if one exists, with a maximum flow algorithm. The Max-Flow Min-Cut Theorem tells us that (3.7) has a solution if and only if every (r, s)-cut of G' has capacity at least $\Sigma(b_q : q \in Q)$. The capacity of a cut $\delta'(A \cup B \cup \{r\})$ where $A \subseteq P$, $B \subseteq Q$, is $\Sigma(a_i : i \in P \backslash A) + \Sigma(b_j : j \in B)$, assuming there is no arc pq for $p \in A$, $q \in Q \backslash B$. (Otherwise the cut capacity is ∞.) This capacity is at least $\Sigma(b_j : j \in Q)$ if and only if $\Sigma(a_i : i \in P \backslash A) \ge \Sigma(b_j : j \in Q \backslash B)$. It is clear that it is enough to check this condition for the sets A such that

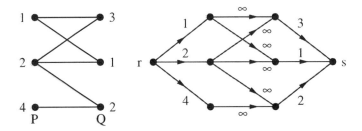

Figure 3.9. Flow network for the transportation problem

every node in $P\backslash A$ is adjacent to a node in $Q\backslash B$, since any node violating this could be added to A. That is, if we define the *neighborset* $N(C)$ of a set C of nodes to be $\{w : vw \in E$ for some $v \in C\}$, then we may assume that $N(Q\backslash B) = P\backslash A$. So the resulting necessary and sufficient condition for the existence of a solution to (3.7) is

$$a(N(C)) \geq b(C), \text{ for all } C \subseteq Q.$$

Notice that, in the context of the transportation problem, the condition is a natural one: For every subset of outlets, the requirement of that subset should not exceed the total production of the factories that can ship to the subset.

A quite general kind of flow feasibility problem asks for $x \in \mathbf{R}^E$ (or $\mathbf{Z}^E$) such that

$$a_v \leq f_x(v) \leq b_v, \text{ for all } v \in V$$
$$\ell_e \leq x_e \leq u_e, \text{ for all } e \in E.$$

Rather than treat this problem directly, we shall discuss a few of its attractive special cases. First, consider the situation in which $\ell = 0$ and $a = b$. That is, given digraph G, $u \in \mathbf{R}_+^E$, and $b \in \mathbf{R}^V$, we want to find, if possible, $x \in \mathbf{R}^E$ such that

$$f_x(v) = b_v, \text{ for all } v \in V \tag{3.8}$$
$$0 \leq x_e \leq u_e, \text{ for all } e \in E.$$

First, we observe that, because of the equality constraints, the total demand $b(V)$ must be 0. (Proof: Just add the equations.) We form a digraph G' with $V' = V \cup \{r, s\}$. Each $vw \in E$ is an arc of G', and has capacity u_{vw}. For each $v \in V$ with $b_v < 0$, there is an arc rv with $u_{rv} = -b_v$. For each $v \in V$ with $b_v > 0$, there is an arc vs with $u_{vs} = b_v$. It is easy to see that there is a solution to (3.8) if and only if there is an (r,s)-flow in G' of value $\Sigma(b_v : v \in V, b_v > 0)$. So this can be tested by a maximum flow calculation. Moreover, by the Max-Flow Min-Cut Theorem, (3.8) has a solution if and only if there exists no $A \subseteq V$ such that $u(\delta'(A \cup \{r\})) < \Sigma(b_v : v \in V, b_v > 0)$. The former quantity is $\Sigma(-b_v : v \notin A,\ b_v < 0) + \Sigma(b_v : v \in A,\ b_v > 0) + u(\delta(A))$, so it is less than $\Sigma(b_v : v \in V,\ b_v > 0)$ if and only if $u(\delta(A)) < \Sigma(b_v : v \notin A)$. So we get the following feasibility theorem, due to Gale [1957].

Theorem 3.15 *There exists a solution to (3.8) if and only if $b(V) = 0$ and, for every $A \subseteq V$, $b(A) \leq u(\delta(\overline{A}))$. If b and u are integral, then (3.8) has an integral solution if and only if the same conditions hold.* ∎

An immediate consequence is the following.

Corollary 3.16 *Given a digraph G and $b \in \mathbf{R}^V$, there exists $x \in \mathbf{R}^E$ with*

$$f_x(v) = b_v,\ \text{for all } v \in V$$
$$x_e \geq 0,\ \text{for all } e \in E$$

if and only if $b(V) = 0$ and for every $A \subseteq V$ with $\delta(\overline{A}) = \emptyset$ we have $b(A) \leq 0$. Moreover, if b is integral, then there exists such an x that is also integral if and only if the same conditions hold. ∎

A *circulation* is a vector $x \in \mathbf{R}^E$ with $f_x(v) = 0$ for all $v \in V$. The following is a useful result of Hoffman [1960].

Theorem 3.17 *(Hoffman's Circulation Theorem) Given a digraph G, $\ell \in (\mathbf{R} \cup \{-\infty\})^E$, and $u \in (\mathbf{R} \cup \{\infty\})^E$, with $\ell \leq u$, there is a circulation x with $\ell \leq x \leq u$ if and only if every $A \subseteq V$ satisfies $u(\delta(\bar{A})) \geq \ell(\delta(A))$. Moreover, if ℓ and u are integral, then there is such a circulation that is also integral if and only if the same condition holds.*

Proof: First, we observe that any $\ell_e = -\infty$ can be replaced by $-M$ where M is integral and very large, without affecting either the existence of the desired circulation or the condition of the theorem. We can deduce the result from Theorem 3.15, by replacing ℓ by 0, u by $u' = u - \ell$, x by $x' = x - \ell$. Since $f_{x'}(v) = f_x(v) - f_\ell(v)$, therefore x is a circulation if and only if $f_{x'}(v) = -f_\ell(v)$. Therefore, we can apply Theorem 3.15 with demands $-f_\ell(v)$ and capacities $u - \ell$. Notice that the sum of the demands is 0. So there exists a feasible circulation if and only if for every $A \subseteq V$ we have

$$(u - \ell)(\delta(\overline{A})) \geq -\Sigma(f_\ell(v) : v \in A).$$

Since the right-hand side is $\ell(\delta(A)) - \ell(\delta(\overline{A}))$, the first statement is proved. If ℓ and u are integral, the same proof using the integral version of Theorem 3.15, yields the second statement. ∎

There is a common generalization of Theorem 3.15 and Theorem 3.17. We state it here and leave the proof for an exercise.

Theorem 3.18 *Given a digraph G, $b \in \mathbf{R}^V$ such that $b(V) = 0$, $\ell \in (\mathbf{R} \cup \{-\infty\})^E$, and $u \in (\mathbf{R} \cup \{\infty\})^E$ with $\ell \leq u$, there exists $x \in \mathbf{R}^E$ such that*

$$f_x(v) = b_v, \text{ for all } v \in V$$
$$\ell_e \leq x_e \leq u_e, \text{ for all } e \in E$$

if and only if every $A \subseteq V$ satisfies

$$u(\delta(\overline{A})) \geq b(A) + \ell(\delta(A)).$$

Moreover, if b, ℓ, and u are integral, there exists such an x that is also integral if and only if the same condition is satisfied. ∎

As a final example, we discuss the problem of finding an (r, s)-flow of minimum value subject to lower bounds. That is, given digraph G with $r, s \in V$ and $\ell \in \mathbf{R}^E$, $\ell \geq 0$

$$\text{Minimize } f_x(s) \tag{3.9}$$
$$\text{subject to}$$
$$f_x(v) = 0, \text{ for all } v \in V \backslash \{r, s\}$$
$$x_e \geq \ell_e, \text{ for all } e \in E.$$

We assume that every arc e having $\ell_e > 0$ is in an (r, s)-dipath (otherwise, there may be no feasible solution), and that there is no (s, r)-dipath (otherwise, there may be feasible solutions for which $f_x(s)$ is arbitrarily small). The latter assumption implies that there exists a set R, $r \in R \subseteq V \backslash \{s\}$, such that $\delta(\overline{R}) = \emptyset$. Any such set R provides a lower bound; namely, for any solution x of (3.9) we have

$$f_x(s) = x(\delta(R)) \geq \ell(\delta(R)).$$

This bound can be made tight.

Theorem 3.19 *(Min-Flow Max-Cut Theorem) There exists a solution to (3.9) having $f_x(s) \leq k$ if and only if there does not exist $R \subseteq V$ with $r \in R$, $s \notin R$, $\delta(\overline{R}) = \emptyset$, and $\ell(\delta(R)) > k$. Moreover, if ℓ is integral, there exists such an x that is also integral if and only if the same condition is satisfied.*

Proof: Add an arc sr to G with $\ell_{sr} = 0$, $u_{sr} = k$ and put $u_e = \infty$ for all $e \in E$. Then the new digraph G' has a feasible circulation if and only if (3.9) has a solution of value at most k. By Theorem 3.17 this is true if and only if there does not exist $A \subseteq V$ with $u(\delta'(A)) < \ell(\delta'(\overline{A}))$. Since $u_e = \infty$ for all $e \in E$, such a set A satisfies $\delta(A) = \emptyset$. Since $\ell(\delta(\overline{A})) > 0$, there exists $e \in \delta(\overline{A})$ with $\ell_e > 0$. Because there must be an (r, s)-dipath using e, it follows that $r \in \overline{A}$, $s \in A$. So $sr \in \delta'(A)$. Therefore, $k = u(\delta'(A)) < \ell(\delta'(\overline{A})) = \ell(\delta(\overline{A}))$. So taking $R = \overline{A}$, we have the desired condition. The second statement of the theorem follows in the same way from the integral version of Theorem 3.17. ∎

Minimum Cuts and Linear Programming

The maximum flow problem is itself a linear-programming problem, and so a good characterization of maximum flows could be given via linear-programming duality. It is natural to wonder what connection there might be between this characterization and the one we have given via cuts. Here we give two linear programming interpretations of the minimum cut problem.

First, let us write explicitly the dual of the maximum flow problem. We have dual variables y_v, $v \in V \backslash \{r, s\}$ (corresponding to the conservation of flow constraints), and z_e, $e \in E$ (corresponding to the inequalities $x_e \leq u_e$). So the dual problem is

$$\text{Minimize } \Sigma(u_e z_e : e \in E) \qquad (3.10)$$

$$\text{subject to}$$
$$-y_v + y_w + z_{vw} \geq 0, \text{ for all } vw \in E, \ v, w \in V \backslash \{r, s\}$$
$$y_w + z_{rw} \geq 0, \text{ for all } rw \in E$$
$$-y_v + z_{vr} \geq 0, \text{ for all } vr \in E$$
$$-y_v + z_{vs} \geq 1, \text{ for all } vs \in E$$
$$y_w + z_{sw} \geq -1, \text{ for all } sw \in E$$
$$z_e \geq 0, \text{ for all } e \in E.$$

We can make (3.10) less ugly by the following device. First, putting $y_r = 0$ and $y_s = -1$, we can unify the constraints (other than nonnegativity) into the form

$$-y_v + y_w + z_{vw} \geq 0, \text{ for all } vw \in E.$$

Then we can add 1 to each y_v without affecting anything. So we can restate (3.10) as

$$\text{Minimize } \Sigma(u_e z_e : e \in E) \qquad (3.11)$$

$$\text{subject to}$$
$$y_r = 1, y_s = 0$$
$$-y_v + y_w + z_{vw} \geq 0, \text{ for all } vw \in E$$
$$z_e \geq 0, \text{ for all } e \in E.$$

In view of the linear programming Duality Theorem, the following result can be viewed as a restatement of the Max-Flow Min-Cut Theorem.

Theorem 3.20 *If (3.11) has an optimal solution, then it has one of the form: For some (r, s)-cut $\delta(R)$, y is the characteristic vector of R and z is the characteristic vector of $\delta(R)$.*

Proof: Choose $\delta(R)$ to be a minimum cut. It is straightforward to check (Exercise 3.40) that (y, z) is feasible to (3.11). Its objective value is $\Sigma(u_e z_e : e \in E) = u(\delta(R))$, which is the maximum flow value by the Max-Flow Min-Cut Theorem. By the linear programming Duality Theorem this is the optimal value of (3.10) and hence of (3.11). ∎

Now we describe a surprisingly different proof of Theorem 3.20, which does not use the Max-Flow Min-Cut Theorem. A similar proof of a related result occurs in D. Bertsimas, C. Teo, and R. Vohra [1995].

Proof: Let (y, z) be an optimal solution of (3.11) and order V as

$$v_1, v_2, \ldots, v_n$$

such that

$$y_{v_1} \geq y_{v_2} \geq \cdots \geq y_{v_n}.$$

Suppose that $v_p = r$, $v_q = s$. Define R_i to be $\{v_1, \ldots, v_i\}$ for $p \leq i \leq q - 1$. Choose the cut $\delta(R)$ by choosing R to be R_i with probability $y_{v_i} - y_{v_{i+1}}$, $p \leq i \leq q - 1$. Then the expected value of $u(\delta(R))$ is

$$\sum_{i=p}^{q-1} (y_{v_i} - y_{v_{i+1}})u(\delta(R_i)).$$

Notice that an arc $v_j v_k$ with $j > k$ does not contribute at all to the above sum, while an arc $v_j v_k$ with $j < k$ contributes at most $(y_{v_j} - y_{v_k})u_{v_j v_k}$. So the expected value of $u(\delta(R))$ is at most

$$\sum_{vw \in E} (y_v - y_w)u_{vw} \leq \sum_{vw \in E} u_{vw} z_{vw}.$$

It follows that at least one R_i has $u(\delta(R_i))$ equal to the optimal value of (3.11), as required. ∎

Next, we give a slightly different interpretation. It allows us to get rid of some variables in (3.10), at the cost of increasing the number of constraints. Consider the problem:

$$\text{Minimize } \Sigma(u_e z_e : e \in E) \tag{3.12}$$

$$\text{subject to}$$

$$\Sigma(z_e : e \in P) \geq 1, \text{ for all arc-sets } P \text{ of simple } (r, s)\text{-dipaths}$$

$$z_e \geq 0, \text{ for all } e \in E.$$

Clearly every (r, s)-cut $\delta(R)$ determines a feasible solution to (3.12) by: $z_e = 1$, $e \in \delta(R)$ and $z_e = 0$ otherwise. We claim that there is always an optimal solution of this form.

Theorem 3.21 *If (3.12) has an optimal solution, then it has one that is the characteristic vector of an (r, s)-cut.*

Proof: Obviously (3.12) has a feasible solution, and it is unbounded if some $u_e < 0$, so we may assume that $u \geq 0$. We choose z' to be the characteristic vector of a minimum cut $\delta(R)$. The dual of (3.12) is

$$\text{Maximize } \Sigma(w_P : P \text{ a simple } (r, s)\text{-dipath}) \qquad (3.13)$$

$$\text{subject to}$$

$$\Sigma(w_P : e \text{ an arc of } P) \leq u_e, \text{ for all } e \in E$$

$$w_P \geq 0, \text{ for all } P \text{ a simple } (r, s)\text{-dipath.}$$

Let x be a maximum flow. We can find a simple (r, s)-dipath P such that $x_e > 0$ for each arc of P, put w_P equal to the minimum value of x_e on P, and subtract w_P from x_e for each arc e of P. We repeat this operation until $f_x(s) = 0$, at which point Σw_P is equal to (the original value of) $f_x(s)$. The procedure will terminate, because at each step one more value of x_e becomes 0. It is easy to see that the resulting w is feasible to (3.13). Since $\Sigma w_P = \Sigma(u_e z'_e)$, z' is optimal to (3.12). ∎

Notice that (3.13) is very much like the initial formulation of the truck-routing problem at the beginning of Section 3.2, except that there we required w_P to be integral. The argument used in the proof is a variant of the one used there.

Exercises

3.16. Find a maximum matching and a minimum cover in the bipartite graph of Figure 3.10.

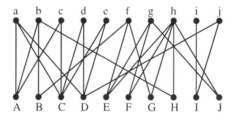

Figure 3.10. Bipartite matching exercise

3.17. A bipartite graph and a weight c_v for each $v \in V$ are given. Give a good algorithm to find a minimum-weight cover of G.

3.18. Given a bipartite graph $G = (V, E)$ and an integer d_v for each node v, does there exist a spanning subgraph H of G such that each node has degree d_v in H? Give a good algorithm to answer this question, and also necessary and sufficient conditions for the existence of such a subgraph.

3.19. Prove that in a matrix the maximum number of nonzero entries, no two in the same line (row or column), is equal to the minimum number of lines that include all the nonzero entries.

3.20. Prove that if G is a bipartite graph having bipartition $\{P, Q\}$ and C, C' are covers of G, then $\{(P \cap C \cap C') \cup (Q \cap (C \cup C'))\}$ is also a cover.

3.21. Prove that in a bipartite graph with bipartition $\{P, Q\}$, there exists a matching of size $|P|$ if and only if for every subset A of P we have $|N(A)| \geq |A|$.

3.22. For a family $(S_1, \ldots, S_k)$ of subsets of a set Q, a *system of distinct representatives (SDR)* is a set $\{q_1, \ldots, q_k\}$ of distinct elements of Q such that $q_i \in S_i$ for $1 \leq i \leq k$. Prove Hall's Theorem, that the family has an SDR if and only if for every subset I of $\{1, \ldots, k\}$ we have $|\cup (S_i : i \in I)| \geq |I|$.

3.23. Prove that a bipartite graph in which every node has degree $k \geq 1$ has a perfect matching. Deduce that such a graph has k disjoint perfect matchings.

3.24. Show that Exercise 3.23 is false when "degree k" is replaced by "degree at least k."

3.25. Let G be a bipartite graph with bipartition $\{P, Q\}$ where $|P| = |Q| = t$. Prove that G has k disjoint perfect matchings if and only if for all $P' \subseteq P, Q' \subseteq Q$ there are at least $k(|P'| + |Q'| - t)$ edges from P' to Q'. (Hint: By Exercise 3.23, G has k disjoint perfect matchings if and only if it has a spanning subgraph in which each node has degree exactly k.)

3.26. For two families $(S_1, \ldots, S_k)$ and $(T_1, \ldots, T_k)$ of subsets of Q, a *common SDR* is a set that is an SDR for both families. Prove that there exists a common SDR if and only if for every pair I, J of subsets of $\{1, \ldots, k\}$ we have

$$|(\cup(S_i : i \in I)) \cap (\cup(T_i : i \in J))| \geq |I| + |J| - k.$$

3.27. Show that every optimal closure problem can be reduced to an optimal closure problem on a digraph that is bipartite.

3.28. Solve the max flow problem of Figure 3.7 and use the solution to find a maximum-weight closure.

3.29. Projects $1, 2, \ldots k$ are available to be undertaken. With each project i is associated a positive revenue r_i. Each project i requires a set S_i of resources to be available, and each resource j, $1 \leq j \leq \ell$, has an associated cost c_j. However, if j is purchased, it is available for any projects for which it is required. Give an algorithm to choose a set of projects so that the associated revenue minus the cost of the required resources is maximized.

3.30. Solve the max flow problem of Figure 3.8 and use the solution to determine if the Buzzards are eliminated.

3.31. Suppose that we want to know whether it is possible for the Buzzards to finish first or second, that is, so that at most one team has more total wins. How can this be done?

3.32. Prove Theorem 3.18.

3.33. Prove that there exists x satisfying $0 \leq x \leq u$ and $f_x(v) \leq b_v$ for all $v \in V$ if and only if for every $A \subseteq V$ we have $b(A) + u(\delta(A)) \geq 0$. Use this result to get a similar one for $f_x(v) \geq b_v$.

3.34. Given an undirected graph G and a nonnegative integer k, we want to choose a direction for each edge of G so that in the resulting digraph G', each node v is the tail of at most k arcs. Give a good algorithm and a good characterization for this property.

3.35. In a tournament of n players in which each player plays each other once, and there are no ties, let w_i denote the number of wins for player i. Given a vector $w = (w_1, \ldots, w_n)$, how can it be determined whether w arises from such a tournament? Give a good algorithm and a good characterization.

3.36. State and solve a problem that yields the solution to both Exercise 3.34 and Exercise 3.35.

3.37. Let G be an acyclic digraph. A set $A \subseteq V$ is *independent* if no two elements of A are joined by a dipath in G. A *path cover* of G is a set $\{P_1, \ldots, P_k\}$ of dipaths such that every node of G is a vertex of at least one P_i. Prove Dilworth's Theorem, that the maximum size of an independent set is equal to the minimum size of a path cover. Hint: Use the Min-Flow Max-Cut Theorem.

3.38. Prove Kőnig's Theorem from Dilworth's Theorem.

3.39. We are given a collection J of jobs and a fixed schedule for completing them, so that the processing of job j must begin at time p_j and end at time q_j. The processing of each job is done on one of a number of identical processors. In addition, there is a transition time t_{ij} required to prepare a processor, which has just completed processing job i, to begin processing job j. (For example, the jobs could be scheduled airline flights, and the processors could be aircraft.) Find a feasible schedule requiring as few processors as possible.

3.40. Let $R \subseteq V \setminus \{s\}$ with $r \in R$ and define $z_e = 1$ for $e \in \delta(R)$, $z_e = 0$ otherwise, and $y_v = 1$ for $v \in R$, $y_v = 0$ otherwise. Show that (y, z) is a feasible solution of (3.11).

3.41. We are given n jobs $J_1, J_2, \ldots, J_n$ and k identical processors. Each job J_i requires total processing time p_i on any one of the processors. Moreover, "preemption" is allowed. That is, J_i need not be processed during an entire interval of length p_i, nor always on the same processor. However, it is processed on at most one processor at a given time, and it is subject to a release date r_i and a due date d_i. That is, the processing of J_i cannot begin before r_i and must be completed by d_i. Show how to find a schedule that satisfies these requirements by solving a maximum flow problem. (Hint: By sorting the r_i and d_i, $1 \leq i \leq n$, into a sequence of distinct dates $t_1 < t_2 < \ldots < t_{m+1}$, where $m + 1 \leq 2n$, we can divide the time period available into m intervals $T_1 = (t_1, t_2)$, $T_2 = (t_2, t_3), \ldots, T_m = (t_m, t_{m+1})$. It will be enough to decide what value of the processing of J_i will be done during T_j for $1 \leq i \leq n$, $1 \leq j \leq m$.)

3.42. Two paths in a digraph are *internally node-disjoint* if the only nodes they have in common are their initial or final nodes. For nodes r, s in a digraph G, a set $S \subseteq V \backslash \{r, s\}$ *separates* r from s if there is no (r, s)-dipath in $G \backslash S$. Prove Menger's Theorem, that the maximum number of internally node-disjoint (r, s)-dipaths is equal to the minimum size of a set that separates r from s.

3.43. Use Exercise 3.42 to state and prove an analogous result for undirected graphs.

3.44. We are given a digraph G, a special node r, and for each arc e a non-negative cost c_e of destroying e. If an attacker destroys a set A of arcs, she receives a benefit b_v for each node v that can no longer be reached from r by a dipath. The attacker wants to choose A to minimize cost minus benefit. Suggest an algorithm.

3.4   PUSH-RELABEL MAXIMUM FLOW ALGORITHMS

In Section 3.2 we modified the augmenting path algorithm for the maximum flow problem to choose a path having as few arcs as possible at each step. We proved that the resulting algorithm requires at most nm augmentations, independent of the capacities. There are further improvements possible to the augmenting path algorithm for the maximum flow problem. In fact, Dinits showed how to implement the shortest augmenting path algorithm in time $O(mn^2)$. However, here we describe another important class of algorithms for the maximum flow problem that is not based on augmenting paths, but on an idea that is, in a sense, simpler.

For purposes of this section it is convenient to take $u_{wv} = x_{wv} = 0$ if $vw \in E$ and $wv \notin E$. (This is just a device that simplifies the description, and does *not* mean that we are going to add a clutter of additional arcs to G.) For a vector x satisfying $0 \leq x \leq u$, we define the auxiliary digraph $G(x)$ just as though x were a feasible flow, except that we do not bother with parallel arcs. That is, we put $vw \in E(G(x))$ if and only if $x_{wv} > 0$ or $x_{vw} < u_{vw}$. If vw is an arc of $G(x)$, then up to $\tilde{u}_{vw} = u_{vw} - x_{vw} + x_{wv}$ units of flow can be "pushed" from v to w, without violating any nonnegativity or capacity restrictions. (We call $\tilde{u}_{vw}$ the *residual capacity* of vw.) Namely, we can increase x_{vw} to u_{vw} and we can decrease x_{wv} to 0. The catch, of course, is that this operation does not preserve conservation of flow. However, it can be refined to preserve a slightly weaker property. We call x a *preflow* if it satisfies $f_x(v) \geq 0$ for all $v \in V \backslash \{r, s\}$. It is a *feasible* preflow if it also satisfies $0 \le x \le u$. If we choose the pair (v, w) so that $\tilde{u}_{vw} > 0$ and $f_x(v) > 0$, then pushing up to $\varepsilon = \min(\tilde{u}_{vw}, f_x(v))$ from v to w will produce a new feasible preflow. This is what we mean by doing a *push* on vw. (There is some ambiguity if both vw and wv are arcs of G and $\varepsilon < \tilde{u}_{vw}$. We can resolve it by decreasing x_{wv} as much as possible. That is, we decrease x_{wv} by $\varepsilon' = \min(\varepsilon, x_{wv})$, and we increase x_{vw} by $\varepsilon - \varepsilon'$.) Figure 3.11 shows a feasible preflow and the effect of pushing 2 units of flow on ab. There are other possibilities, including pushing 2 units on ba, or 1 unit on bs.

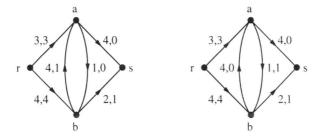

Figure 3.11. A push operation

We call a node $v \in V \backslash \{r, s\}$ *active* if it has positive net flow, that is, if $f_x(v) > 0$. (So a preflow is a flow precisely if there are no active nodes.) The basic step of the *push-relabel* algorithm for the maximum flow problem is to choose an active node v, then choose an arc vw of $G(x)$, and do a push on vw. However, we need to specify more carefully the choice of vw. Otherwise, for example, one could easily have an infinite loop consisting of a push on vw, then a push on wv, then a push on $vw, \ldots$. The additional restriction comes from the idea that we want, as much as possible, to push flow toward the sink, s. However, it is quite possible that we reach a point where no more flow can be pushed toward the sink, but there are still active nodes. In this case the only way to restore conservation of flow is to push the excess back toward the source, r. The device that allows us to make decisions about the direction of pushes is an estimate of distances in $G(x)$. We say that a vector $d \in (\mathbf{Z}_+ \cup \{\infty\})^V$ is a *valid labelling* with respect to a preflow x if

$$d(r) = n, \quad d(s) = 0 \tag{3.14}$$

for every arc vw of $G(x)$, $d(v) \le d(w) + 1$.

Notice that $d(v) = d_x(v, s)$ satisfies all of these conditions except $d(r) = n$. (Recall that $d_x(v, w)$ denotes the number of arcs in a shortest (v, w)-dipath in $G(x)$.) So we can "almost" get a valid labelling for any feasible preflow. In fact, however, it is not true that every feasible preflow admits a valid labelling, as we see below. But it is easy to construct *some* feasible preflow and a valid labelling for it, using the following procedure, which we call *initialize* x,d. Put $x_e = u_e$ for all arcs e having tail r, and put $x_e = 0$ for all other arcs $e \in E$. Put $d(r) = n$ and $d(v) = 0$ for all other nodes $v \in V$. It is easy to check that x and d do satisfy (3.14). (Remark: We must assume that each $u_{rv} \neq \infty$. Exercise 3.47 deals with this small difficulty.) Figure 3.12 shows the initial values of x and d for the problem treated in Section 3.2. The existence of

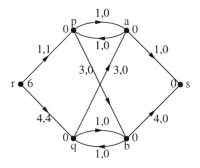

Figure 3.12. Initializing x,d

a valid labelling for a preflow implies an important property of the preflow, that it "saturates a cut."

Proposition 3.22 *If x is a feasible preflow and d is a valid labelling for x, then there exists an (r, s)-cut $\delta(R)$ such that $x_{vw} = u_{vw}$ for all $vw \in \delta(R)$ and $x_{vw} = 0$ for all $vw \in \delta(\overline{R})$.*

Proof: Since there are n nodes, there exists a value k, $0 < k < n$, such that $d(v) \neq k$ for all $v \in V$. Take $R = \{v \in V : d(v) > k\}$. Then $r \in R$ and $s \notin R$. Clearly, (3.14) implies that no arc of $G(x)$ leaves R, which implies the statement of the proposition. ∎

Corollary 3.23 *If a feasible flow x has a valid labelling, then x is a maximum flow.* ∎

The corollary gives a possible termination condition for a maximum flow algorithm. A push-relabel algorithm maintains a feasible preflow and a valid labelling (and thus by Proposition 3.22, a saturated cut) and terminates when the preflow becomes a flow. So there is a certain duality with an augmenting path algorithm, which maintains a feasible flow and terminates when a cut becomes saturated.

Next we show in what sense a valid labelling gives an approximation to distances in $G(x)$.

Lemma 3.24 *For any feasible preflow x, and any valid labelling d for x, we have*

$$d_x(v,w) \geq d(v) - d(w), \text{ for all } v, w \in V.$$

Proof: If $d_x(v,w) = \infty$ this is certainly true, so suppose $d_x(v,w)$ is finite, and consider any shortest (v,w) dipath in $G(x)$. Adding up the inequality $d(p) - d(q) \leq 1$ on the arcs pq of the dipath gives the result. ∎

In particular, it follows from Lemma 3.24 that $d(v)$ is a lower bound on $d_x(v,s)$, and $d(v) - n$ is a lower bound on $d_x(v,r)$. Notice that if $d(v) \geq n$, this means that $d_x(v,s) = \infty$, and excess flow at v should be moved toward the source r. Whether $d(v)$ is large or small, we try to move flow toward nodes w having $d(w) < d(v)$, since such nodes are estimated to be closer to the ultimate destination. Moreover, by the definition of valid labelling, $d(w) < d(v)$ and vw an arc of $G(x)$ implies that $d(w) = d(v) - 1$. Therefore, *push* is applied only to arcs vw of $G(x)$ such that v is active and $d(v) = d(w) + 1$. Such arcs are called *admissible*. Notice that d is still a valid labelling for the new preflow, since the only (possibly) new restriction arises if wv becomes an arc of $G(x)$; this would require $d(w) \leq d(v) + 1$, which is already satisfied.

Now suppose that v is active but there is no arc vw of $G(x)$ with $d(v) = d(w) + 1$. Then we can increase $d(v)$ to $\min(d(w) + 1 : vw \in E(G(x)))$, without violating the validity of the labelling. This is the *relabel* operation. There is a story that goes with the procedure. Imagine that flow really is liquid, that it always runs downhill, and that $d(v)$ represents the height of node v. Then the procedure finds a node that has positive net flow and an arc leading downhill from it, and allows some of the excess to run down along the arc. But if there is positive net flow at the node and no downhill arc from it, then the node is raised until at least one such arc is created.

Although many of the results we are going to give also apply to more general versions of the algorithm, we have chosen to simplify the description by restricting the order in which the operations are performed. Namely, once an active node v is chosen, we continue to perform push operations on admissible arcs vw of $G(x)$ until v either becomes inactive or is relabelled. Notice that this is possible, because if there are no admissible arcs vw and v is still active, then v can be relabelled. To perform this sequence of operations is to *process* v.

Process v

While there exists an admissible arc vw
    Push on vw;
If v is active
    Relabel v.

Now we can state the *push-relabel* maximum flow algorithm quite simply. It is due to Goldberg [1985] and to Goldberg and Tarjan [1988].

Push-Relabel Algorithm

Initialize x,d;
While x is not a flow
    Choose an active node v;
    Process v.

As an example, we apply the push-relabel algorithm to the example of Figure 3.12. So that the order of operations on this example can be followed easily, we use a rule for choosing the next active node v: We choose v to be an active node whose distance label $d(v)$ is maximum. Actually, we shall see that this *maximum distance* version of the algorithm has some very nice properties. Notice that, with this rule, once we choose an active node v, we stay with v until it becomes inactive, since relabelling increases $d(v)$. In addition, for this example, we break ties in the choice of v by alphabetical order. (*This rule cannot be especially recommended in general!*) So the first step is to choose p, to increase $d(p)$ to 1 and then to increase x_{pa} to 1. (x_{pb} could have been increased instead.) After this, the steps of the algorithm happen to be completely determined, and we encourage the reader to work through them. Active nodes are chosen in the sequence $p, a, q, a, q, a, q, a, q, a, q, p, b$ and pushes are on arcs $pa, as, qb, qa, ap, aq, qa, aq, qa, aq, qr, pb, bs$. The final pair (x, d) is shown in Figure 3.13. We have found the same maximum flow as with the augmenting path algorithm. Note also that there is no node v with $d(v) = 5$, so $R = \{r, q, a\}$ determines a minimum cut, as suggested by the proof of Proposition 3.22.

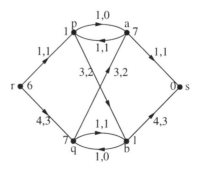

Figure 3.13. Final values of x,d

The push-relabel algorithm does maintain a feasible preflow and a valid labelling. (It is obvious that a push produces a feasible preflow, and that a relabel produces a valid labelling, and we have already checked that a push keeps the labelling valid.) Therefore, by Corollary 3.23, the algorithm terminates with a maximum flow, if it terminates. It remains to analyze its complexity. First, let us state two main results of that analysis.

Theorem 3.25 *The push-relabel algorithm performs* $O(n^2)$ *relabels and* $O(mn^2)$ *pushes.*

Theorem 3.26 *The maximum distance push-relabel algorithm performs* $O(n^2)$ *relabels and* $O(n^3)$ *pushes.*

Most of the proof applies to the general algorithm, and even to more general ones that are not based on processing nodes. Only Lemma 3.31 refers specifically to the maximum distance algorithm.

Lemma 3.27 *If x is a preflow and w is an active node, then there is a (w, r)-dipath in $G(x)$.*

Proof: Let R denote the set of nodes v for which there is a (v, r)-dipath in $G(x)$. Then no arc leaves $\overline{R}$ in $G(x)$, so $x(\delta(R)) = 0$. But suppose that we add the inequalities $f_x(v) \geq 0$ for $v \in \overline{R}$. Then we get $x(\delta(R)) - x(\delta(\overline{R})) \geq 0$. With $x(\delta(R)) = 0$, this implies $x(\delta(\overline{R})) = 0$. So the sum of the inequalities holds with equality, which means that each of them does. That is, there is no active node in $\overline{R}$, so $w \in R$, as required. $\blacksquare$

Lemma 3.28 *At every stage of the push-relabel algorithm, for every $v \in V$, we have $d(v) \leq 2n - 1$. Each node is relabelled at most $2n - 1$ times, and there are $O(n^2)$ relabels in all.*

Proof: Since each relabel of v increases $d(v)$ by at least 1, the second statement follows from the first. Since only active nodes are relabelled, it is enough to prove the first statement for v active. By Lemma 3.27, $d_x(v, r) \leq n - 1$. By Lemma 3.24, $d_x(v, r) \geq d(v) - n$. Combining these two inequalities gives $d(v) \leq 2n - 1$. $\blacksquare$

It is useful to divide the push operations into two kinds. A push on vw is *saturating* if $\tilde{u}_{vw} \leq f_x(v)$, so that the value pushed is $\tilde{u}_{vw}$, and arc vw leaves $G(x)$. Otherwise, the push is *nonsaturating*, and in this case v is no longer active.

Lemma 3.29 *The number of saturating pushes performed by the push-relabel algorithm is at most $2mn$.*

Proof: Consider a fixed pair (v, w) of nodes, such that $vw \in E$ or $wv \in E$. Between two saturating pushes on vw, there must be a push on wv, since otherwise vw is not an arc of $G(x)$. But since $d(v) = d(w) + 1$ for a push on vw, and $d(w) = d(v) + 1$ for a push on wv, and since $d(v)$ never decreases, there must be a relabel of w before there can be a push on wv. Hence between any two saturating pushes on vw, $d(w)$ increases by at least 2, and this can happen, by Lemma 3.28, at most $n-1$ times. So the total number of saturating pushes on vw is at most n. Therefore, the total number of saturating pushes associated with an arc $vw \in E$ (that is, on vw or wv) is at most $2n$, and the total for all arcs is at most $2mn$. ∎

Lemma 3.30 *The number of nonsaturating pushes performed by the push-relabel algorithm is $O(mn^2)$.*

Proof: Let A be the set of active nodes with respect to the preflow x, and let $D = \Sigma(d(v) : v \in A)$. Observe that D is initially 0 and is never negative. Each relabel increases D. A saturating push on vw may increase $d(A)$ by as much as $2n - 1$, since w could enter A and v could remain in A. Each nonsaturating push on vw decreases D, either by $d(v)$ if w was already in A, or by $d(v) - d(w) = 1$ if w enters A. Therefore, all increases in D during the algorithm are due to relabels and saturating pushes, and by Lemma 3.28 and Lemma 3.29 this increase is at most $(n - 2)(2n - 1) + 2mn(2n - 1) = O(mn^2)$. Every nonsaturating push decreases D by at least 1. Since the total decrease in D is at most the total increase, there are $O(mn^2)$ nonsaturating pushes. ∎

Theorem 3.25 now follows from Lemmas 3.28, 3.29, and 3.30. Theorem 3.26 follows from Lemmas 3.28 and 3.29 and the following easy result.

Lemma 3.31 *The maximum distance push-relabel algorithm performs $O(n^3)$ nonsaturating pushes.*

Proof: Any nonsaturating push from a node v makes v inactive, and v cannot become active again before there is a relabel, since all active nodes w have $d(w) \leq d(v)$. Hence, if there are n nonsaturating pushes with no relabel, there is no active node and the algorithm terminates. Therefore, the number of nonsaturating pushes is less than n times the number of relabels, so by Lemma 3.28, it is $O(n^3)$. ∎

We remark that there are other choice rules that give an $O(n^3)$ bound for the number of pushes. However, the maximum distance algorithm has actually been proved by Tunçel [1994] to require only $O(n^2\sqrt{m})$ pushes. Earlier, Cheriyan and Maheshwari [1989] had proved this bound under some additional hypotheses.

Implementation of Push-Relabel Algorithms

We have proved quite good bounds on the numbers of basic steps of the push-relabel algorithms. In order to convert these into statements about running times, we need to give some details of implementation. For each node v we keep a list L_v of the pairs vw such that vw or wv (or both) is an arc of G. We may refer to these as arcs; actually, they are the possible arcs of $G(x)$. With each element vw of L_v, we keep $\tilde{u}_{vw}$. We also keep links between the pairs $vw \in L_v$ and $wv \in L_w$, so that after a push on vw, we can update both of the affected residual capacities in constant time. The order of L_v is fixed. In addition, we keep with each node v the values $d(v)$ and $f_x(v)$. We discuss the storage of the active nodes later.

It is clear that we can do each relabel of node v in time $O(|L_v|)$, and each push in constant time, once we know that these particular actions are justified. How much time must be spent looking for admissible arcs and deciding whether it is time to relabel? The following observation provides the key.

Lemma 3.32 *Suppose that $v \in V$ is active and the arc vw is not admissible. Then before vw becomes admissible, there will be a relabel of v.*

Proof: If vw is not admissible, then $d(v) \leq d(w)$ or $\tilde{u}_{vw} = 0$. If the latter is the case, then for this to change, there must be a push on wv, which requires $d(w) = d(v) + 1$. So in either case, before vw becomes admissible, $d(v) \leq d(w)$ will hold, and only a relabel of v can change it. ∎

Naturally, we process a vertex v by traversing L_v, performing pushes where applicable. The processing of v can be ended by a relabelling of v or by a push that makes v inactive. In the latter case, the next time that v is processed, there is no point in looking again at arcs that were inadmissible when they were considered before (either because they were already inadmissible or were made inadmissible by a push): Lemma 3.32 tells us that they will still be inadmissible. This suggests that when v is chosen to be processed, the traversal of L_v should resume where it left off the last time. (This can easily be implemented by remembering for each v a *current* element of L_v.) Moreover, this idea leads to an even more important observation. If we reach the end of L_v having done no relabel of v, it will be possible to relabel v, because all elements of L_v were inadmissible when the traversal left them. Hence we relabel v when, and only when, a traversal of L_v is completed. This implies not only a simplification of the algorithm, but a valuable fact about its complexity: Since v is relabelled at most $2n - 1$ times, there will be only $O(n)$ traversals of L_v.

In summary, the total time spent looking for admissible arcs is

$$O(\Sigma(n|L_v| : v \in V)) = O(nm).$$

The total time spent on relabels is the same, since each relabel of v takes time $O(|L_v|)$. The total time spent on pushes is $O(N)$, where N is the number of pushes. Finally, we must consider the time spent choosing the next active node to process. The number of times that this operation is performed is $O(N+n^2)$, since each one leads to a push or a relabel. For the general algorithm, it is easy to do each such operation in constant time. Namely, since any active node can be chosen, we just keep the active nodes in a list. Then the total time spent on this operation is $O(n^2 m)$, by Theorem 3.25. Hence we get the following result on the running time of the general algorithm.

Theorem 3.33 *The push-relabel maximum flow algorithm can be implemented to run in time $O(n^2 m)$.* ∎

In the case of the maximum distance version of the algorithm, it is not completely obvious how to maintain the set of active nodes. We want to get a running time that is $O(N)$, which means that, on the average, we need to be able to choose the next active node in constant time. If we just use a list, the time would be $O(n)$ per step. However, there is a better way. Suppose we keep all of the nodes v for which $d(v) = k$ in a list D_k. Given k, we can access D_k in constant time. This list is doubly linked, allowing insertion and deletion of elements in constant time. Moreover, for each v we keep a pointer that allows us to access it in its list (if it is active) in constant time.

Each time a node is relabelled, we remove it from its list and insert it into its new list. When a push on vw makes w active, we insert w into the correct list, and when it makes v inactive, we delete it from its list. So none of these additional operations increases the running time of the algorithm. But why is it easy to find an active node v having $d(v)$ maximum? After v is relabelled, it still has maximum distance, so this case is easy. Suppose that a push makes v inactive. Then if $d(v) = k$, we first look to see whether D_k is still nonempty. If so, then we have found our next node to process in constant time. If not, then we look at D_{k-1}. It will almost always contain a node, because the last push from v was on an arc vw satisfying $d(w) = k - 1$. Such a node w must be active, *unless $w = r$!* So we can find the next node to process in constant time, except for one very special case in which we have just finished deleting a node from D_{n+1} and then we find that both D_{n+1} and D_n are empty. But it is not hard to see (Exercise 3.50) that this can happen at most n times, so even if we have to look at all the values of k less than n in this case, it will contribute only $O(n^2)$ to the total time. (There are other ways to get around this difficulty.) Hence we get the desired running time. Notice also that the $O(n^2 \sqrt{m})$ bound on the number of pushes leads directly to an even better running time.

Theorem 3.34 *The maximum distance push-relabel maximum flow algorithm can be implemented to run in time $O(n^3)$.* ∎

The push-relabel algorithms work extremely well in practice. We want to mention an idea that has no effect on the theoretical running time, but makes a substantial difference to observed computation times. It is to periodically (say, after a sequence of $n/2$ node processing steps) compute a largest valid labelling for the current preflow. This idea is considered in Exercise 3.48.

Exercises

3.45. Solve the problem of Figure 3.12 using the same rules as in the text, but making the first push on pb.

3.46. Find a maximum flow and minimum cut for the problem of Figure 3.4 using the push-relabel algorithm.

3.47. Suppose that $u_{rv} = \infty$ for some $rv \in E$. Show how a maximum flow can be constructed from the solution of a maximum flow problem on a smaller digraph.

3.48. Prove that, if x is a feasible preflow and there exists a valid labelling for x, then the labelling defined by $d(v) = \min(d_x(v,s), n + d_x(v,r))$ is valid. In what sense is this labelling "best"? How fast can it be computed?

3.49. Suppose that we want to find a minimum cut, but not necessarily a maximum flow. Once $d(v)$ reaches n, we know something about how it relates to the minimum cut we will eventually find. How can we use this observation to modify the algorithm to find a minimum cut (but not a maximum flow) more efficiently?

3.50. In the implementation of the maximum distance algorithm, show that if we encounter the situation that D_{n+1} has just become empty and D_n is also empty, then the next time this happens, at least one more label will have exceeded n (and hence this can happen at most n times).

3.51. Suggest a way, involving keeping slightly different data structures, to ensure that the maximum distance active node can be identified in constant time in every case.

3.5   MINIMUM CUTS IN UNDIRECTED GRAPHS

Cut problems in undirected graphs are closely related to the minimum cut problems that we have discussed thus far. In several cases, however, the structure of the undirected problems allows us to solve them much more efficiently than we could by a direct application of maximum-flow methods. This is the subject of this section.

3.5.1   Global Minimum Cuts

We begin with a discussion of the "global minimum cut" problem for undirected graphs. We are given an undirected graph $G = (V, E)$ and a capacity function $u \in \mathbf{R}^E$; we suppose that $u_e > 0$ for each $e \in E$. An example is shown in Figure 3.14. We want to find a set $S \subseteq V$ with $\emptyset \neq S \neq V$ such that

$u(\delta(S))$ is minimized. In this section we use *minimum cut* to mean an optimal solution to this problem. Notice that we can assume that G is connected, for otherwise the solution is trivial. The minimum cut problem includes as

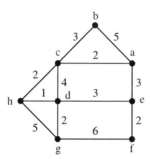

Figure 3.14. A sample input for the minimum cut problem

a special case a fundamental problem of graph theory. Namely, when each $u_e = 1$, the minimum is the "edge-connectivity" of G. Another important application, related to the traveling salesman problem, is discussed in Chapter 7. The problem can be solved by the methods of previous sections, but we present here two specialized methods which, interestingly, seem to have nothing to do with solving flow problems.

Minimum Cut Problem

Given a connected, undirected graph $G = (V, E)$ and $u_e > 0$ for all $e \in E$, find a set $A = \delta(S)$ such that $\emptyset \subset S \subset V$ and $u(A)$ is minimum.

We say that nodes v, w of G are *separated* by a cut $\delta(S)$, or that $\delta(S)$ is a (v, w)-cut, if exactly one of v, w is in S. Of course, the problem of finding a minimum (v, w)-cut is solvable by solving one maximum flow problem. (We can replace each undirected edge by a pair of oppositely directed arcs and give them the same capacity as the edge.) Now if we fix one node r of G, every cut of G is an (r, s)-cut for some node s. Therefore, we can solve the minimum cut problem by solving $n - 1$ maximum flow problems, and taking the best of the cuts found. In particular, it follows that we can solve the minimum cut problem in time $O(n^4)$. (Actually, for a fixed node r, Hao and Orlin [1992] have given a push-relabel algorithm that finds a minimum cut separating r from s for all $s \neq r$, and whose running time is $O(n^3)$. Their method, moreover, appears to work very well in practice — see the study of Chekuri, Goldberg, Karger, Levine, and C. Stein [1997].) It is convenient to introduce the notation $\lambda(G)$ for the capacity of a minimum cut of G, and

$\lambda(G; v, w)$ for the capacity of a minimum (v, w)-cut of G. Of course, both of these values depend also on u.

The two methods that we will present have a common feature. They both rely on the following simple operation. Let v, w be distinct nodes. Then G_{vw} is obtained by *identifying* v with w. Namely, we put $V(G_{vw}) = (V \setminus \{v, w\}) \cup \{x\}$, where x is a new node, and $E(G_{vw}) = E \setminus \gamma(\{v, w\})$; for each edge $e \in E$ and end p of e in G, p is an end of e in G_{vw} if $p \neq v, w$, and otherwise x is an end of e in G_{vw}. The edges of G_{vw} have the same capacities as they had in G. Figure 3.15 shows the effect of identifying nodes f, g in the example of Figure 3.14.

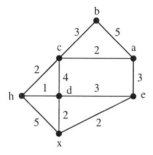

Figure 3.15. Node identification

Note that the operation may create multiple edges. (It does not create loops, even if G already contains multiple edges.) For purposes of finding minimum cuts, of course, we could replace multiple edges by a new edge with capacity equal to the sum of the capacities of the edges replaced. However, for purposes of describing the ideas of the algorithms, it is often more convenient not to do this.

The main observation we will use is a simple one: Every cut of G_{vw} is a cut of G. We can say more.

Proposition 3.35 *Every cut of G_{vw} is a cut of G. Every cut of G that does not separate v from w is a cut of G_{vw}.* ∎

Suppose that we do a sequence of node-identifications beginning with G. Then we get a graph G' each of whose cuts is a cut of G. So a minimum cut of G' is a candidate to be a minimum cut of G, but certain cuts of G have been "lost," that is, they are not cuts of G'. The two algorithms we describe pursue two natural approaches to handling the problem of the lost cuts. Roughly speaking, the first idea is to identify nodes in a way that allows knowing the capacity of the best lost cut, and the second is to choose the nodes to identify so that the best cut is not too likely to be lost. These ideas build on the work of Padberg and Rinaldi [1990], who considered node identification as a practical means of speeding up the $n-1$ maximum flow approach for finding minimum cuts.

Node Identification Algorithm

It follows from the observations above that

$$\lambda(G) = \min(\lambda(G_{vw}), \lambda(G; v, w)).$$

This tells us that we could solve the minimum cut problem by the following process: Choose a pair of nodes v, w, compute a minimum (v, w)-cut in G, and replace G by G_{vw}. After $n-2$ steps, we will have a graph for which the minimum cut problem is trivial. (In fact, this graph has only one cut!) So we will have accumulated $n-1$ cuts in various graphs obtained from G by node-identification, and the best of these cuts is a minimum cut of G. If we do this in a straightforward way, will have to solve $n-2$ minimum (r, s)-cut problems, which is an improvement on the network flow solution we mentioned at the outset, because many of the problems are on smaller graphs. However, it is not a very significant improvement, and it still relies on flows.

A crucial observation is that the choice of v, w is up to us, and we may be able to choose them so that a minimum (v, w)-cut is easy to find. Nagamochi and Ibaraki [1992] gave a fast method to find such a pair. Later, Stoer and Wagner [1994] gave a simpler description and justification of their algorithm. Our presentation follows a description told to us by A. Frank.

A *legal ordering* of G is an ordering $v_1, v_2, \ldots, v_n$ of the nodes of G such that, where V_i denotes $\{v_1, \ldots, v_i\}$,

$$u(\delta(V_{i-1}) \cap \delta(v_i)) \geq u(\delta(V_{i-1}) \cap \delta(v_j)) \text{ for } 2 \leq i < j \leq n.$$

To say this more algorithmically, we choose any node to be v_1 and at step i we choose v_i to be the node that has the largest total capacity of edges joining it to the previously chosen nodes. For example, in Figure 3.14, a legal ordering beginning with a is a, b, c, d, e, h, g, f. It is not too difficult to find a legal ordering in time $O(n^2)$. In fact, the complexity of finding a legal ordering is no more than that of Prim's or Dijkstra's Algorithm. (See Exercise 3.53.) The key result is now the following.

Theorem 3.36 *If $v_1, \ldots, v_n$ is a legal ordering of G, then $\delta(v_n)$ is a minimum (v_n, v_{n-1})-cut of G.*

Notice that, indeed, in the example of Figure 3.14, $\delta(f)$ is a minimum (f, g)-cut. Hence a minimum cut of this graph is $\delta(f)$ or a minimum cut of the graph of Figure 3.15. Theorem 3.36 implies that the following algorithm returns a minimum cut. Moreover, it runs in time $O(n^3)$, or n times the running time of Prim's Algorithm.

Node Identification Minimum Cut Algorithm

Initialize M to be ∞, A to be undefined;
While G has more than one node
    Find a legal ordering $v_1, v_2, \ldots, v_n$ of G;
    If $u(\delta(v_n)) < M$
        Replace M by $u(\delta(v_n))$, A by $\delta(v_n)$;
    Replace G by $G_{v_{n-1}v_n}$;
Return A.

Although we have assumed that G is connected, it is convenient to prove Theorem 3.36 without this assumption. (Note that the definition of "legal ordering" does not require the connectivity of G.) First we need the following simple observation.

Lemma 3.37 *If $p, q, r \in V$, then $\lambda(G; p, q) \geq \min(\lambda(G; r, q), \lambda(G; p, r))$.*

Proof: Consider the minimum (p, q)-cut $\delta(S)$, with $p \in S$. If $r \in S$, then $\delta(S)$ is an (r, q)-cut and so $u(\delta(S)) \geq \lambda(G; q, r)$. Otherwise, $\delta(S)$ is a (p, r)-cut and so $u(\delta(S)) \geq \lambda(G; p, r)$. The result follows. ∎

Proof of Theorem 3.36: Since $\delta(v_n)$ is a (v_n, v_{n-1})-cut, we just need to show that $u(\delta(v_n)) \leq \lambda(G; v_{n-1}, v_n)$. The proof is by induction on the number of edges and nodes. (Notice that the statement is trivial if $|V| = 2$ or $|E| = 0$.) In the proof we use δ' to refer to δ on the graph G'.

First, suppose that $v_n v_{n-1}$ is an edge e of G. Let G' denote G with edge e deleted. It is easy to see that $v_1, \ldots, v_n$ is still a legal ordering of G'. We have

$$u(\delta(v_n)) = u(\delta'(v_n)) + u_e = \lambda(G'; v_{n-1}, v_n) + u_e = \lambda(G; v_{n-1}, v_n).$$

Here, the middle equation follows by the induction hypothesis. So the result holds in this case.

Now suppose that v_{n-1}, v_n are not adjacent in G. In view of Lemma 3.37, it will be enough to show that

$$u(\delta(v_n)) \leq \lambda(G; v_{n-2}, v_n)$$

and that

$$u(\delta(v_n)) \leq \lambda(G; v_{n-2}, v_{n-1}).$$

To do the former, we apply induction to $G' = G \setminus v_{n-1}$. It is easy to see that $v_1, \ldots, v_{n-2}, v_n$ is a legal ordering of G'. Now

$$u(\delta(v_n)) = u(\delta'(v_n)) = \lambda(G'; v_{n-2}, v_n) \leq \lambda(G; v_{n-2}, v_n).$$

Again, the middle equation follows from the induction hypothesis. To prove the second inequality, we apply induction to $G' = G \setminus v_n$. Again, it is easy to see that $v_1, \ldots, v_{n-1}$ is a legal ordering of G'. Now

$$
\begin{aligned}
u(\delta(v_n)) \leq u(\delta(v_{n-1})) &= u(\delta'(v_{n-1})) \\
&= \lambda(G'; v_{n-2}, v_{n-1}) \leq \lambda(G; v_{n-2}, v_{n-1}).
\end{aligned}
$$

∎

Random Contraction Algorithm

The second minimum cut algorithm based on node identification is simpler, in a sense, than the earlier one. First, it identifies a pair of nodes only if they are the ends of some edge, which we refer to as *contracting* the edge. Second, it outputs only the cut from the last (two-node) graph that appears at the end of the sequence of edge-contractions. (That is, it does not keep any information relative to the cuts lost through the contractions.) Finally, it makes the choice of the edge to contract at random, with probability proportional to its capacity. This algorithm is due to Karger [1993].

Random Contraction Algorithm

While G has more than two nodes
  Choose an edge e of G with probability $u_e / u(E)$;
  Where $e = vw$, replace G by G_{vw};
Return A, the unique cut of G.

The idea of an algorithm making random choices is not one that we have encountered before, and at first glance it may seem a bit suspect. Of course, it means that the algorithm can do almost anything, and so it will not be possible to claim that the algorithm will return a minimum cut. What we will show is that it returns a minimum cut with at least a certain probability. This probability can be made higher by running the algorithm repeatedly and keeping track of the best cuts that are returned. (This idea would be silly for "ordinary" algorithms. Running such an algorithm again would give the same output.) Knowing that the probability of success in one run is positive, is not by itself a useful fact. For example, the algorithm that chooses any proper nonempty subset S of V with equal probability and then returns $\delta(S)$ has a positive probability of giving a minimum cut, but this algorithm is not of much use. One way to quantify its uselessness is to observe that we would have to run it exponentially many times to have a probability of even $1/2$ of finding a minimum cut. In fact, the probability that the random contraction algorithm returns a minimum cut is much higher than this naive algorithm, and therefore we will be able to show that it need not be run so many times.

Theorem 3.38 *Let A be a minimum cut of G. Then the random contraction algorithm returns A with probability at least $2/n(n-1)$.*

Proof: The algorithm will return A provided that none of its edges is chosen to be contracted. (Note that this observation is a little trickier than it looks. An edge can be lost because a parallel one is contracted. However, parallel edges are in the same cuts, so if an edge of A is lost, then some edge of A has been contracted.) Suppose that i edges have been chosen, none from A. Denote the current graph by $G' = (V', E')$. Then $|V'| = n - i$. Now A is a minimum cut of G', so its capacity is at most the average of the capacities of all the cuts of the form $u(\delta'(v))$ over all $v \in V'$. (Here, as before, δ' is δ applied to G'.) Thus $u(A) \le 2u(E')/(n - i)$. Therefore, the probability that an edge from A is chosen at step $i + 1$ is

$$\frac{u(A)}{u(E')} \le \frac{2u(E')}{(n-i)u(E')} = \frac{2}{n-i}.$$

Thus the probability that no edge of A is chosen at step $i + 1$ is at least

$$1 - 2/(n - i) = (n - i - 2)/(n - i),$$

and so the probability that no edge of A is *ever* chosen is at least

$$\frac{n-2}{n} \cdot \frac{n-3}{n-1} \cdot \frac{n-4}{n-2} \cdots \frac{3}{5} \cdot \frac{2}{4} \cdot \frac{1}{3} = \frac{2}{n(n-1)}.$$

∎

Since the probability that the algorithm fails to find the minimum cut A after k runs of the algorithm is q^k, where q is the probability that it fails to find A in one run, we can calculate a bound on how many runs are needed to obtain any desired probability of success. To provide a simple bound, we use the inequality from calculus that we saw earlier (Exercise 2.31): $1 - x \le e^{-x}$.

Corollary 3.39 *Let A be a minimum cut of G and let k be a positive integer. The probability that the random contraction algorithm does not return A in one of kn^2 runs is at most e^{-2k}.*

Proof: The desired probability is, by Theorem 3.38, at most

$$\left(1 - \frac{2}{n(n-1)}\right)^{kn^2} \le \left(1 - \frac{2}{n^2}\right)^{kn^2} \le \left(e^{-\frac{2}{n^2}}\right)^{kn^2} = e^{-2k}.$$

∎

So, for example, running the algorithm $10n^2$ times, we get any given minimum cut A with probability at least $1 - e^{-20}$, which is more than .99999999. If we run it $n^2 \log n$ times, the probability of *not* getting A at least one time is at most $1/n^2$.

The random contraction algorithm is very simple and attractive, but some care needs to be taken in order to obtain a practical method. There are two questions to address: how to implement the individual runs, and whether one can decrease the number of runs. Karger and Stein [1993] show that the random contraction algorithm can be made to run in time $O(n^2 \log n)$. They also exploit the fact that the algorithm is much more likely to contract a minimum-cut edge after G becomes small than near the beginning. (For example, the first step chooses a minimum cut edge with probability at most $2/n$, the last with probability at most $2/3$.) This suggests modifying the algorithm so that after some number $f(n)$ of contractions, an exact algorithm (that is, one that is guaranteed to find a minimum cut) is applied to the current G. Then the failure probability is lower and the number of runs required will be lower. A slightly different idea turns out to be even better. We could run the random contraction algorithm itself *twice* on the current graph after $f(n)$ contractions. This idea can be applied again to each of *these* runs, and so on; this leads to a recursive algorithm (an algorithm that uses itself as a subroutine). This algorithm also runs in time $O(n^2 \log n)$, but it finds a given minimum cut with probability at least about $1/\log n$. Therefore, running it $O(\log n)$ times yields a constant probability that the given cut is found. Some of the above ideas are explored in the exercises.

There is one more aspect of the random contraction algorithm that is important in certain applications, namely that *all* the minimum cuts can be generated with high probability. This idea is investigated in the exercises.

3.5.2   Cut-Trees

In this section we discuss the "multiterminal" case, where we have a set K of terminal nodes and we wish to find minimum (r, s)-cuts for each pair of nodes $r, s \in K$. Such problems arise, for example, in the design of communications networks.

Multiterminal Cut Problem

Given an undirected graph $G = (V, E)$, $u_e \geq 0$ for all $e \in E$, and a set of terminals $K \subseteq V$, find a minimum (r, s)-cut for each pair of nodes $r, s \in K$.

If $|K|$ is rather large then it would appear to be a daunting task merely to write down the collection of cuts, let alone find them. A very elegant solution, however, was developed by Gomory and Hu [1961]. Their method involves the solution of only $|K| - 1$ maximum flow problems, on graphs that are no larger than G (and for the most part, much smaller). Moreover, the output of their procedure is a tree structure that makes it a simple matter to pick out the (r, s)-cut corresponding to any pair of terminal nodes. As Ford and Fulkerson [1962] write: "Thus one could hardly ask for anything better."

It is perhaps easiest to describe the Gomory-Hu algorithm with the help of a series of figures, to avoid some unnecessary notation. We begin by choosing some pair of terminal nodes $r, s \in K$ and find a minimum (r, s)-cut that splits the nodes into sets R and S. We represent this cut as an edge of a tree, T, having ends corresponding to R and S. This initial tree is depicted in Figure 3.16. We label the edge of the tree with the capacity of the cut.

Figure 3.16. First step of Gomory-Hu

If both R and S contain exactly one node from K, then we stop. Otherwise, we choose a tree node, say R, that contains two terminal nodes, p and q. We then build the graph G_R from G by identifying all of the nodes in S, and find a minimum (p, q)-cut $\delta(X)$ in G_R. The cut splits R into set sets Q and P. We form a new tree by replacing R with Q and P, joining these two new nodes by an edge labeled with the capacity of the cut $\delta(X)$. We join S to one of the nodes, using the following rule: If the P side of the cut contains the identified node corresponding to S then S is joined to P; otherwise S is joined to Q. The label on the edge is not altered during the operation. The new tree may look like the one given in Figure 3.17.

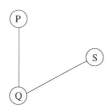

Figure 3.17. Second step of Gomory-Hu

Skipping ahead a few steps to a general stage of the algorithm, we might have the tree given in Figure 3.18. The sets corresponding to the nodes of the tree form a partition of V. Suppose that some set A contains terminal nodes $y, z \in K$. We form the graph G_A from G, by identifying the nodes in the components of the forest obtained by deleting A from the tree, that is, we identify, in turn, the nodes in $B1 \cup B2 \cup B3$, $C1$, $D1 \cup D2$, and $E1 \cup E2 \cup E3$. We then split A into sets Y and Z, by finding a minimum (y, z)-cut in G_A. We form a new tree by replacing the tree node A by nodes Y and Z, joining

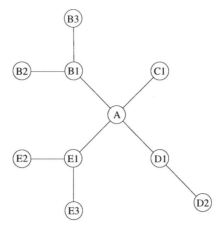

Figure 3.18. General stage of Gomory-Hu

these new nodes by an edge labeled with the capacity of the cut. Each of the former tree neighbors of A, namely $B1, C1, D1$, and $E1$, is joined to either Y or Z, according to which side of the cut contains its corresponding identified node. Again, the labels on the edges in the tree are not altered during the operation. The new tree may look like the one given in Figure 3.19.

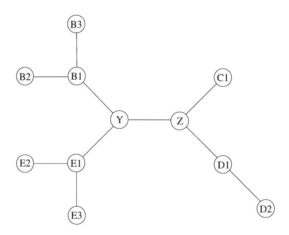

Figure 3.19. General step of Gomory-Hu

The procedure terminates when all of the tree nodes in T correspond to sets containing exactly one terminal node. For each edge e in the tree, let f_e be its assigned label.

Theorem 3.40 *For any pair of terminal nodes $r, s \in K$, the capacity of a minimum (r, s)-cut in G is equal to the minimum label f_e among the edges e in* *the path from r to s in the tree T. Moreover, if edge e* achieves this minimum, then a minimum (r, s)-cut is given by the bipartition of V corresponding to the two-tree forest obtained by deleting e* from T.*

Any capacitated tree possessing the first property in the theorem is said to be *K-flow-equivalent* to G. If the tree also possesses the second property, then it is called a *Gomory-Hu K-cut-tree*. If K happens to be the entire set of nodes V, then we drop the "K" from the notation. An example of a graph G and cut-tree T is given in Figure 3.20. It should be noted that, for a given graph

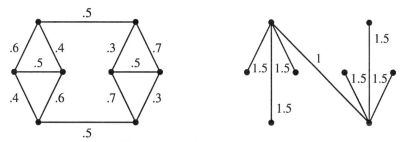

Figure 3.20. Gomory-Hu cut-tree

G, flow-equivalent trees are not necessarily cut-trees. (See Exercise 3.68.)

Our proof of Theorem 3.40 follows the original argument used by Gomory and Hu [1961]. We begin with the following structural result.

Lemma 3.41 *Let $\delta(S)$ be a minimum (r, s)-cut for some nodes $r, s \in V$, and let $v, w \in S$. Then there exists a minimum (v, w)-cut $\delta(T)$ such that $T \subset S$.*

Proof: Let $\delta(X)$ be a minimum (v, w)-cut, and suppose that $X \cap S \neq \emptyset$ and $(V \setminus X) \cap S \neq \emptyset$. We may assume that $s \in X$. There are two cases to consider, depending on whether $r \in X$ or $r \in V \setminus X$. Suppose, first, that $r \in X$. By the "submodular inequality" for cuts (see Exercise 3.66), we have

$$u(\delta(S \cap (V \setminus X))) + u(\delta(S \cup (V \setminus X))) \leq u(\delta(S)) + u(\delta(V \setminus X)). \quad (3.15)$$

Now, since $\delta(S \cup (V \setminus X))$ is an (r, s)-cut, we have

$$u(\delta(S \cup (V \setminus X))) \geq u(\delta(S)). \quad (3.16)$$

Combining (3.15) and (3.16), we have

$$u(\delta(S \cap (V \setminus X))) \leq u(\delta(V \setminus X)) = u(\delta(X)).$$

So $\delta(S \cap (V \setminus X))$ is a minimum (v, w)-cut.

Now suppose that $r \in V \setminus X$. In this case, using a similar argument, we can show that $S \cap X$ is a minimum (v, w)-cut. ∎

For each pair of nodes $v, w \in V$, let $f(v, w)$ denote the capacity of a minimum (v, w)-cut in G. Let T be a tree produced at some stage of the Gomory-Hu procedure, and let e be a tree edge joining the sets R and S. We say that terminal nodes $r \in R$ and $s \in S$ are *representatives* for e if $f_e = f(r, s)$.

Lemma 3.42 *At every stage of the Gomory-Hu procedure, there exist representatives for each edge of the tree T.*

Proof: This is obviously true for the initial, two-node tree. We will show that the property is maintained by a general step of the algorithm.

Suppose we split a tree node A into sets X and Y, based on an (x, y)-cut, for some terminal nodes $x \in X$ and $y \in Y$. By repeated application of Lemma 3.41, we know that the capacity of a minimum (x, y)-cut in G_A is equal to $f(x, y)$. So x and y are representatives for the new tree edge joining the sets X and Y.

What about the remaining edges in the tree? The only tree edges whose sets are altered in the operation are those that were previously joined to A. Let h be such an edge, joining sets A and B, and suppose that h is joined to X in the new tree. (The other case, when h is joined to Y, can be handled in the same manner.) By assumption, there exist terminal nodes $a \in A$ and $b \in B$ such that $f_h = f(a, b)$, since h had representatives in the previous tree. If $a \in X$, then a and b can continue to be the representatives for h. Suppose, on the other hand, $a \in Y$. We show, in this case, that x and b are representatives for h in the new tree.

The cut that was used originally to obtain the edge h separates x and b, since these nodes are on opposite sides of the bipartition we obtain by deleting h from the tree. It follows that

$$f(x, b) \leq f(a, b).$$

We need to show that the reverse inequality holds.

In our argument, we work with a new graph G', obtained from G by identifying all nodes in Y. We let $f'(v, w)$ denote the capacity of a minimum (v, w)-cut in G' for all pairs of nodes v, w in G' (including the node v_Y that corresponds to Y). By Lemma 3.41, we know that

$$f(x, b) = f'(x, b).$$

Furthermore, by a simple property of minimum cuts (see Exercise 3.69), we have
$$f'(x, b) \geq \text{minimum } (f'(x, v_Y), f'(v_Y, b)).$$

Now, since $a \in Y$, we have

$$f'(v_Y, b) \geq f(a, b).$$

Also,

$$f'(x, v_Y) \geq f(x, y) \geq f(a, b),$$

since $y \in Y$ and since the (x, y)-cut that splits A into X and Y also separates a and b. Putting these pieces together, we have

$$f(x, b) \geq f(a, b), \tag{3.17}$$

which is what we wanted to show.   ∎

Proof of Theorem 3.40: Notice that, by the construction of T, the second assertion in the theorem follows from the first assertion. Indeed, each edge e in T corresponds to a cut in G specified by the bipartition of V obtained by deleting e from T, and this cut has capacity f_e.

We will prove the first assertion. Let $V_0, e_1, V_1, \ldots, e_k, V_k$ be the path in T that joins r and s. We need to show that

$$f(r, s) = \text{minimum}(f_{e_1}, \ldots, f_{e_k}).$$

The inequality

$$f(r, s) \leq \text{minimum}(f_{e_1}, \ldots, f_{e_k})$$

is easy, since each of the cuts corresponding to the edges $e_1, \ldots, e_k$ separates r and s. To see the reverse inequality, for each $i = 0, \ldots, k$, let v_i be the unique terminal node in V_i. It follows that $v_0 = r$ and $v_k = s$. By Lemma 3.42, we have

$$f_{e_i} = f(v_{i-1}, v_i)$$

for all $i = 1, \ldots, k$. Thus, by Exercise 3.69, we have

$$f(r, s) \geq \text{minimum}(f_{e_1}, \ldots, f_{e_k}),$$

completing the proof.   ∎

The Gomory-Hu algorithm works well in practice, although it does begin to exhibit its underlying complexity on problem instances with greater than several thousand nodes. Note that the algorithm does not need to do as many shrinking and unshrinking steps as our description suggests, since once a set is shrunk we do not need to unshrink it until it is chosen as the next set to be split. Nonetheless, the overhead in doing the shrink operations partially offsets the benefits we gain by working on graphs that are smaller than the original graph G. This, together with the programming complexity involved in making a good implementation of Gomory-Hu, motivated Gusfield [1990] to develop a version of the algorithm that avoids shrinking altogether. His method is surprisingly easy, although the arguments to establish its correctness are considerably more involved than in the original Gomory-Hu algorithm. One interesting feature of Gusfield's approach is that it can operate without explicit knowledge of the underlying graph G, using only a series of $|V| - 1$ calls to a routine that returns a minimum (r, s)-cut when given as input the nodes r and s.

Exercises

3.52. Complete the solution of the minimum cut problem of Figure 3.14 using the node identification algorithm.

3.53. Explain the connection between the implementation of an algorithm to find a legal ordering, and the implementation of Prim's and Dijkstra's Algorithms.

3.54. If $p, q, r \in V$, show that the smallest two of $\lambda(G; p, q)$, $\lambda(G; p, r)$, and $\lambda(G; q, r)$ are equal.

3.55. According to the bound of Corollary 3.39, how many times must you run the random contraction algorithm on the graph of Figure 3.14 to have a probability of at least .5 of finding the minimum cut? Now try an experiment; run the algorithm until it returns the minimum cut. (Note: This *may* take forever, but if it does, it is probably because you are doing something wrong.)

3.56. A cut of G is *minimal* if there is no cut of G properly contained in it. Prove that the random contraction algorithm returns only minimal cuts.

3.57. Estimate the probability that a given minimum cut is still a cut of the current graph after i contractions.

3.58. Suppose that the random contraction algorithm is modified as follows. The old algorithm is run for i iterations and then an algorithm guaranteed to get an exact minimum cut is applied to the current graph. How many runs of this new version are needed to guarantee that a minimum cut is found with probability at least .5?

3.59. Suppose that the random contraction algorithm is modified as follows. The old algorithm is run for i iterations and then two runs of the old algorithm are applied to the current graph. How many runs of this new version are needed to guarantee that a minimum cut is found with probability at least .5?

3.60. Use Theorem 3.38 to prove that a graph on n nodes has at most $(n(n-1))/2$ minimum cuts.

3.61. Characterize the graphs G and capacity vectors u such that there are exactly $n(n-1)/2$ minimum cuts.

3.62. Prove that if the random contraction algorithm is run $3n^2 \log n$ times, it will produce *all* the minimum cuts with probability at least $1 - \frac{1}{n}$. Hint: First prove that a given minimum cut is generated with probability at least $1 - \frac{1}{n^3}$.

3.63. Let α be a positive integer. A cut A is α-*approximate* if $u(A) \leq \alpha\lambda(G)$. Show that after $n - 2\alpha$ iterations of the random contraction algorithm, any α-approximate cut A is still present with probability $\binom{n}{2\alpha}^{-1}$.

3.64. Consider the following algorithm for computing α-approximate cuts. Do $n - 2\alpha$ iterations of the random contraction algorithm, and then output any cut of the current graph with equal probability. Give a lower bound on the probability that this algorithm outputs a given α-approximate cut.

3.65. Combining the ideas of Exercise 3.62 and Exercise 3.64, show that if the random contraction algorithm is run $(2\alpha + 1)n^{2\alpha} \log n$ times, it will produce *all* α-approximate cuts with probability $1 - \frac{1}{n}$.

3.66. Let $G = (V, E)$ be a graph and $u_e \geq 0$ for all $e \in E$, and let $A \subseteq V$ and $B \subseteq V$. Show that

$$u(\delta(A)) + u(\delta(B)) \geq u(\delta(A \cup B)) + u(\delta(A \cap B)).$$

3.67. We say that two sets $A, B \subseteq V$ *cross* if all four sets $A \cap B, \bar{A} \cap B, A \cap \bar{B}, \bar{A} \cap \bar{B}$ are nonempty. Let $\delta(S)$ be a minimum (r, s)-cut for two nodes $r, s \in V$. Show that for any pair of nodes $v, w \in V$, there exists a minimum (v, w)-cut $\delta(T)$ such that S and T do not cross.

3.68. Find an example of a tree $G = (V, E)$, with nonnegative capacity function $u \in R^E$, and a flow-equivalent tree T for G such that T is not a cut-tree for G.

3.69. Let $G = (V, E)$ be a graph with a nonnegative capacity function $u \in R^E$. For each pair of nodes $r, s \in V$ let $f(r, s)$ denote the capacity of a minimum (r, s)-cut. Show that f satisfies

$$f(v_0, v_k) \geq \text{minimum}(f(v_0, v_1), f(v_1, v_2), \ldots, f(v_{k-1}, v_k)) \qquad (3.18)$$

for any choice of nodes $v_0, v_1, \ldots, v_k$. (Hint: Show the result for $k = 2$, then use induction on k.)

3.70. (R. Gomory and T.C. Hu) Suppose that f is a nonnegative, symmetric function on the pairs of elements of some finite set V. Show that if f satisfies (3.18) for all choices of nodes $v_0, v_1, \ldots, v_k$, then there exists a graph $G = (V, E)$ and a nonnegative capacity function $u \in R^E$ such that for any pair of nodes $u, v \in V$, the capacity of a minimum (u, v)-cut is $f(u, v)$.

3.71. Find a Gomory-Hu cut-tree for the graph given in Figure 3.15.

3.72. (A. Frank) Let $G = (V, E)$ be a graph such that each node $v \in V$ meets at least k edges, for some integer k. Show that there exists a pair of nodes $r, s \in V$ such that a minimum (r, s)-cut has cardinality at least k.

3.6 MULTICOMMODITY FLOWS

We have seen that the problem of finding a maximum flow from a source r to a sink s is highly tractable. Often in practice however, one is interested in connecting several pairs of sources and sinks simultaneously. Consider, for example, a large communication or transportation network, where several messages or goods must be transmitted at the same time over the same network, between different pairs of terminals; or the design of *very large-scale integrated* (VLSI) circuits, where several pairs of pins must be interconnected by wires on a chip, in such a way that the wires follow given "channels" and the wires connecting different pairs of pins do not intersect one another.

These types of problems provide motivation for the following *(fractional) multicommodity flow problem*:

Multicommodity Flow Problem

Input: A directed graph $G = (V, E)$, pairs $(r_1, s_1), \ldots, (r_k, s_k)$ of nodes of G, nonnegative arc capacities $u \in \mathbf{R}^E$, and nonnegative demands $d_1, \ldots, d_k \in \mathbf{R}$.
Objective: For each $i = 1, \ldots, k$, find an (r_i, s_i)-flow x^i of value d_i such that for each arc $e \in E$ we have $\sum(x_e^i : i = 1, \ldots, k) \leq u_e$.

The pairs (r_i, s_i) are called *commodities*. (Throughout the chapter, we will assume that $r_i \neq s_i$ for all $i = 1, \ldots, k$.)

This problem has a natural analogue for undirected graphs G. We replace each undirected edge $e = vw$ by two opposite arcs vw and wv and ask for flows $x^1, \ldots, x^k$ of values $d_1, \ldots, d_k$, respectively, so that for each edge $e = vw$ of G we have $\sum(x_{vw}^i + x_{wv}^i : i = 1, \ldots, k) \leq u_e$.

For both directed and undirected graphs, if we require each x^i to be an integral flow, then the resulting problem is called an *integer multicommodity flow problem*. Unfortunately, this integral problem is known to be as difficult as solving the traveling salesman problem, even in the special case with $k = 2$ commodities and all capacities equal to 1. Nonetheless, there is a rich literature on solving integer multicommodity flow problems, concentrating on important special case results and on approximation algorithms. We refer the interested reader to the collection of papers in Korte, Lovász, Prömel, and Schrijver [1990] for an in-depth treatment of the area.

In this section, we will study an algorithm of Ford and Fulkerson [1958] for solving the fractional multicommodity flow problem.

The fractional problem asks for flows $x^1, \ldots, x^k$ of given values $d_1, \ldots, d_k$ such that the total value of flow through any arc e does not exceed the capacity of e. So it amounts to finding a solution to the following system of linear inequalities in the $k|E|$ variables x_e^i $(i = 1, \ldots, k; \ e \in E)$:

$$f_{x^i}(v) = 0, \text{ for all } i = 1, \ldots, k, \ v \in V \setminus \{r_i, s_i\} \tag{3.19}$$
$$f_{x^i}(s_i) = d_i, \text{ for all } i = 1, \ldots, k$$
$$\sum(x_e^i : i = 1, \ldots, k) \leq u_e, \text{ for all } e \in E$$
$$x_e^i \geq 0, \text{ for all } i = 1, \ldots, k, \ e \in E.$$

(We are using the f_x notation introduced on page 38.) Thus any linear-programming method can solve the fractional multicommodity flow problem. In particular, the problem is solvable in polynomial time.

If k is large, however, the above method may not be practical since (3.19) will have a very large number of variables. To get around this, we will describe an alternative linear formulation that is much larger still!

If there exists a nonnegative (r_i, s_i)-flow of value d_i, then there exists one, y^i, such that $x^i \geq y^i$ and y^i is the nonnegative combination of (r_i, s_i)-path flows. Therefore, for each fixed $i = 1, \ldots k$, we may assume that there exist (r_i, s_i)-dipaths $P_{i1}, \ldots, P_{in_i}$ and nonnegative reals $z_{i1}, \ldots, z_{in_i}$ satisfying

$$\sum(z_{ij}\mathcal{X}_e^{P_{ij}} : j = 1, \ldots, n_i) = x_e^i, \text{ for all } e \in E$$
$$\sum(z_{ij} : j = 1, \ldots n_i) = d_i,$$

where $\mathcal{X}^P$ denotes the characteristic vector of $E(P)$ in $\mathbf{R}^E$, that is, $\mathcal{X}^P(e) = 1$ if $e \in E(P)$ traverses e, and is 0 otherwise.

Hence the fractional multicommodity flow problem amounts to finding dipaths P_{ij} and nonnegative reals z_{ij}, where P_{ij} is an (r_i, s_i)-dipath, such that

$$\sum(z_{ij} : j = 1, \ldots, n_i) = d_i, \text{ for all } i = 1, \ldots, k \qquad (3.20)$$
$$\sum(z_{ij}\mathcal{X}_e^{P_{ij}} : i = 1, \ldots, k, \ j = 1, \ldots, n_i) \leq u_e, \text{ for all } e \in E. \quad (3.21)$$

This formulation applies to both the directed and undirected problems.

Solving (3.20), (3.21) again amounts to solving a system of linear inequalities, albeit with an enormous number of variables.

This does not appear to be a practical alternative to solving (3.19) directly. Indeed, for any moderately large problem instance, we could not even write down the system of inequalities. The trick, however, is that we will *not* write down the system. Instead, we will take advantage of the form of the P_{ij} variables, to work explicitly with only a small subset of them during the algorithm. The approach is known as *column generation*, since we only "generate" the variables (or columns of the matrix) as they are needed. This is an important general technique in combinatorial optimization. In the case of solving (3.20), (3.21), the underlying solution method is the simplex algorithm—see Section 3 of Appendix A, but column generation can be used in other contexts as well, as we will see in Chapters 5 and 7.

First convert the problem to a maximization problem. To this end, we add, for each $i = 1, \ldots, k$, a node r_i' and an arc $r_i'r_i$, with capacity equal to d_i. Then we can delete the constraint (3.20), and maximize $\sum(z_{ij} : i = 1, \ldots, k, j = 1, \ldots, n_i)$ over the remaining constraints (replacing r_i by r_i'). If the maximum value is equal to $\sum(d_i : i = 1, \ldots, k)$ then we have a solution to (3.20), (3.21). Otherwise, we can conclude that (3.20), (3.21) has no nonnegative solution z_{ij}.

Using this reduction, the problem is equivalent to the following linear-programming problem, where $\mathcal{P}$ is the collection of all (r_i, s_i)-dipaths for all

$i = 1, \ldots, k.$

$$\text{Maximize} \quad \sum(x_P \; : \; P \in \mathcal{P}) \tag{3.22}$$
$$\text{subject to}$$
$$\sum(x_P \mathcal{X}_e^P \; : \; P \in \mathcal{P}) \leq u_e, \text{ for all } e \in E$$
$$x_P \geq 0, \text{ for all } P \in \mathcal{P}.$$

We next add a slack variable x_e for each $e \in E$. Thus if A denotes the $E \times \mathcal{P}$-matrix with the characteristic vectors of all dipaths in $\mathcal{P}$ as its columns (in some order) and w is the vector in $R^{\mathcal{P}} \times R^E$ with $w_P = 1$ ($P \in \mathcal{P}$) and $w_e = 0$ ($e \in E$), we will solve the linear-programming problem

$$\text{Maximize } w^T x \tag{3.23}$$
$$\text{subject to}$$
$$[A \; I]x = u$$
$$x \geq 0.$$

Suppose that we have subsets $\mathcal{P}'$ of $\mathcal{P}$ and E' of E such that $|\mathcal{P}'| + |E'| = |E|$. The pair $\mathcal{P}', E'$ is a basis if the matrix B consisting of the columns of $[A \; I]$ corresponding to $\mathcal{P}'$ and E' has full row rank. We let x_B denote the variables associated with $\mathcal{P}', E'$, and let w_B denote the corresponding part of the vector w. The basic solution corresponding to B is defined as the unique solution to the system $Bx_B = u$, setting all other variables to 0.

The form of (3.23) allows us to initialize the simplex algorithm by simply letting $B = I$, that is $\mathcal{P}' = \emptyset$ and $E' = E$, so that $x_P = 0$ for all $P \in \mathcal{P}$ and $x_e = u_e$ for all $e \in E$.

Now the first step of the core algorithm is to check if the current basis is optimal. To do this, we compute the corresponding solution y to the dual linear-programming problem by solving the linear system $y^T B = w_B$. With y in hand, we must check whether or not it is a feasible solution to the dual problem. There are two types of dual constraints, those corresponding to arc variables and those corresponding to dipath variables. The constraints corresponding to arcs are easy to check. For each $e \in E$, the constraint states that y_e should be nonnegative. So we can check these directly. If for some e we have $y_e < 0$, then we choose e as the index to enter the basis.

So suppose that y is nonnegative. We still need to check whether the dipath constraints are satisfied. This is the part that is challenging, since there are an enormous number of these inequalities.

For each $P \in \mathcal{P}$, the dipath constraint is

$$y^T \mathcal{X}^P \geq 1. \tag{3.24}$$

If we give each $e \in E$ the cost y_e, then this inequality means that the dipath P must have cost at least 1. Thus, if all dipaths $P \in \mathcal{P}$ have cost at least 1, then y satisfies all dipath constraints. This is the key to the algorithm. Using the fact that y is nonnegative, we can find a shortest (r_i, s_i)-dipath for each $i = 1, \ldots, k$ with Dijkstra's Algorithm. If each of these k dipaths has cost at least 1, then we can conclude that y is a feasible dual solution and terminate the algorithm. Otherwise, we choose some dipath P of cost less than 1 to be the index to enter the basis.

We can now need to select an index j to leave the basis. If the entering index is an arc e, we solve the system $Bz = 1_e$, where 1_e denotes the eth unit vector in $\mathbf{R}^E$. Otherwise, some dipath P is the entering index and we solve the system $Bx = \mathcal{X}^P$. Once we have z, we can select the leaving index with the standard ratio test. That is, we select an element j in $\mathcal{P}' \cup E'$ for which the quotient x_j / z_j has positive denominator and is as small as possible.

This completes the description of the algorithm. Like many variations of the simplex method, this algorithm is not guaranteed to be finite since it may be possible to construct an example where the algorithm will cycle through a series of bases, returning to the same basis that started the cycle. Although in practice it may not be necessary, we could avoid the cycling problem by applying a "lexicographic rule" for selecting the index to leave the basis. (See Chvátal [1983] for a description of this pivot rule for general linear-programming problems.)

Exercises

3.73. Modify the column generation technique to solve the following problem: Given a directed graph $G = (V, E)$, nonnegative capacities $u \in \mathbf{R}^E$, commodities $(r_1, s_1), \ldots, (r_k, s_k)$, and positive "profits" $p_1, \ldots, p_k \in \mathbf{R}$, find vectors $x^1, \ldots, x^k$ in $\mathbf{R}^E$ and numbers $d_1, \ldots, d_k$ in $\mathbf{R}$ so that:

$$x^i \text{ is an } (r_i, s_i)\text{-flow of value } d_i \text{ for } i = 1, \ldots, k,$$
$$\sum(x_e^i \; : \; i = 1, \ldots, k) \leq u_e \text{ for all } e \in E,$$
$$\sum(p_i d_i \; : \; i = 1, \ldots, k) \text{ is as large as possible.}$$

3.74. Using Farkas' Lemma, show that the fractional multicommodity flow problem has a solution, if and only if for each nonnegative "length" vector $\ell \in \mathbf{R}^E$:

$$\sum_{i=1}^{k} d_i \cdot \text{dist}_\ell(r_i, s_i) \leq \sum_{e \in E} \ell(e). \tag{3.25}$$

(Here $\text{dist}_\ell(r, s)$ denotes the length of a shortest (r, s)-dipath with respect to ℓ.)

3.75. Show that if the fractional multicommodity flow problem has no solution, Ford and Fulkerson's column generation technique yields a length function ℓ violating (3.25).

CHAPTER 4

Minimum-Cost Flow Problems

4.1 MINIMUM-COST FLOW PROBLEMS

A *minimum-cost flow problem* is one of finding a feasible flow x of minimum cost $\sum(c_e x_e : e \in E)$. The definition of feasibility is with respect to any of a number of equivalent models, several of which were mentioned in Section 3.3. We take as our standard minimum-cost flow model the following. (Later in this section we describe some equivalent models.)

$$\text{Minimize } \sum(c_e x_e : e \in E) \tag{4.1}$$
$$\text{subject to}$$
$$f_x(v) = b_v, \text{ for all } v \in V$$
$$0 \leq x_e \leq u_e, \text{ for all } e \in E.$$

The *minimum-cost integral flow* problem is just (4.1) with the additional requirement that x be integer-valued. We shall see that, when b and the finite components of u are integer-valued, this restriction causes no additional difficulty. Here we assume $u \in (\mathbf{R} \cup \{\infty\})^E$ is nonnegative, and that $b(V) = 0$. The numbers b_v are called *demands*. (A negative demand can be thought of as a "supply.") We point out some important special cases. When $c = 0$, every feasible solution to (4.1) is optimal, so the problem becomes one of finding a feasible solution, which is essentially a maximum flow problem. (Another way in which (4.1) generalizes the maximum flow problem is the subject of Exercise 4.7.) If for some $r, s \in V$, $b_r = -1$, $b_s = 1$, and $b_v = 0$ for $v \in V \backslash \{r, s\}$, and in addition each $u_e = \infty$, then (4.1) becomes the linear-programming problem (2.14) that we associated with the shortest path problem. Another

91

class of minimum-cost flow problems arises from *transportation problems*, mentioned at the beginning of Chapter 3. Here G is bipartite with bipartition $\{P, Q\}$ and we are given positive numbers a_p, $p \in P$ and b_q, $q \in Q$, as well as costs c_{pq}, $pq \in E$. The problem is

$$\text{Minimize } \sum(c_{pq}x_{pq} : pq \in E) \tag{4.2}$$

subject to

$$\sum(x_{pq} : p \in P, \ pq \in E) = b_q, \text{ for all } q \in Q$$
$$\sum(x_{pq} : q \in Q, \ pq \in E) = a_p, \text{ for all } p \in P$$
$$x_{pq} \geq 0, \text{ for all } pq \in E.$$

If we allow also constraints of the form $x_{pq} \leq u_{pq}$ with u_{pq} finite, we have a *capacitated transportation problem*. In either case, multiplying each of the second group of equations by -1, and putting $b_v = -a_v$ for $v \in P$, we get a problem of type (4.1). Finally, let us mention the important special case of (4.1) in which each u_e is ∞. Such a problem is called a *transshipment problem*.

We can obtain conditions for optimality of a feasible solution of (4.1) using linear programming. (Later we shall show how these conditions are related to shortest path problems and to maximum flow problems.) The dual of (4.1), obtained by first writing the constraints $x_{vw} \leq u_{vw}$ as $-x_{vw} \geq -u_{vw}$, is

$$\text{Maximize } \sum(b_v y_v : v \in V) - \sum(u_{vw}z_{vw} : \quad vw \in E) \tag{4.3}$$

subject to

$$-y_v + y_w - z_{vw} \leq c_{vw}, \text{ for all } vw \in E$$
$$z_{vw} \geq 0, \text{ for all } vw \in E.$$

Here it is understood that, if $u_{vw} = \infty$, then there is no dual variable z_{vw}, since the corresponding constraint $x_{vw} \leq u_{vw}$ is missing from (4.1). For a vector $(y_v : v \in V)$, it is convenient to denote $c_{vw} + y_v - y_w$ by $\bar{c}_{vw}$, and call it the *reduced cost* of vw. Hence a feasible solution (y, z) of (4.3) satisfies $\bar{c}_e \geq 0$ if $u_e = \infty$. Moreover, if $u_e \neq \infty$, then z_e must satisfy only $z_e \geq -\bar{c}_e$ and $z_e \geq 0$, so the best choice of z_e is just $\max(0, -\bar{c}_e)$. In some sense, then, the variables z_e are unnecessary. We may refer to y alone as a feasible or optimal dual solution, if it extends in this way to a feasible or optimal solution of (4.3). In terms of y the complementary slackness conditions for the pair (4.1), (4.3) of problems can be written:

$$x_e > 0 \text{ implies } -\bar{c}_e = \max(0, -\bar{c}_e), \text{ that is, } x_e > 0 \text{ implies } \bar{c}_e \leq 0;$$

$$z_e > 0 \text{ implies } x_e = u_e, \text{ that is, } -\bar{c}_e > 0 \text{ implies } x_e = u_e.$$

The Complementary Slackness Theorem then gives the desired characterization of optimality.

Theorem 4.1 *A feasible solution x of (4.1) is optimal if and only if there exists a vector $(y_v : v \in V)$ satisfying for each $e \in E$*

$$\bar{c}_e < 0 \text{ implies } x_e = u_e(\neq \infty)$$
$$\bar{c}_e > 0 \text{ implies } x_e = 0.$$

Moreover, every pair (x, y) of optimal primal and dual solutions satisfies these conditions. ∎

Given a path P in G, not necessarily directed, we define the cost (with respect to c) of P to be $\sum(c_e : e \text{ forward in } P) - \sum(c_e : e \text{ reverse in } P)$. The reason for this definition can be explained as follows. Suppose we send ε units of flow on an x-incrementing path P in G. That is, x_e is raised by ε on forward arcs of P and lowered by ε on reverse arcs of P. Then $c^T x$ is increased by $\sum(\varepsilon c_e : vw \text{ forward in } P) - \sum(\varepsilon c_e : e \text{ reverse in } P)$, that is, by ε times the cost of P. In particular, if x is a feasible solution of (4.1), then an x-incrementing circuit of negative cost gives a solution of lower cost.

For a vector $x = (x_e : e \in E)$ satisfying $0 \leq x \leq u$, we define the auxiliary digraph $G(x)$ as we did in the context of maximum flows, except that we include a notion of arc cost. Namely, for each $vw \in E$ such that $x_{vw} < u_{vw}$, we include vw in $E(G(x))$ with cost $c'_{vw} = c_{vw}$. Also, for each $vw \in E$ such that $x_{vw} > 0$, we include wv in $E(G(x))$ with cost $c'_{wv} = -c_{vw}$. Then every x-incrementing path in G corresponds to a dipath in $G(x)$ having the same cost. In particular, a negative-cost x-incrementing circuit of G corresponds to a negative-cost dicircuit of $G(x)$. Figure 4.1 illustrates some of these ideas. On the left is G; numbers beside arc e are c_e, u_e, x_e. On the right is $G(x)$. Notice that $G(x)$ has a dicircuit of cost -2, having node-sequence p, b, q, p. The corresponding circuit of G can be used to produce a feasible solution whose cost is lower by 6.

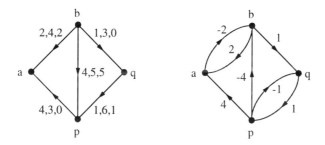

Figure 4.1. (G, c, u, x) and its auxiliary digraph

We can use shortest path methods to determine whether $G(x)$ has a dicircuit of negative cost. This leads to the following characterization of optimality for (4.1). It is the basis for the "primal" minimum-cost flow algorithms—those that maintain a feasible solution to (4.1) and attempt at each step to find a better one.

Theorem 4.2 *A feasible solution x of* (4.1) *is optimal if and only if there is no x-incrementing circuit having negative c-cost.*

Proof: Add to $G(x)$ a node r and, for each $v \in V$, an arc rv with $c'_{rv} = 0$. If we solve the shortest path problem in this new digraph G', we get either a negative-cost dicircuit or a feasible potential. A negative-cost dicircuit cannot use r, and hence corresponds to a negative-cost x-incrementing circuit of G. On the other hand, a feasible potential y satisfies

$$y_v + c'_{vw} \geq y_w \text{ for all } vw \in E(G(x)).$$

But this is equivalent to:

$$y_v + c_{vw} \geq y_w \quad \text{if } x_{vw} < u_{vw}$$
$$y_v - c_{wv} \geq y_w \quad \text{if } x_{wv} > 0$$

and these conditions are equivalent to those of Theorem 4.1. ∎

An immediate consequence of the method of proof, since shortest path costs are integral if arc-costs are integral, is the following.

Theorem 4.3 *Suppose that* (4.1) *has an optimal solution and c is integral. Then y in Theorem 4.1 can be chosen integral; equivalently, the dual linear-programming problem of* (4.1) *has an optimal solution that is integral.* ∎

Using the same technique we can also characterize the instances of (4.1) that have an optimal solution.

Theorem 4.4 *Suppose that* (4.1) *has a feasible solution. Then it has an optimal solution if and only if there exists no negative-cost dicircuit of G, each of whose arcs has infinite capacity.*

Proof: It is easy to see that, if such a dicircuit exists, then (4.1) could not have an optimal solution. Now suppose that no such dicircuit exists. It will be enough to show that the dual linear-programming problem (4.3) has a feasible solution (y, z). Since we can choose z_e to be $\max(0, -\bar{c}_e)$ if $u_e \neq \infty$, it is enough to find y such that $\bar{c}_e \geq 0$ for all vw with $u_e = \infty$. We form a digraph G' by deleting from G all arcs of finite capacity, adding a new node r and an arc rv of cost 0 for each $v \in V$. The resulting digraph has no negative-cost dicircuit and hence has a feasible potential y, which has the desired property. ∎

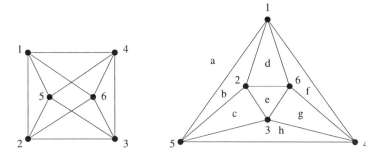

Figure 4.2. G and a plane embedding of G

We have shown how, given a feasible solution x of (4.1), to find if possible a y such that (x, y) satisfies the optimality conditions of Theorem 4.1. Now let us consider the converse problem, that is, given y to find x. We want x to satisfy

$$f_x(v) = b_v, \text{ for all } v \in V$$
$$0 \leq x \leq u$$
$$x_e = 0 \text{ if } \bar{c}_e > 0$$
$$x_e = u_e \text{ if } \bar{c}_e < 0.$$

Let us define u' by $u'_e = 0$ if $\bar{c}_e > 0$ and $u'_e = u_e$ otherwise, and ℓ' by $\ell'_e = u_e$ if $\bar{c}_e < 0$ and $\ell'_e = 0$ otherwise. Then the above conditions are equivalent to the flow feasibility conditions

$$f_x(v) = b_v, \text{ for all } v \in V$$
$$\ell' \leq x \leq u'.$$

Of course, we can find such an x, or determine that none exists, by solving a single maximum flow problem. Moreover, if one exists, and in addition b and u are integral, then x can be chosen integral. This immediately gives a companion to Theorem 4.3.

Theorem 4.5 *If b and u are integral and (4.1) has an optimal solution, then it has an optimal solution that is integral.* ∎

Application to Rectilinear Graph Drawing

For a graph $G = (V, E)$, a *plane embedding* of G is a drawing of G in the plane so that two edges touch only at a node to which they are both incident. A graph is *planar* if it has a plane embedding. (An example of a graph that is not planar is the complete graph on five nodes.) See Figure 4.2 for an example of a planar graph G and a plane embedding of it.

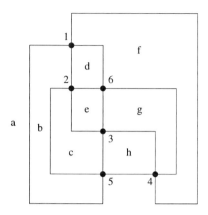

Figure 4.3. A rectilinear embedding

A plane embedding of G divides its complement in the plane into a number of simply connected regions, called *faces*. Exactly one of the faces, called the *external face,* is unbounded. In Figure 4.2, the faces are labeled with letters, and the one labeled a is the external face. With a plane embedding of G and a face f of it, we can associate the nodes and edges of the boundary of f. Under the assumption (which we will make) that G is connected, has at least three nodes, and has no cut node, one can show that these are the edges and nodes of some circuit of G, which we call the *boundary circuit* of f.

A *rectilinear embedding* of G is a plane embedding for which each of the edges is represented by a polygonal chain consisting of vertical and horizontal line segments. In Figure 4.3, we show a rectilinear embedding of the graph in Figure 4.2.

Suppose that we are given a plane embedding of a graph G. A *rectilinear layout* of G is a rectilinear embedding of G having the same boundary circuits as the given embedding, and such that the boundary circuits of the external faces of the two embeddings are the same. So Figure 4.3 shows a rectilinear layout of the plane embedding of Figure 4.2. It is easy to see that if there is any node of degree more than four in G, then G cannot have a rectilinear layout. It is possible to prove that this is the only difficulty, that is, that any plane embedding of a graph having maximum degree at most four has a rectilinear layout. Our problem is not to find such a layout, but to find a "nice" one. The edges of rectilinear embeddings will usually have some bends in them. For example, in Figure 4.3 there are a total of 14 bends. Our problem is to find a rectilinear layout such that the total number of bends is minimized. The solution we will describe is due to Tamassia [1987]. We will slide over some details, which can be found in that paper.

Suppose that we have such an optimal layout. Suppose that we know, for each edge uv of G, the number of left bends and the number of right bends in uv as we move from u to v, and that we also know, for each pair of edges on a face incident with a node v, the angle that they make at v. It is possible to prove that we can construct from this information a layout that is optimal. In other words, determining these numbers is really the problem we want to solve. Here are two conditions that they must satisfy. The first is that the sum of the angles at any node must be 4 right angles. The second is that, for any face f, the number of inside turns minus the number of outside turns on the boundary of f equals 4 unless f is external, when it equals -4. These turns are of two kinds—those that arise from bends in edges, and those that arise from two boundary edges meeting at a node at right angles. (For example, in Figure 4.3 face b has 6 inside and 2 outside turns, while face a has 3 inside and 7 outside turns.)

Now we define some variables to represent the unknown numbers mentioned above, and describe the relations among them with some linear equations. Given an edge e that is in the boundary circuits of faces f and g, let x_{fg} denote the number of inside bends in e from f's point of view. Therefore, x_{gf} is the number of outside bends. (This notation is ambiguous, since there may be more than one edge in the boundary circuits of both f and g.) Given a face f and a node v on the boundary of f, let

$$x_{vf} = \begin{cases} 1 & \text{if the boundary of } f \text{ takes an inside turn at } v \\ -1 & \text{if the boundary of } f \text{ takes an outside turn at } v \\ 0 & \text{otherwise.} \end{cases}$$

(For example, in the layout of Figure 4.3, we have $x_{vf} = 1$ for all vf, $x_{eg} = 0$, $x_{df} = 1$, $x_{fa} = 4$, $x_{fg} = 0, \ldots$.) Then for every face f we have

$$\sum_v x_{vf} - \sum_g x_{gf} + \sum_g x_{fg} = \begin{cases} -4 & \text{if } f \text{ is external} \\ 4 & \text{otherwise.} \end{cases}$$

Also, for each node v, we see (after some thought) that

$$\sum_f x_{vf} = \begin{cases} 0 & \text{if } v \text{ has degree 2} \\ 2 & \text{if } v \text{ has degree 3} \\ 4 & \text{if } v \text{ has degree 4.} \end{cases}$$

We also require, for each fg, that x_{fg} be a nonnegative integer and, for each vf, that x_{vf} be an integer between -1 and 1. Suppose that we multiply the second set of equality constraints by -1. Then it is easy to check that every variable occurs exactly twice, once with a coefficient of 1 and once with a coefficient of -1. Therefore, together with the bound constraints, they define the constraints of a minimum-cost flow problem. We want to minimize the number of bends, which is $\sum x_{fg}$, so the problem reduces to a minimum-cost flow problem. It is easily converted to the form (4.1) by replacing the variables x_{vf} by $x_{vf} + 1$. The fact that we require an integral solution does not pose a difficulty, by Theorem 4.5.

Reductions

There are several forms in which minimum-cost flow problems can appear. As is also the case for the various flow feasibility theorems of Section 3.3, it is fairly easy to move from one form to another. Here we indicate how some more general models can be reduced to our standard model (4.1). We also show that (4.1) can be transformed into a transshipment problem. Further transformations are treated in the exercises.

Let us begin with the general (looking!) problem

$$\text{Minimize } c^T x \tag{4.4}$$

$$\text{subject to}$$

$$a_v \leq f_x(v) \leq b_v, \text{ for all } v \in V$$

$$\ell_e \leq x_e \leq u_e, \text{ for all } e \in E.$$

Here we allow $\ell_e = -\infty$ and/or $u_e = \infty$. We proceed through a sequence of reductions.

We may assume that $a_v = b_v$ for each node v. $\tag{4.5}$

To accomplish this, we add a new vertex r and an arc rv for each node v, assigning rv cost 0, lower bound $\ell_{rv} = 0$ and capacity $u_{rv} = b_v - a_v$. We put $b_r = -\sum(b_v : v \in V)$, and replace each a_v by b_v. Let us verify that this works. (We shall leave other verifications to the reader.) The new problem requires $x_{rv} + f_x(v) = b_v$ for each $v \in V$. Therefore $x_{rv} = b_v - f_x(v)$, and so $a_v \leq f_x(v) \leq b_v$ if and only if $0 \leq x_{rv} \leq b_v - a_v$. Thus, the feasible solutions of the new problem, restricted to the old arcs, are precisely the feasible solutions of the old problem. Moreover, since the new arcs have cost zero, corresponding solutions of the two problems have the same objective function values, so the problems really are equivalent. Notice also that if the new problem has m' arcs and n' nodes, then $m' = O(m)$ and $n' = O(n)$, so an algorithm can solve the old problem by solving the new problem with no essential loss of efficiency.

We may assume that for each $e \in E$, at least one of ℓ_e, u_e is finite. $\tag{4.6}$

In fact, if $\ell_e = -\infty$, $u_e = \infty$, where $e = vw$, we can actually reduce the problem to a smaller one. The idea is essentially that in this case as long as the net flow into $\{v, w\}$ is $b_v + b_w$, the value of x_e can be adjusted to ensure that $f_x(v) = b_v$ and $f_x(w) = b_w$. (In general, this is not true, because the "right" value of x_e might violate a bound constraint $x_e \leq u_e$ or $\ell_e \leq x_e$.) So we can contract $e = vw$ into a single node s. However, we must adjust the costs of some arcs to account for the contribution of $c_e x_e$ to the objective function. This is the subject of Exercise 4.1.

We may assume that, for each $e \in E$, ℓ_e is finite. $\tag{4.7}$

By (4.6), if $\ell_e = -\infty$, then $u_e \neq \infty$. We replace $e = vw$ by an arc $e' = wv$ with $c_{e'} = -c_e$, $\ell_{e'} = -u_e$, $u_{e'} = \infty$.

$$\text{We may assume that, for each } e \in E, \ \ell_e = 0. \tag{4.8}$$

This is proved by a standard trick, replacing ℓ_e by 0, u_e by $u_e - \ell_e$, and adjusting the demands at the ends of v. Namely, if $e = vw$, then b_v becomes $b_v + \ell_e$ and b_w becomes $b_w - \ell_e$.

We have completed the reduction of problem (4.4) to problem (4.1). Now we show that the latter problem can be further reduced to a transshipment problem. We emphasize that this last reduction is quite expensive in practice; in particular, it can triple the number of arcs. However, the simplicity of the transshipment model is so attractive that we occasionally take advantage of this reduction to explain the basic idea underlying an algorithm, or to prove conveniently that an algorithm solves the minimum-cost flow problem in polynomial time. The reduction is simple; see Figure 4.4. Given an arc $e = vw$ with $u_e \neq \infty$, replace e by two nodes p, q and three arcs vp, qp, qw with $c_{vp} = c_e$, $c_{qp} = c_{qw} = 0$, $b_p = u_e$, $b_q = -u_e$; the new arcs have infinite capacity. Notice that the new problem has $O(m + n)$ nodes and $O(m)$ arcs.

Figure 4.4. Getting rid of capacities

Exercises

4.1. In the reduction of (4.6) how should the arc costs be defined?

4.2. Consider a minimum-cost flow problem (4.1) in which each u_e is finite. Write each constraint $x_e \leq u_e$ as $x_e + z_e = u_e$, where z_e is a nonnegative slack variable. Now for each $v \in V$ subtract from the constraint $f_x(v) = b_v$, all of the equations $x_{wv} + z_{wv} = u_{wv}$. The resulting linear-programming problem is equivalent to (4.1). Is it also a minimum-cost flow problem? In what way is it special?

4.3. Formulate the problem of finding a minimum-cost perfect matching in a bipartite graph as a minimum-cost flow problem.

4.4. A *directed Euler tour* of a digraph G is a closed dipath that uses every node at least once and every arc exactly once. Show that G has a directed Euler tour if and only if G is connected and every node v has $|\delta(v)| = |\delta(\bar{v})|$ ("indegree" = "outdegree").

4.5. Suppose that we are given a connected digraph G in which $|\delta(v)| + |\delta(\bar{v})|$ is even for each node v, and for each $vw \in E$ a cost c_{vw} of "reversing vw," that is, replacing vw by wv. We want to choose an optimal subset of arcs to reverse, so that the resulting digraph has a directed Euler tour. Formulate this problem as a minimum-cost flow problem.

4.6. Suppose that we are given a connected digraph G and a cost for every $e \in E$ of "duplicating e," that is, of making a new arc parallel to e. (This can be done more than once, if desired.) We want to find a minimum cost way of duplicating arcs to make the new digraph have a directed Euler tour. Formulate this problem as a minimum-cost flow problem.

4.7. Suppose that we are given an instance of the maximum flow problem, with source r and sink s. We add an arc sr and assign $u_{sr} = \infty$, $c_{sr} = -1$, and $c_e = 0$ for all other arcs e. Notice that the resulting minimum-cost circulation problem is equivalent to the original maximum flow problem. What facts about the maximum flow problem can be derived from Theorems 4.2, 4.3, 4.4, and 4.5?

4.8. For the minimum-cost flow problem of Figure 4.5 determine whether the indicated vector x is optimal. (Numbers on arcs are in the order c_e, u_e, x_e. Numbers at nodes are demands.) If it is optimal, find a vector y that certifies its optimality.

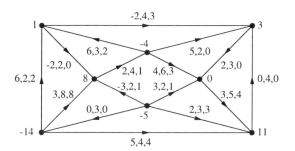

Figure 4.5. Given x, find y

4.9. For the minimum-cost flow problem of Figure 4.6, determine whether there exists a flow x satisfying the conditions of Theorem 4.1 with the indicated vector y. (Numbers on arcs are (c_e, u_e); numbers at nodes are (b_v, y_v).)

4.10. State and prove an analogue of Theorem 4.2, giving a necessary and sufficient condition for the existence of a feasible flow x satisfying the optimality conditions of Theorem 4.1 with a given $y \in \mathbf{R}^V$.

4.11. Suppose we have an instance of (4.1) that we know has an optimal solution. For each arc e with $u_e = \infty$, we want to impose a finite capacity K, so that any optimal solution of the new problem will be optimal for the

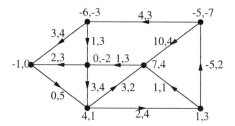

Figure 4.6. Given y, find x

old. Show that $K = \sum(u_e : e \in E, \ u_e \neq \infty) + \sum(b_v : v \in V, \ b_v > 0)$ has this property.

4.12. Show, directly from their definitions, that the sum of the demands in the minimum-cost flow formulation of the rectilinear layout problem is zero.

4.13. Find a layout of the plane graph of Figure 4.2 having 12 bends.

4.14. Write the minimum-cost flow problem for finding an optimal layout of the plane graph of Figure 4.2. Convert it into the form (4.1).

4.15. Find the feasible solution of the problem of Exercise 4.14 corresponding to the layout of Exercise 4.13. Show that it is optimal by finding an appropriate vector y, as in Theorem 4.1.

4.16. Suppose we allow $a_v = -\infty$ and/or $b_v = \infty$ in the general model (4.4). How can this be reduced to (4.1)?

4.17. Convert the minimum-cost flow problems of Exercises 4.8 and 4.9 into transshipment problems.

4.2   PRIMAL MINIMUM-COST FLOW ALGORITHMS

By Theorem 4.2, a feasible solution x of the minimum-cost flow problem is optimal if and only if there is no x-incrementing circuit of negative cost. (We may use "augmenting circuit" to describe such a circuit.) Moreover, if x is not optimal, an augmenting circuit provides an easy way to obtain a better feasible solution. This idea forms the basis for the *primal* minimum-cost flow algorithms. The name comes from the fact that at every step such an algorithm has a feasible solution to the "primal" linear-programming problem, that is, the minimum-cost flow problem. The basic algorithm of this type was apparently first proposed by Kantorovich [1942], although certain versions of it are older.

Augmenting Circuit Algorithm for the Min Cost Flow Problem

     Find a feasible solution x;
    While there exists an augmenting circuit
        Find an augmenting circuit C;
        If C has no reverse arc and no forward arc of finite
           capacity, then stop;
        Augment x on C.

In the most straightforward implementation of this approach (proposed by Kantorovich), we test for the existence of an augmenting circuit by trying to find a negative-cost dicircuit in $G(x)$. This requires the use of an algorithm, such as the Ford-Bellman Algorithm, having running time O(nm).

To get a feeling for the number of flow changes that can occur in the augmenting circuit algorithm and its refinements, it is helpful to consider the maximum flow problem as a special kind of minimum-cost circulation problem. (Recall that this was also the subject of Exercise 4.7.) Namely, add an arc sr with $c_{sr} = -1$ and $u_{sr} = \infty$, give all original arcs a cost of zero, and put all demands equal to zero. This construction, applied to a familiar maximum flow example, results in the instance depicted in Figure 4.7. Any feasible (r, s)-flow

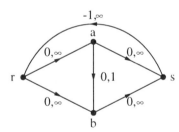

Figure 4.7. A bad example for the Augmenting Circuit Algorithm

x in the original problem will yield a feasible flow x' in the new one, and the value of x will be the negative of the cost of x'. Moreover, each augmenting circuit in the primal algorithm will include sr as a forward arc, and so will correspond to an augmenting path in the maximum flow problem. Since we know that the number of steps in the basic augmenting path algorithm can be unacceptably high, we get the same conclusion for the Augmenting Circuit Algorithm.

Therefore, it is tempting to put some extra effort into finding "good" augmenting circuits. One idea is to find a most-negative augmenting circuit at each step. However, in the above maximum flow example, *every* negative cost augmenting circuit is most negative, since each has cost -1, so these may not provide a good choice. (There is another argument against finding most negative augmenting circuits; see Exercise 4.18.) A better idea is to find an augmenting circuit whose "average arc cost" is small. The *mean cost* of a circuit C of k arcs is its cost divided by k. Notice that in the maximum flow setting, a minimum-mean-cost circuit *is* a good choice—it corresponds to a shortest augmenting path! More generally, it is true that an augmenting circuit algorithm based on minimum mean circuits leads to a polynomial-time algorithm. This result can be found in Goldberg and Tarjan [1989].

From a practical point of view, it seems that many versions of the Augmenting Circuit Algorithm require relatively few iterations on typical problems occurring in applications. But it is also important for a primal algorithm to require very little work per iteration. (Performing a complete $O(nm)$ computation every time is quite expensive.) So the algorithms that we have mentioned so far are seldom used in practice. The one that is currently most used in actual computation is the Network Simplex Method.

The Network Simplex Method

The *Network Simplex Method*, as the name suggests, is an interpretation of the linear-programming simplex method applied to the minimum cost flow problem. The connection can be made through Proposition 2.15 and linear algebra. However, we shall present the method as another specialization of the Augmenting Circuit Algorithm. We assume that the digraph G on which the problem is defined is connected. Otherwise, the problem can be solved by solving its restriction to each of the components of G. Therefore, G has a spanning tree. Spanning trees are so pervasive in the simplex method that we take the liberty of using only "tree" to mean "spanning tree," and of referring to a tree T when we really mean just its set of arcs. It is convenient to present the algorithm first for the special case of the transshipment problem, that is, to assume that $u_e = \infty$ for every arc e. A *tree solution* of the transshipment problem is a vector $x \in \mathbf{R}^E$ satisfying, for some tree T,

$$f_x(v) = b_v, \text{ for all } v \in V$$
$$x_e = 0 \text{ for all } e \notin T.$$

We begin by showing that the tree T uniquely determines its associated tree solution. In the proof, and elsewhere, we use the following fundamental fact about trees, whose proof is left as an exercise.

Proposition 4.6 *Let v, w be nodes of a tree T. Then there is a unique simple path from v to w in T.* ∎

It is convenient to identify a node r and to think of trees as being "rooted at r." The choice of r is arbitrary, but it is fixed. A tree T and an arc $h = pq$ of

T determine a partition of the nodes into two sets, $R(T, h)$ and $V \backslash R(T, h)$, as follows. $R(T, h)$ is the set of those nodes v such that the simple path in T from r to v does not use h. Obviously, $r \in R(T, h)$. This situation is illustrated in Figure 4.8. Obviously, h is the only arc of T having one end in $R(T, h)$ and

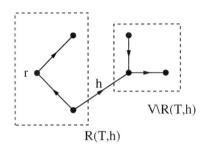

Figure 4.8. Partition induced by a tree and an arc

one end not in $R(T, h)$. Thus in any tree solution x associated with T the net flow into $R(T, f)$ must be entirely carried by h. That is, $x_h = b(R(T, h))$, if $q \in R(T, h)$, and $x_h = -b(R(T, h))$, otherwise. This observation will be useful. Its first consequence is the following.

Proposition 4.7 *A tree T uniquely determines its tree solution.*   ∎

We remark that the converse of Proposition 4.7, that a tree solution uniquely determines its tree, is far from being true. To take an extreme example, if all of the demands are zero, then *every* tree determines the same tree solution, namely $x = 0$.

Next we show that the restriction to tree solutions does not affect the existence of feasible or optimal solutions. Notice that, since G is assumed to be connected, a feasible solution is a tree solution if and only if there is no circuit each of whose arcs has positive flow.

Theorem 4.8 *If (G, b) has a feasible solution, then it has a feasible tree solution. If it has an optimal solution, then it has an optimal tree solution.*

Proof: Let x be a feasible solution. If x is not a tree solution, then there is a circuit C, each of whose arcs carries positive flow. We may assume that C has at least one reverse arc, since otherwise we can replace C by the circuit defined by the same sequence taken in reverse. Now let $\varepsilon = \min(x_e : e$ a reverse arc of $C)$. We replace x_e by $x_e + \varepsilon$ if e is a forward arc of C, and by $x_e - \varepsilon$ if e is a reverse arc of C. The new x is feasible and has fewer arcs carrying positive flow. Continuing this procedure, we get a feasible tree solution.

Now suppose that x is an optimal solution, and each arc of C carries positive flow. From Theorem 4.2 we can conclude that C has cost zero, since neither it nor its reverse can have negative cost. Then the same construction used in the first part of the proof results in a new optimal solution in which fewer arcs carry positive flow, and the result follows.   ∎

The Network Simplex Method maintains feasible tree solutions and looks for negative-cost circuits of a special kind. For each arc $e = vw \notin T$, there is a unique circuit $C(T, e)$ having the following properties:

(a) Each arc of $C(T, e)$ is an element of $T \cup \{e\}$;

(b) e is a forward arc of $C(T, e)$;

(c) The initial node s of $C(T, e)$ is the first common node of the simple paths in T from v and w to r.

The situation is illustrated in Figure 4.9.

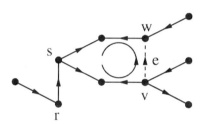

Figure 4.9. Example of $C(T, e)$

To see the advantage of this special kind of circuit, consider the vector $y = y^T \in \mathbf{R}^V$, where y_v is defined to be the cost of the simple path in T from r to v. Notice that, for any two nodes v, w, the cost of the simple path in T from v to w is just $y_w - y_v$. (See Exercise 4.19.) Hence the reduced costs $\bar{c}_{vw}$ defined by $\bar{c}_{vw} = c_{vw} + y_v - y_w$ satisfy

$$\bar{c}_{vw} = 0 \text{ for all } vw \in T;$$
$$\bar{c}_{vw} \text{ is the cost of } C(T, e) \text{ for all } vw \in E \backslash T.$$

It follows immediately from this that, if every $C(T, e)$ has nonnegative cost, then the tree solution x determined by T satisfies the conditions of Theorem 4.1, so we have the following result.

Proposition 4.9 *If the tree T determines the feasible tree solution x and $C(T, e)$ has nonnegative cost for every $e \notin T$, then x is optimal.*   ∎

Testing whether T satisfies this optimality condition is relatively easy. We can compute y in time $O(n)$, and then $\bar{c}$ can be computed in time $O(m)$. This straightforward method is certainly faster than solving a shortest path problem to find a negative-cost augmenting circuit; later we shall see further improvements. However, there is also a catch. We cannot be sure that a circuit of this form is actually an augmenting circuit, because it may have a reverse arc having zero flow. This difficulty is illustrated in Figure 4.10. Here

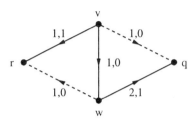

Figure 4.10. No augmenting circuit of type $C(T, e)$

the numbers at the nodes are the demands, the pair c_e, x_e is on arc e, and the tree arcs are the unbroken ones. Then $C(T, wr)$ has positive cost, and $C(T, vq)$ has a reverse arc vw having zero flow. But x is not optimal—sending one unit of flow on the circuit with node-sequence r, v, q, w, r gives a better solution.

The Network Simplex Method handles such cases, roughly speaking, by pretending that they never happen! Although a negative-cost circuit $C(T, e)$ has a reverse arc h having zero flow, it is possible to make use of $C(T, e)$ to find a different tree $\hat{T}$ having the same tree solution x. (In the example of Figure 4.10, we would obtain the tree $\hat{T} = \{vr, wq, vq\}$.) Although this kind of step does not improve the cost of the flow, in fact does not change the flow, it makes different circuits available for further steps. (In the example, $C(\hat{T}, wr)$ has negative cost, and positive flow can be sent on it. In fact, this is the augmenting circuit whose existence we observed before, but it is not of the form $C(T, e)$ for any e.) We can now state a preliminary form of the algorithm we have in mind.

Network Simplex Method for the Transshipment Problem

Find a tree T whose associated flow x is feasible;
Compute y_v, the (r, v) path cost in T, for each node v;
While there exists an arc $e = vw$ such
$\qquad$ that $\bar{c}_e = c_e + y_v - y_w < 0$
$\quad$ Find such an arc e;
$\quad$ If $C(T, e)$ has no reverse arc, then stop;
$\quad$ Compute $\theta = \min(x_j : j$ a reverse arc of $C(T, e))$;
$\quad$ Find a reverse arc h of $C(T, e)$ with $x_h = \theta$;
$\quad$ Augment x by θ on $C(T, e)$;
$\quad$ Replace T by $(T \cup \{e\}) \setminus \{h\}$;
$\quad$ Update y.

One question that arises immediately is how to find the initial tree and flow. The answer is simple. We use the tree T whose arcs are $\{rv : v \in V \setminus \{r\}, \ b_v \geq 0\} \cup \{vr : v \in V \setminus \{r\} \ b_v < 0\}$. For each such arc that does not exist in G, we add it to G, but assign it a large enough cost so that it will not carry positive flow in any optimal solution (unless there is no feasible flow for the original problem). These extra arcs are sometimes called *artificial* arcs. Exercise 4.21 considers the cost of artificial arcs.

There are many implementation issues that we will not discuss (see, for example, Chvátal [1983]), but we do include one of them here because it is important for other reasons. This is the way in which y changes at each step.

Proposition 4.10 *In an iteration of the Network Simplex Method, let T be the old tree, $\hat{T} = (T \cup \{e\}) \setminus \{h\}$ be the new tree, y be the old path costs, and $\hat{y}$ be the new path costs. Then, where $e = vw$,*

$$\hat{y}_q = y_q, \text{ for all } q \in R(T, h)$$
$$\hat{y}_q = y_q + \bar{c}_e, \text{ for all } q \notin R(T, h) \text{ if } v \in R(T, h)$$
$$\hat{y}_q = y_q - \bar{c}_e, \text{ for all } q \notin R(T, h) \text{ if } w \in R(T, h).$$

Proof: If $q \in R(T, h)$ then the (r, q)-path in the new tree is the same as that in the old tree, so the cost is the same. Now suppose that $q \notin R(T, h)$, and that $v \in R(T, h)$. Then, where $e = vw$, the (r, q)-path in $\hat{T}$ consists of the (r, v)-path in T, together with e, together with the (w, q)-path in T. (See Figure 4.11.) Therefore, $\hat{y}_q = y_v + c_e + (y_q - y_w) = y_q + \bar{c}_e$. The proof for the last case is similar, and is left for Exercise 4.23. ∎

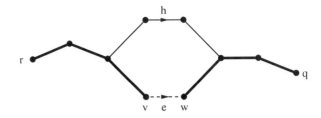

Figure 4.11. The (r, q)-path in $\hat{T}$

Finiteness of the Network Simplex Method

If a tree T determines a flow x such that at least one arc of T carries zero flow, then we say that x is *degenerate*. An iteration of the algorithm that changes the tree but not the flow can occur only if the flow is degenerate. We call such an iteration *degenerate*. The existence of such iterations leads to an important theoretical question. Can the algorithm return to the same tree after a sequence of degenerate iterations? Such behavior, called *cycling*, could cause a failure to terminate. On the other hand, if cycling never occurs, then the algorithm *will* terminate, because the number of different trees is finite.

Cycling *can* occur (although it has never occurred in the solution of a practical problem, so far as we know). See Exercise 4.24 for an example. However, there is a very simple modification of the algorithm, introduced in Cunningham [1976], for which cycling *cannot* happen.

Let $h = pq$ be an arc of T. We say that h is *away from r in T* if $p \in R(T, h)$. Otherwise, h is *toward r in T*. In Figure 4.8 arc h is away from r. A tree T is said to be *strongly feasible* if it determines a feasible flow x such that, for each arc h of T such that $x_h = 0$, h is away from r in T. Notice that if the flow is not degenerate then the tree trivially satisfies this condition. Also notice that the tree we used to initialize the algorithm is strongly feasible. Now suppose that we start the algorithm with a strongly feasible tree, and at each step of the algorithm we choose the arc h according to the following rule.

Leaving arc rule: Choose h to be the *first* reverse arc of $C(T, e)$ with $x_h = 0$.

Proposition 4.11 *If T is strongly feasible and $\hat{T}$ is obtained from T using the leaving arc rule, then $\hat{T}$ is strongly feasible.*

Proof: Let e be the entering arc, h the leaving arc, x the old flow, and $\hat{x}$ the new flow. It is easy to see that an arc g of $\hat{T}$ that is not an arc of $C(T, e)$ is away from r in $\hat{T}$ if and only if it is away from r in T. Moreover, such arcs satisfy $\hat{x}_g = x_g$. So we can restrict attention to arcs of $C(T, e)$.

Suppose first that $\theta > 0$. Then the arcs g of $C(T, e)$ having $\hat{x}_g = 0$, are precisely the reverse arcs of $C(T, e)$ such that $x_g = \theta$. Since h is chosen to be the first of these, it is clear that the others will be away from r in $\hat{T}$.

Now suppose that $\theta = 0$. Let $e = vw$ and let s be the first node of $C(T, e)$. The arcs g of $C(T, e)$ for which $\hat{x}_g = 0$ are those for which $x_g = 0$. (This includes e.) But because T is strongly feasible, those other than e are all forward arcs of the paths in T from s to v and from s to w. The rule chooses h to be the last such arc of the second of these paths (there must be one, since $\theta = 0$), so the others will all be away from r in $\hat{T}$. Moreover, e, being a forward arc of $C(T, e)$, is also away from r in $\hat{T}$. ∎

Theorem 4.12 *The Network Simplex Method, started with a strongly feasible tree and using the leaving arc rule, terminates finitely.*

Proof: It is enough to prove that cycling cannot occur. Suppose that a degenerate iteration takes T to $\hat{T}$, with associated path cost vectors y and $\hat{y}$. Then the leaving arc h is a forward arc of the path in T from the first node s of $C(T, e)$ to the head of the entering arc e (since h would violate strong feasibility of T). It follows that the head w of the entering arc e is not in $R(T, h)$, and so by Proposition 4.10 $\hat{y}_a \le y_a$ for all $a \in V$, while $\hat{y}_w < y_w$. Therefore, $\sum_{a \in V} \hat{y}_a < \sum_{a \in V} y_a$. Hence any sequence of degenerate iterations will generate a strict decrease in the sum of the path costs, so no tree can be repeated. ∎

The Network Simplex Method for the Minimum-Cost Flow Problem

Here we summarize the Network Simplex Method for the solution of the minimum-cost flow problem (4.1). For the most part, the algorithm is a straightforward extension of the version already developed for the transshipment problem. It can also be viewed as an interpretation of the bounded variable simplex method of linear programming.

Again we assume that the underlying digraph G is connected. A *tree solution* for (4.1) is a vector $x \in \mathbf{R}^E$ such that for some tree T and partition (L,U) of $E \backslash T$ we have

$$f_x(v) = b_v, \text{ for all } v \in V$$
$$x_e = 0 \text{ for all } e \in L;$$
$$x_e = u_e \text{ for all } e \in U.$$

So whereas before we required that there be no circuit on which each edge carries positive flow, now we require that there be no circuit on which each edge carries flow strictly between its bounds. Notice that it follows from the definition that u_e is finite for each $e \in U$. (Thus, in particular, if (4.1) is actually a transshipment problem, then $U = \emptyset$, so this definition corresponds with the original one.) One can show again that (T, L, U) uniquely determines x. Also, there is no loss in restricting attention to such tree solutions, as the following result shows. We leave the proofs to the exercises.

Theorem 4.13 *If (G, b, u) has a feasible solution, then it has a feasible tree solution. If it has an optimal solution, then it has an optimal tree solution.* ∎

The Network Simplex Method will move from tree solution to tree solution, again using only circuits formed by adding a single arc to T. However, if the nontree arc e being considered satisfies $x_e = u_e$, then we consider sending flow in the opposite direction, that is, we form a circuit of which e is a reverse arc. This suggests the following definition. Given (T, L, U) and $e \in L \cup U$, $C(T, L, U, e)$ denotes the circuit satisfying

(a) Each arc of $C(T, L, U, e)$ is an element of $T \cup \{e\}$;

(b) e is a forward arc of $C(T, L, U, e)$ if $e \in L$, and otherwise is a reverse arc;

(c) The initial node s of $C(T, L, U, e)$ is the first common node of the simple paths in T from v and w to r.

Now if we define the node numbers y_v, $v \in V$ for T just as before, then the cost of $C(T, L, U, e)$ is $\bar{c}_e$ if $e \in L$ and is $-\bar{c}_e$ if $e \in U$. Now from the optimality conditions for the minimum-cost flow problem we get the following extension of Proposition 4.9.

Proposition 4.14 *If the tree (T, L, U) determines the feasible tree solution x and $C(T, L, U, e)$ has nonnegative cost for every $e \notin T$, then x is optimal.* ∎

It is now easy to state the algorithm.

Network Simplex Method for the Minimum-Cost Flow Problem

Find a triple (T, L, U) whose associated flow x is feasible;
Compute y_v, the (r, v) path cost in T, for each node v;
While there exists an arc $e = vw \in L$ such that $\bar{c}_e < 0$ or
    $e \in U$ with $\bar{c}_e > 0$
Find such an arc e;
If $C(T, L, U, e)$ has no reverse arc and no forward arc of
    finite capacity, then stop;
Compute $\theta_1 = \min(x_j : j$ a reverse arc of $C(T, L, U, e))$;
Compute $\theta_2 = \min(u_j - x_j : j$ a forward arc
    of $C(T, L, U, e))$;
Let $\theta = \min(\theta_1, \theta_2)$;
Find an arc h of $C(T, L, U, e)$ with $x_h = \theta$ and h reverse or
    $u_h - x_h = \theta$ and h forward;
Augment x by θ on $C(T, L, U, e)$;
Replace T by $(T \cup \{e\}) \setminus \{h\}$;
Update L, U by deleting h and adding e as appropriate;
Update y.

Finally, we outline the extension of the method for preventing cycling to this situation. Again the main thing is to extend the definitions appropriately. We call a triple (T, L, U) defining a feasible tree solution x *strongly feasible* if each $e \in T$ with $x_e = 0$ is directed away from r in T *and* each $e \in T$ with $x_e = u_e$ is directed toward r in T. We begin the algorithm with a strongly feasible triple and choose the leaving arc to be the first arc of $C(T, L, U, e)$ that satisfies the requirements in the statement of the algorithm. The finiteness proof given for the transshipment problem now can be carried through in almost the same way.

Exercises

4.18. Prove that the existence of an efficient algorithm to find a most negative augmenting circuit would imply the existence of an efficient algorithm to solve the traveling salesman problem.

4.19. Prove Proposition 4.6. (Hint: First show that if there are two different such paths, then one of them uses an arc that the other one does not.) Then apply Lemma 2.1.

4.20. If we make a different choice of the root node r, then y^T will change. What is the relationship between the two vectors?

4.21. Suppose that each artificial arc is given cost equal to $1 + n(\max\{|c_e| : e \in E\})$. Show that if the Network Simplex Method terminates with some artificial arc carrying positive flow, then the original problem does not have a feasible solution.

4.22. Solve the transshipment problem of Figure 4.12 with the Network Simplex Method, beginning with the tree indicated with solid arcs.

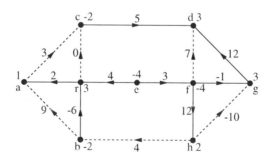

Figure 4.12. Exercise for Network Simplex Method

4.23. Complete the proof of Proposition 4.10.

4.24. Consider the transshipment problem of Figure 4.13. All demands are zero and the arc-costs are as indicated. The problem has no optimal solution. Show that the Network Simplex Method, beginning with the tree

$T = \{g, h\}$, can cycle. (The cycle visits twelve different trees.) The example is from Cunningham [1979].

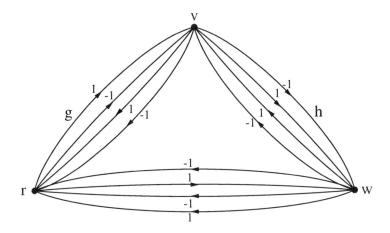

Figure 4.13. Example of cycling

4.25. Given an instance of the transshipment problem on digraph G with demand vector b, we define a new demand vector b' by $b'_v = b_v + \varepsilon$ if $v \neq r$, and $b'_r = b_r - (n-1)\varepsilon$. Prove that, for ε positive and sufficiently small, a tree T of G is feasible for b' if and only if it is strongly feasible for b.

4.26. Prove that a given triple (T, L, U) in the minimum-cost flow problem uniquely determines its tree solution x.

4.27. Prove Theorem 4.13.

4.28. Prove Proposition 4.14.

4.29. Consider an iteration of the Network Simplex Method for the minimum-cost flow problem with $\theta = 0$. Show that, if both the old and the new triple are strongly feasible, then $\sum(y_v : v \in V)$ strictly decreases.

4.3 DUAL MINIMUM-COST FLOW ALGORITHMS

In this section, we describe a different kind of minimum-cost flow algorithm. These *dual* algorithms maintain the optimality conditions of Theorem 4.1, and hence (at least implicitly) a feasible solution to the dual linear-programming problem (4.3), and work toward obtaining a feasible solution to the minimum-cost flow problem (4.1).

The Primal-Dual Algorithm

Suppose that we are given an instance of the minimum-cost flow problem (4.1). We will keep $x \in \mathbf{R}^E$ and $y \in \mathbf{R}^V$ satisfying $0 \le x \le u$ and the optimality conditions of Theorem 4.1, and work toward satisfying the net flow conditions $f_x(v) = b_v$. So the requirements on x, y can be written as

$$x_e = u_e, \text{ for each arc } e \text{ with } \bar{c}_e < 0 \tag{4.9}$$

$$x_e = 0, \text{ for each arc } e \text{ with } \bar{c}_e > 0 \tag{4.10}$$

$$0 \le x_e \le u_e, \text{ for each arc } e. \tag{4.11}$$

If $c \ge 0$, then clearly choosing $x = 0$, $y = 0$ satisfies these conditions. In the general case, we can solve a shortest path problem as in the proof of Theorem 4.4; we get (either that (4.1) has no optimal solution or) a y such that $\bar{c}_e \ge 0$ whenever $u_e = \infty$. Then we can take $x_e = u_e$ if $\bar{c}_e < 0$ and $x_e = 0$ otherwise. So finding initial solutions satisfying (4.9)–(4.11) involves, at most, solving a single shortest path problem. For convenience we assume *throughout this section* that the instances we deal with have already been checked for the existence of an optimal solution. We remind the reader that in general this requires solving one maximum flow problem and one shortest path problem.

Let us call node v an *x-source* if $b_v < f_x(v)$, and an *x-sink* if $b_v > f_x(v)$. We shall follow the example in Figure 4.14, where v, b_v, y_v are indicated with each node v, and c_e, u_e, x_e are indicated with each arc e. Then x, y satisfy

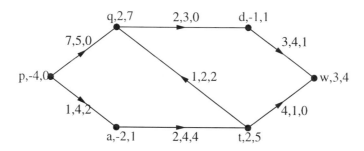

Figure 4.14. An example for the Primal-Dual Algorithm

(4.9),(4.10). Moreover, p is the only x-source and w is the only x-sink. Clearly, we can come closer to satisfying $f_x(v) = b_v$ if we can find an *x-augmenting path*, that is, an x-incrementing path P from an x-source r to an x-sink s. Then we could augment on P by an amount $\varepsilon > 0$. Notice that there is no point in turning an x-source into an x-sink or vice versa. Therefore, ε should be chosen to satisfy $\varepsilon \le b_s - f_x(s)$, $f_x(r) - b_r$; in addition, of course, it should not exceed the width of P. However, a moment's thought shows that such a flow augmentation will keep the conditions (4.9),(4.10) only if every arc e of P satisfies $\bar{c}_e = 0$. (We call such an arc an *equality* arc.) For example, if $\bar{c}_e > 0$, then necessarily $x_e = 0$, so e will be a forward arc of P, and augmentation will increase x_e, violating (4.10). (In Figure 4.14, if we try to augment on the path having node sequence p, q, d, w, we will have this problem with arc qd.) Therefore, we search for an x-incrementing path from an x-source to an x-sink *using only equality arcs.*

What if no such path exists? Then a by-now-familiar argument shows that there exists a set R of nodes, such that R contains every x-source and no x-sink and

$$e \in \delta(R) \text{ implies } \bar{c}_e \neq 0 \text{ or } x_e = u_e$$
$$e \in \delta(\overline{R}) \text{ implies } \bar{c}_e \neq 0 \text{ or } x_e = 0.$$

(Namely, R can be chosen to consist of all the nodes reachable from an x-source by an x-incrementing path of equality arcs. In our example, $R = \{p, q, a\}$.) Combining this observation with the restrictions imposed by (4.9),(4.10), we see that the arcs in $\delta(R) \cup \delta(\overline{R})$ partition into just four sets:

$$
\begin{aligned}
T_1 &= \{e \in \delta(R) : \bar{c}_e \leq 0, x_e = u_e\} \\
T_2 &= \{e \in \delta(R) : \bar{c}_e > 0, x_e = 0\} \\
T_3 &= \{e \in \delta(\overline{R}) : \bar{c}_e \geq 0, x_e = 0\} \\
T_4 &= \{e \in \delta(\overline{R}) : \bar{c}_e < 0, x_e = u_e\}.
\end{aligned}
$$

(In our example, at is the only element of T_1, qd is the only element of T_2, tq is the only element of T_4, and T_3 is empty.) In this situation we can increase y_v, for each $v \in \overline{R}$ by a small amount $\sigma > 0$. Notice that such a "dual change" does not affect $\bar{c}_e$ for arcs $e \in \gamma(R) \cup \gamma(\overline{R})$, it decreases $\bar{c}_e$ by σ for $e \in \delta(R)$, and it increases $\bar{c}_e$ by σ for $e \in \delta(\overline{R})$. Therefore, conditions (4.9),(4.10) will be preserved, provided we choose σ small enough so that $\sigma \leq \bar{c}_e$ for every $e \in T_2$ and $\sigma \leq -\bar{c}_e$ for every $e \in T_4$. Moreover, if we choose σ as large as possible so that these conditions are satisfied, then after the change some arc formerly in $T_2 \cup T_4$ will now satisfy $\bar{c}_e = 0$. That is, after the change R will no longer block the search for an augmenting path of equality arcs. Thus we should choose $\sigma = \min(\sigma_1, \sigma_2)$, where $\sigma_1 = \min(\bar{c}_e : e \in \delta(R), \bar{c}_e > 0)$ and $\sigma_2 = \min(-\bar{c}_e : e \in \delta(\overline{R}), \bar{c}_e < 0)$. (In the example, $\sigma = \min(\bar{c}_{qd}, -\bar{c}_{tq}) = \min(8, 1) = 1$, and tq becomes an equality arc.) One small difficulty must be handled here: What if $\sigma = \infty$, that is, $T_2 = T_4 = \emptyset$? Actually, this is impossible, because (4.1) would have no feasible solution! Namely, since $\overline{R}$ contains at least one x-sink and no x-sources, we have

$$
\begin{aligned}
b(\overline{R}) &> \sum (f_x(v) : v \in \overline{R}) \\
&= x(\delta(R)) - x(\delta(\overline{R})) \\
&= u(\delta(R)).
\end{aligned}
$$

So a necessary (and, by Theorem 3.15, sufficient) condition for the existence of a feasible solution is violated. In summary, we can state the algorithm as follows.

Primal-Dual Algorithm for the Minimum-Cost Flow Problem

Find x, y satisfying (4.9)–(4.11).
While x is not a feasible solution
    If there is an x-augmenting path of equality arcs
        Find such a path P and augment x on P.
    Else
        Find a set R blocking all such paths and change
        y at R.

For completeness, the definition of *augment x* on P is: Given an augmenting path P of equality arcs from an x-source r to an x-sink s, compute $\varepsilon = \min(\varepsilon_1, \varepsilon_2, f_x(r) - b_r, \ b_s - f_x(s))$, where $\varepsilon_1 = \min(u_e - x_e : e$ forward in $P)$, and $\varepsilon_2 = \min(x_e : e$ reverse in $P)$, and increase x_e by ε on forward arcs of P and decrease x_e by ε on reverse arcs of P. The definition of *change y at R* is: Given a set $R \subseteq V$ containing all x-sources and no x-sinks and such that $\bar{c}_e \neq 0$ or $x_e = u_e$ for all $e \in \delta(R)$ and $\bar{c}_e \neq 0$ or $x_e > 0$ for all $e \in \delta(\overline{R})$, compute $\sigma = \min(\sigma_1, \ \sigma_2)$ where $\sigma_1 = \min(\bar{c}_e : e \in \delta(R), \ \bar{c}_e > 0)$ and $\sigma_2 = \min(-\bar{c}_e : e \in \delta(\overline{R}), \ \bar{c}_e < 0)$, and increase y_v by σ for every $v \in \overline{R}$.

We do not go into detail on the implementation and analysis of the Primal-Dual Algorithm. (Implementation of the above version of the algorithm is the subject of Exercise 4.34. Later we will give a different interpretation that also allows an efficient implementation.) However, it should be clear that one gets an augmentation after at most $n - 1$ successive dual variable changes, and that the number of augmentations is bounded by $\sum(b_v - f_{x'}(v) : v \in V$, v an x'-sink) if u and b are integer-valued, and x' is the initial (integral) flow. If there is no assumption about integrality of the data, one can prove only the following. (See Exercises 4.36 and 4.37.)

Theorem 4.15 *If the Primal-Dual Algorithm always chooses, among augmenting paths of equality arcs, one having as few arcs as possible, then it terminates finitely.*   ■

Least-Cost Augmenting Paths

The Primal-Dual Algorithm was introduced by Ford and Fulkerson [1957] (and Kuhn [1955] for a special case). The general approach is important; in Chapter 5 we shall use primal-dual algorithms to solve optimal matching problems. In this subsection we describe a nice interpretation of the algorithm that suggests also a way to implement it using a shortest path subroutine.

Let x, y satisfy the conditions required in the Primal-Dual Algorithm, let r, $s \in V$, and let P be an x-incrementing path from r to s. Then, using conditions (4.9),(4.10), we have

$$
\begin{aligned}
c(P) &= \sum(c_{vw} : vw \text{ forward in } P) - \sum(c_{vw} : vw \text{ reverse in } P) \\
&\geq \sum(y_w - y_v : vw \text{ forward in } P) - \sum(y_w - y_v : vw \text{ reverse in } P) \\
&= y_s - y_r.
\end{aligned}
$$

(This calculation should look familiar.) Moreover, if there exist y and P such that every arc of P is an equality arc, then $c(P) = y_s - y_r$, so P is a least-cost x-incrementing path. This suggests that we should find successive augmentations simply by solving shortest path problems. (Actually, this approach was proposed in Busacker and Gowen [1961], apparently independently of the Primal-Dual Algorithm.) In the example of Figure 4.14, the path having node-sequence p, q, t, w is the unique least-cost augmenting path. The reader can check that this is exactly the path found by the Primal-Dual Algorithm after two dual changes.

Notice that we would have to use an $O(mn)$ shortest path algorithm, even if $c \geq 0$, because the actual shortest paths are calculated on $G(x)$, where certain arcs vw are assigned cost $-c_{wv}$. But there is a way to use Dijkstra's Algorithm, via a trick mentioned in Section 2.2. Namely, the cost of an (r, s)-augmenting path with respect to costs $\bar{c}_e$ differs from that with respect to costs c_e only by a constant. Moreover, if we use $\bar{c}$, then the digraph $G(x)$ will, by (4.9),(4.10), have nonnegative costs. So this works fine, *provided we have* y. That is, we need a way to compute a new y that satisfies (4.9),(4.10) with the new x. Let us look again at the result of the shortest path calculation. For each node v, let σ_v be the least cost (with respect to $\bar{c}$) of an x-incrementing path from an x-source to v; $\sigma_v = \infty$ if none exists. (Recall that to get these values we need a single shortest path computation, after adding to the auxiliary digraph one node and arcs of cost zero from it to each x-source.) If we choose the (r, s)-path P for augmentation, having computed it using $\bar{c}$ as arc-costs, then we have $\sigma_v + \bar{c}_{vw} = \sigma_w$ for each forward arc vw of P. So $c_{vw} + (y_v + \sigma_v) - (y_w + \sigma_w) = 0$; similarly this holds for reverse arcs. That is, with respect to y' defined by $y'_v = y_v + \sigma_v$ for each node v, P is a path of equality arcs. Therefore, x' (the new x) and y' satisfy (4.9),(4.10) for arcs of P, but this also needs to be checked for other arcs. (We encourage the reader to check that, for the example of Figure 4.14, the dual solution y' that is current when an augmentation is first performed satisfies $y' = y + \sigma$.) We also need to handle the nodes v for which $\sigma_v = \infty$.

Primal-Dual Algorithm with Least-Cost Augmenting Paths

Find x, y satisfying (4.9)–(4.11);
While x is not a feasible solution
    For each $v \in V$, find a least-cost (with respect to $\bar{c}$)
        x-incrementing path P_v from an x-source to v;
    Let σ_v be the cost of P_v (∞ if P_v does not exist);
    Choose an x-sink s such that σ_s is minimum;
    Replace y_v by $y_v + \min(\sigma_v, \sigma_s)$ for each $v \in V$;
    Augment x on P_s.

Lemma 4.16 *The least-cost augmenting path version of the Primal-Dual Algorithm maintains x, y satisfying (4.9)–(4.11).*

Proof: Let x, y and x', y' denote successive pairs in the algorithm, and suppose that x, y satisfies the conditions. We want to show that x', y' does also. The condition (4.11) is satisfied, by the definition of augmentation. The conditions (4.9),(4.10) are satisfied for arcs e of P_s, because P_s consists of equality arcs with respect to y', as we have seen. Now consider an arc vw not on P_s, so $x'_{vw} = x_{vw}$. We suppose that $x_{vw} < u_{vw}$ and show that $c_{vw} + y'_v - y'_w \geq 0$. (A similar method shows that, if $x_{vw} > 0$, then $c_{vw} + y'_v - y'_w \leq 0$.) We will use the fact that $\sigma_v + \bar{c}_{vw} \geq \sigma_w$.

Case 1: $\sigma_v \leq \sigma_s$. Then

$$
\begin{aligned}
c_{vw} + y'_v - y'_w &= \bar{c}_{vw} + \sigma_v - \min(\sigma_w, \ \sigma_s) \\
&\geq \bar{c}_{vw} + \sigma_v - \sigma_w \geq 0.
\end{aligned}
$$

Case 2: $\sigma_v \geq \sigma_s$. Then

$$
c_{vw} + y'_v - y'_w = \bar{c}_{vw} + \sigma_s - \min(\sigma_w, \ \sigma_s) \geq \bar{c}_{vw} \geq 0.
$$

∎

As a consequence of results of this subsection we can make the following statement about the efficiency of the Primal-Dual Algorithm. We use B_x to denote $\sum(b_v - f_x(v) : v \in V, \ v$ an x-sink$)$.

Theorem 4.17 *If b and u are integral and x is the initial (integral) flow, then the Primal-Dual Algorithm solves the minimum-cost flow problem in time $O(S(n, m)B_x)$.* ∎

In the important special case of the transshipment problem, we know that we can always take the initial flow x to be 0, if the problem has an optimal solution. In this case we get $B_x = B$, which is defined to be $\sum(b_v : v \in V, b_v > 0)$.

Corollary 4.18 *If b is integer-valued, then the Primal-Dual Algorithm will solve the transshipment problem in time $O(S(n, m)B)$.*   ∎

The next section describes modifications of the Primal-Dual Algorithm that improve its theoretical running time.

Exercises

4.30. Finish the solution of the problem of Figure 4.14 with the Primal-Dual Algorithm.

4.31. The running times for the algorithms of this section should really include a term of $O(n^3)$, because we have assumed that each instance has already been checked for the existence of an optimal solution. Suppose that b is integral and small, for example, that $\sum(b_v : b_v > 0)$ is $O(n)$. Then it seems that the work of the preprocessing step actually increases the running time. Is this really true?

4.32. Solve the minimum-cost flow problems of Exercises 4.8 and 4.9 with the Primal-Dual Algorithm.

4.33. We can find a minimum-cost perfect matching in a bipartite graph by applying the Primal-Dual Algorithm to the formulation of Exercise 4.3. How fast is the resulting method for solving the matching problem?

4.34. Discuss the direct implementation of the Primal-Dual Algorithm, without least-cost augmenting paths. How could each dual change be implemented in time $O(n)$?

4.35. Suppose that P and P' are paths on which augmentations are performed in the Primal-Dual Algorithm, and that P is used before P'. Prove that $c(P) \leq c(P')$.

4.36. Show that the number of different values of y occurring in the execution of the Primal-Dual Algorithm is bounded by the number of different costs of simple paths in G. In particular, show that the number of dual changes is finite.

4.37. Show that the sequence of flow augmentations occurring in the execution of the Primal-Dual Algorithm while y remains the same, corresponds to the sequence of flow augmentations in the solution of a certain maximum flow problem by the augmenting path algorithm. Use this observation and the result of the previous exercise to obtain a proof of Theorem 4.15.

4.38. Suppose that each cost is 0 or 1. Use the first result of Exercise 4.36 to obtain a bound on the running time of the Primal-Dual Algorithm.

4.39. Consider the transshipment instances in Figure 4.15. Only nonzero demands are indicated, and the numbers beside arcs are c_e, u_e. Check that the Primal-Dual Algorithm requires three and eight augmentations, respectively, to solve them. The first three augmentations in the second problem correspond to those of the first problem, and the last three iterations of the second "undo" the first three. Define a problem with ten nodes requiring $18 = 8 + 2 + 8$ augmentations. This idea (from Zadeh [1973]) can be used to construct a problem having $2n + 2$ nodes and requiring $2^n + 2^{n-2} - 2$ augmentations.

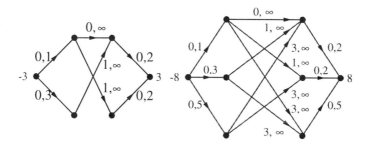

Figure 4.15. Bad examples for the Primal-Dual Algorithm

4.40. Modify the Primal-Dual Algorithm so that it maintains a flow x such that there is no circuit for which every arc has flow strictly between 0 and its capacity (and hence the algorithm terminates with an optimal tree solution). (Hint: This essentially requires modifying Dijkstra's Algorithm for finding a least-cost augmenting path.)

4.4 DUAL SCALING ALGORITHMS

The computation bound for the Primal-Dual Algorithm presented in Theorem 4.17 suffers from an illness that we have seen before, notably in the basic augmenting path algorithm for the maximum flow problem. Namely, the bound depends linearly on the input numbers. Moreover, Exercise 4.39 shows that this bound can really be achieved. In this section we show how to reduce, and eventually eliminate, this dependence. Both of the algorithms we describe use the Primal-Dual Algorithm as a subroutine, and both use a simple technique called *scaling*. We have seen forms of scaling previously in exercises (Exercise 2.31 for shortest paths, Exercise 3.12 for maximum flows).

It is more convenient to present these faster algorithms only for transshipment problems, that is, we assume that each $u_e = \infty$. As was explained in Section 4.1, any minimum-cost flow problem can be transformed into a transshipment problem. The faster algorithms can then be applied to the resulting problem. As we shall see, the computation bounds obtained from this process are quite good. We also assume that the demands b_v are integers, and (as before) that the input instance (G, c, b) has an optimal solution. We make one more technical assumption which makes life easier. We assume that there exists a node $r \in V$ such that $b_r = 0$ and such that $rv \notin E$, $vr \in E$ and $c_{vr} = 0$ for each node $v \neq r$. We can arrange this simply by adding r and the desired arcs to the digraph. Clearly $x_{vr} = 0$ for every feasible solution x, so this change has no real effect.

Scaling the instance (G, c, b) by the factor $\Delta > 0$ means replacing b by b', defined as follows: $b'_v = \lfloor b_v / \Delta \rfloor$ for $v \neq r$, and $b'_r = -b'(V \setminus \{r\})$. The latter definition merely ensures that $b'(V) = 0$. (Although we work with integral demands here, this definition and the lemma below apply to nonintegral demands too. We will need this fact later.) Figure 4.16 shows a transshipment instance and the result of scaling it by a factor of 4; we indicate v, b_v beside node v.

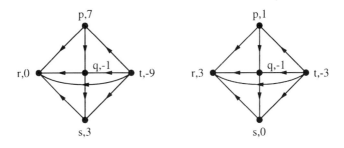

Figure 4.16. An example of scaling

Lemma 4.19 *If (G, c, b) has an optimal solution, then each scaled instance also has an optimal solution.*

Proof: Let (G, c, b') be obtained from (G, c, b) by scaling by Δ. Since the costs have not changed, there is still no negative-cost dicircuit, so the new instance cannot be unbounded, and we need only prove that it has a feasible solution. By Corollary 3.16, we must show that every $A \subseteq V$ with $\delta(\overline{A}) = \emptyset$ satisfies $b'(A) \leq 0$. If $A = V$, this is trivially satisfied. Otherwise, $r \notin A$, so

$$b'(A) = \sum_{v \in A} \left\lfloor \frac{b_v}{\Delta} \right\rfloor \leq \sum_{v \in A} \frac{b_v}{\Delta} = \frac{b(A)}{\Delta}.$$

But $b(A) \leq 0$, since (G, c, b) has a feasible solution, and we are done.   ∎

The Successive-Scaling Algorithm

The *Successive-Scaling Algorithm* of Edmonds and Karp [1972] was the first polynomial-time minimum-cost flow algorithm. For a nonnegative integer k, let (G, c, b^k) denote the instance obtained by scaling (G, c, b) by 2^k. (Notice that, if $b_v > 0$, then b_v^k is obtained by deleting the k lowest-order digits in the binary expression for b_v.) Of course, $b^0 = b$. Suppose that we have optimal primal and dual solutions x', y' for (G, c, b^{k+1}). Then $x = 2x'$, $y = y'$ satisfy conditions (4.9)–(4.11) for (G, c, b^k). Moreover, $b_v^k = 2b_v^{k+1}$ or $2b_v^{k+1} + 1$ for $v \neq r$, and $b_r^k \leq 2b_r^{k+1}$. Hence $\sum(b_v^k - f_x(v) : b_v^k > f_x(v)) \leq n - 1$. Thus if we start the Primal-Dual Algorithm with (x, y), it will solve (G, c, b^k) by solving at most $n - 1$ nonnegative-cost shortest path problems.

Successive-Scaling Algorithm for the Transshipment Problem

Let K be the least integer such that $|b_v| \leq 2^K$ for each node v.
Find $y \in \mathbf{R}^V$ such that $\bar{c}_e \geq 0$ for each arc e.
Set $x = 0$.
For $k = K, K - 1, \ldots, 1, 0$
  Solve (G, c, b^k) with the Primal-Dual Algorithm beginning
  with $(2x, y)$.
  Let (x, y) denote the optimal primal and dual solutions.

Theorem 4.20 *Let (G, c, b) be a transshipment problem instance with $b \in \mathbf{Z}^V$ and having an optimal solution. Then the Successive-Scaling Algorithm solves (G, c, b) in time $O(nS(n, m)(1 + \log B))$, where $B = \sum(b_v : v \in V, b_v > 0)$.*

Proof: We have already explained that the Primal-Dual Algorithm solves (G, c, b^k) for $k < K$ by solving $O(n)$ nonnegative-cost shortest path problems. This is also true for (G, c, b^K), because $\sum(b_v^K : b_v^K > 0) = -\sum(b_v^K : b_v^K < 0) = |\{v : b_v < 0\}|$. Hence the algorithm solves at most $(K + 1)n$ shortest path problems in total. Since $K + 1 \leq 2 + \log(\max |b_v|) \leq 2 + \log B$, the proof is finished. ∎

The Scale-and-Shrink Algorithm

Although the Successive-Scaling Algorithm is a polynomial-time algorithm, we can see a qualitative distinction between its computation bound and that for, say, the shortest augmenting path algorithm for the maximum flow problem. Namely, in the Successive-Scaling Algorithm the number of main steps depends on the size of the input numbers, whereas the running time of the maximum flow algorithm depends on the input numbers, but only in the sense that the amount of work required to do arithmetic operations depends on the sizes of the numbers involved. (The number of augmentations is bounded by nm no matter what are the capacities.) Such algorithms are sometimes called "strongly polynomial." In their 1972 paper, Edmonds and Karp asked whether there exists a strongly polynomial algorithm for the minimum-cost flow problem. Such an algorithm was finally discovered by Tardos [1985]. Her work spawned a number of new algorithms for the problem. The one we describe here is most closely related to algorithms of Orlin [1985] and Fujishige [1986].

We remark that the existence of strongly polynomial algorithms is in doubt only because we allow the input numbers to grow quite fast relative to the size of the input digraph. It may seem reasonable to assume, for example, that the demands have at most $\log n$ digits, or at most n digits. If we do so, then we already have a strongly polynomial algorithm. The number of steps of the Primal-Dual Algorithm is bounded by a polynomial in n whenever the input is a transshipment problem whose demands are integers bounded by a polynomial in n. The number of steps of the Successive-Scaling Algorithm is similarly bounded whenever the number of digits in each demand is bounded by a polynomial in n.

We make the usual assumptions about the input: no finite capacities, existence of an optimal solution, existence of the special node r, *except* that we do not assume that the demands are integral. The algorithm, like the Successive-Scaling Algorithm, is based on solving a sequence of scaled instances. However, the resulting information is used in a different way. The solution to each scaled instance allows us to reduce the number of nodes by one. Moreover, we will be able to solve each scaled instance fast, not because we start with a flow that is close to optimal, but because the positive demands are integers bounded by n^3.

Suppose that we want to solve the transshipment instance $I = (G, c, b)$, and we know that a certain arc f carries positive flow in some optimal solution. (We worry later about how f might be identified.) By Theorem 4.1, it follows that every optimal dual solution y satisfies $\bar{c}_f = 0$. That is, where $f = pq$, the dual constraint $y_q - y_p \leq c_f$ is satisfied with equality for every optimal dual solution. We call f *tight* for I. Then we can replace the dual linear-programming problem

$$\text{Maximize } \sum_{v \in V} b_v y_v$$
$$\text{subject to}$$
$$y_w - y_v \leq c_{vw}, \text{ for all arcs } vw$$

by substituting $y_p + c_f$ for y_q. This changes the objective by replacing $b_p y_p + b_q y_q$ by $(b_p + b_q) y_p$. It changes each constraint $y_q - y_v \leq c_{vq}$ to $y_p - y_v \leq c_{vq} - c_f$, and it changes each constraint $y_v - y_q \leq c_{qv}$ to $y_v - y_p \leq c_{qv} - c_f$. That is, the new problem is the dual of the problem obtained from I by contracting f, as defined on page 98. We denote this new instance by I/f. To get a new optimal dual solution for I from an optimal dual solution y' for I/f, we take $y_v = y'_v$ for $v \neq q$, and $y_q = y'_p + c_f$. Notice the importance of duality in all of this. We cannot conclude that every optimal flow for I/f determines an optimal flow for I. For that to be true we would need to know that f carries positive flow in *every* (not some) optimal flow of I. But duality allows us to conclude that every optimal dual solution for I/f determines an optimal dual solution for I.

Notice that contracting a tight arc does preserve the existence of an optimal solution, since we have seen that there is a correspondence between optimal dual solutions of the two problems. The process of contracting may create loops and parallel arcs. If a loop arises, then it has nonnegative cost, because of the existence of an optimal solution, so the loop may be deleted. Similarly, we may delete the more expensive of any pair of parallel arcs that arises. So we may assume that the digraphs that we work with remain simple. In particular, this implies that each contraction reduces the number of nodes by one, so there will be at most n iterations of a routine to identify tight arcs. Finally, as we saw in Section 4.1, finding an optimal flow is easy once an optimal dual solution is known.

Scale-and-Shrink Algorithm for the Transshipment Problem

Set $I_1 = (G, c, b)$;
Set $k = 1$;
While $b \neq 0$
    Find a tight arc f of I_k and set $I_{k+1} = I_k/f$;
    Replace k by $k + 1$;
Find a feasible potential y for I_k;
While $k \neq 1$
    Extend y to an optimal dual solution of I_{k-1};
    Replace k by $k - 1$;
Find a feasible flow x of (G, c, b) such that
    $\bar{c}_e = 0$ whenever $x_e > 0$.

We remark that, if $b = 0$, then 0 is an optimal flow, so any feasible potential is an optimal dual solution. Also, since any arc vr has zero flow in any feasible solution, such an arc will never be tight, and so we keep the special node r with its required properties throughout the procedure. All that remains is to explain how to identify a tight arc of a transshipment instance (G, c, b) with $b \neq 0$. As the name of the algorithm suggests, this routine uses scaling.

Finding a Tight Arc

Find a feasible tree solution x for (G, c, b);
Scale (G, c, b) by $\Delta = \max(x_e : e \in E)/n(n-1)$
     to get (G, c, b');
Find an optimal tree solution x' for (G, c, b');
Find $f \in E$ such that $x'_f \geq n - 1$.

Lemma 4.21 *The above algorithm finds a tight arc.*

Proof: First we show that there does exist an arc f with $x'_f \geq n - 1$. Let j be an arc such that x_j is maximum. Notice that $x_j \neq 0$, because $b \neq 0$. Since x is a tree solution, there is a set $S \subset V$ such that $b(S) = x_j$. Now $b'(S) \leq x'(\delta(\overline{S}))$, and x' is a tree solution, so there is an arc $f \in \delta(\overline{S})$ such that

$$x'_f \geq \frac{b'(S)}{n-1} \geq \frac{1}{n-1}\left(\frac{n(n-1)b(S)}{x_j} - |S|\right) = n - \frac{|S|}{n-1} \geq n - 1.$$

It remains to show that f is tight. Consider the instance (G, c, b''), where $b''_v = b_v/\Delta$ for each node v. This instance is equivalent to (G, c, b), since there has been no rounding. Suppose we were to solve (G, c, b'') with the Primal-Dual Algorithm, beginning with (x', y'), where y' is an optimal dual solution for (G, c, b'). The algorithm terminates, by Theorem 4.15. We cannot say much about the *number* of augmentations, but since each node v with $b''_v > f_{x'}(v)$ satisfies $b''_v - f_{x'}(v) < 1$, the total *value* of all augmentations is $< n - 1$. Therefore, since $x'_f \geq n - 1$, the flow on f will remain positive. So it is positive in some optimal solution of (G, c, b''), and hence also in some optimal solution of (G, c, b). ∎

We remark that it is a consequence of the proof that any arc having x'-flow at least $n - 1$ is tight. The lemma guarantees that there will be at least one, but if there are several, they can all be contracted.

Theorem 4.22 *The Scale-and-Shrink Algorithm solves the transshipment problem in time $O(n^2 \log n S(n, m))$.*

Proof: The main work of the algorithm is the time involved in the subroutine for identifying tight arcs, which is executed at most n times. Each time we work on a graph having at most n nodes and at most m arcs. The first step is to find a feasible tree solution of (G, c, b). The first time we do this, for the original instance, it requires solving one maximum flow problem, which can be done in time $O(n^3)$. In each subsequent iteration corresponding to instance I, we already have a feasible tree solution x determined by a tree T for an instance I' such that $I = I'/f$ for some arc f. It is easy to see that x, restricted to the remaining arcs, is a feasible solution for I/f. It is a tree solution if $f \in T$. Otherwise, it is turned into a tree solution with a single pivot. So the total amount of work for this step, over all iterations of the tight arc subroutine, is bounded by $O(n^3)$.

It remains to estimate the time required to solve the transshipment instance (G, c, b'). Let v be any node different from r such that $b'_v > 0$. There exist at most $n - 1$ arcs e having head v and satisfying $x_e > 0$, so one of them has $x_e \geq b_v/(n - 1)$. It follows that $\Delta \geq b_v/n(n - 1)^2$, so $b'_v \leq b_v/\Delta \leq n(n - 1)^2$. Moreover, $b'_r \leq n - 1$. It follows that $\sum(b'_v : b'_v > 0) = O(n^4)$. Therefore the Successive-Scaling Algorithm will solve (G, c, b') in time $O(n \log nS(n, m))$. (The additional time required to turn an optimal solution into an optimal tree solution does not affect the running time. Alternatively there is a way to modify the algorithms of this section so that they automatically produce tree solutions. See Exercise 4.40.) Since this step is executed $O(n)$ times, the proof is finished.   ∎

We remark that the Scale-and-Shrink Algorithm uses scaling in two different ways. It scales to produce an instance whose solution will identify a tight arc, and it uses the successive scaling algorithm to solve that instance. However, it is the first application of scaling that is crucial to the strong polynomiality of the method. See Exercise 4.43.

Finite Capacities Revisited

So far in this section we have dealt only with transshipment problems, that is, we have not considered finite capacities. Suppose that we are given an instance of the more general problem (4.1) to solve. We can apply the transformation at the end of Section 4.1 to convert it to an instance of the transshipment problem and then solve that instance. It is straightforward to evaluate the resulting running time for either the Successive-Scaling Algorithm or the Scale-and-Shrink Algorithm. There is, however, one observation that allows an improvement. It is that the time to solve a shortest path problem on the new graph is the same as for the old one. This fact is closely related to an observation made earlier (Exercise 2.23) concerning the effect on the complexity of shortest path calculations, of nodes adjacent to just two nodes. For an instance (G, c, b, u) of the minimum-cost flow problem, we use U to denote $\sum(u_e : u_e \neq \infty)$.

Theorem 4.23 *Let (G, c, b, u) be an instance of the minimum-cost flow problem with $b \in \mathbf{Z}^V$, $u \in (\mathbf{Z} \cup \{\infty\})^E$ and having an optimal solution. Then the Successive-Scaling Algorithm solves (G, c, b, u) in time $O(nS(n, m)(1 + \log(\max(B, U))))$.*   ∎

Theorem 4.24   *The Scale-and-Shrink Algorithm solves the minimum-cost flow problem in time $O((m_0 + n)n \log nS(n,m))$, where m_0 is the number of arcs having finite capacity.* ∎

We remark that Orlin has discovered an improved version of the Scale-and-Shrink Algorithm, having a running time of $O(m \log nS(n,m))$ for the minimum-cost flow problem; see Orlin [1988].

Exercises

4.41. Show that contracting an arc of a transshipment problem does not, in general, preserve the existence of an optimal solution.

4.42. Solve the transshipment problems of Exercise 4.17 with the Successive-Scaling Algorithm.

4.43. If the Successive-Scaling Algorithm were replaced by the Primal-Dual Algorithm in the statement of the Scale-and-Shrink Algorithm, what would be the running time of the new algorithm?

CHAPTER 5

Optimal Matchings

5.1 MATCHINGS AND ALTERNATING PATHS

We have seen some examples of matching problems in Sections 1.1 and 3.3. In this chapter we develop the theory, algorithms, and further applications of optimal matching.

We begin by introducing and reviewing some terminology. A *matching* in a graph $G = (V, E)$ is a set M of edges such that no node of G is incident with more than one edge in M. Given a matching M, we say that M *covers* a node v (or that v is *M-covered*) if some edge of M is incident with v. Otherwise, v is *M-exposed*. Note that the number of nodes covered by M is exactly $2|M|$, and that the number of M-exposed nodes is $|V| - 2|M|$. A *maximum* matching is one of maximum cardinality, or equivalently, one that has the fewest exposed nodes. We denote by $\nu(G)$ the cardinality of a maximum matching of G, and by $\mathrm{def}(G)$ (the "deficiency of G") the minimum number of exposed nodes for any matching of G. So $\mathrm{def}(G) = |V| - 2\nu(G)$. A *perfect* matching is one that covers all the nodes. See Figure 5.1.

A basic problem is to decide whether a graph has a perfect matching. A slightly more general one is to find a maximum matching. Finally, we might want a perfect matching having minimum weight with respect to some given edge-weights. The last problem includes as a special case the matching problem we introduced in Chapter 1, arising from the minimization of plotter pen motion.

A graph G is *bipartite* if its nodes can be partitioned into two sets V_1, V_2 so that every edge joins a node in V_1 to a node in V_2. We have seen that matching problems in bipartite graphs can be solved by network flow methods. In Section 3.3 we reduced the problem of finding a maximum matching in a

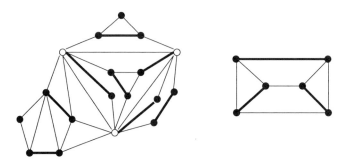

Figure 5.1. A maximum matching and a perfect matching

bipartite graph to an integral maximum flow problem. Similarly, the problem of finding a minimum-weight perfect matching in a bipartite graph can be solved by solving a minimum-cost flow problem. (See Exercises 4.3 and 4.33.) So we already know how to solve matching problems for bipartite graphs. In this chapter we seldom rely on network flow methods, instead treating matching problems directly. We do frequently point out the special case when the graph is bipartite, because it offers an important and rather easy contrast to the general case.

In the remainder of this section we give a few more definitions and basic results. Given a matching M of G, a path P is M-*alternating* if its edges are alternately in and not in M. (Note that we do not specify the type of the first and last edges of P.) If, in addition, the end nodes of P are distinct and are both M-exposed, then P is an M-*augmenting* path. Figure 5.2 shows a matching M and an M-augmenting path. The reader should be able to verify the connection between this use of the term and that in Chapter 3. Namely, if x is the integral flow corresponding to M in the formulation of the maximum bipartite matching problem as a maximum integral flow problem, then x-augmenting paths correspond to M-augmenting paths. This justifies the use of similar terminology. It is a consequence of the Augmenting Path Theorem of maximum flows that a matching M of a bipartite graph is maximum if and only if there is no M-augmenting path. It is a fundamental fact, due to Berge [1957], that this holds for all graphs.

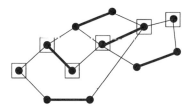

Figure 5.2. Augmenting path

Theorem 5.1 *(Augmenting Path Theorem of Matchings) A matching M in a graph $G = (V, E)$ is maximum if and only if there is no M-augmenting path.*

For sets S and T, we let $S \triangle T$ denote their *symmetric difference*, that is, $S \triangle T$ is the set of elements belonging to one, but not both, of S, T.

Proof of the Augmenting Path Theorem of Matchings: Suppose that there exists an M-augmenting path P joining nodes v and w. Then $N = M \triangle E(P)$ is a matching that covers all nodes covered by M, plus v and w. Therefore, M is not maximum.

Conversely, suppose that M is not maximum and so some other matching N satisfies $|N| > |M|$. Let $J = N \triangle M$. Each node of G is incident with at most two edges of J, so J is the edge-set of some node disjoint paths and circuits of G. For each such path or circuit, the edges alternately belong to M or N. Therefore, all circuits are even, and contain the same number of edges of M and N. Since $|N| > |M|$, there must be at least one path with more edges of N than M. This path is an M-augmenting path. (See Figure 5.3.) ∎

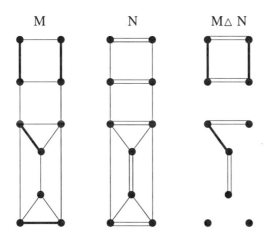

Figure 5.3. Augmenting paths and larger matchings

The Augmenting Path Theorem of Matchings suggests a possible approach to constructing a maximum matching, analogous to the augmenting path algorithm for the maximum flow problem. Namely, we repeatedly find an augmenting path and obtain a new matching using the path, until we discover a matching for which there is no augmenting path. This is the approach that we shall take. It requires having a way of recognizing when a matching is maximum, so that the search for augmenting paths can be abandoned.

Odd Circuits and the Tutte-Berge Formula

Recall that a *cover* of G is a set A of nodes such that every edge has at least one end in A. The cardinality of a cover provides an upper bound on the size of any matching. In Section 3.3 we discussed Kőnig's Theorem, which states that, if G is bipartite, there is a matching and a cover of equal cardinality. Therefore, in that special case, one can prove that a matching is maximum by exhibiting an appropriate cover. In general, however, this is not always possible; for example, when G consists of a circuit of odd length $2k+1$, the maximum size of a matching is k and the minimum size of a cover is $k+1$. Therefore, we need a stronger upper bound on the size of a maximum matching.

A bound arises from the following ideas. Let A be a subset of nodes for which $G \backslash A$ has k components $H_1, \ldots, H_k$ having an odd number of nodes. (See Figure 5.4.) Let M be a matching of G; for each i, either H_i has an

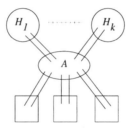

Figure 5.4. An upper bound for the size of a maximum matching

M-exposed node, or M contains an edge having just one end in $V(H_i)$. All such edges have their other ends in A, and, since M is a matching, all these ends must be distinct. Therefore, there can be at most $|A|$ such edges, and so the number of M-exposed nodes must be at least $k - |A|$. This gives the following general upper bound on the size of a matching of G. (We denote by $\mathrm{oc}(H)$ the number of odd components of a graph H.)

$$\text{For any } A \subseteq V,\ \nu(G) \leq \tfrac{1}{2}(|V| - \mathrm{oc}(G \backslash A) + |A|). \tag{5.1}$$

If we choose A in (5.1) to be a cover of G, then there are $|V| - |A|$ odd components of $G \backslash A$ (each having a single node), so the right-hand side reduces to $|A|$. In other words, the bound provided by (5.1) is at least as strong as that provided by covers. In fact, it is better, for example, in the case of a circuit of length $2k + 1$. Then we can choose $A = \emptyset$ to show that there is no matching of size bigger than k, whereas the smallest cover has size $k+1$. Less trivially, one can see that choosing A to be the white nodes in the first graph of Figure 5.1 gives $\mathrm{oc}(G \backslash A) - |A| = 2$, showing that every matching will leave at least two nodes exposed and hence proving that the indicated matching, of

size 8, is maximum. (The smallest cover has size 11.) The central result of the
theory of maximum matching states that this upper bound can be achieved.

Theorem 5.2 *(Tutte-Berge Formula) For a graph $G = (V, E)$ we have*

$$\max\{|M| : \ M \text{ a matching}\} \ = \ \min\{\tfrac{1}{2}(|V| - \text{oc}(G\backslash A) + |A|) : \ A \subseteq V\}.$$

It immediately implies a condition for the existence of a perfect matching.

Theorem 5.3 *(Tutte's Matching Theorem) A graph $G = (V, E)$ has a perfect
matching if and only if for every subset A of nodes we have $\text{oc}(G\backslash A) \leq |A|$.*

Theorem 5.3 was proved by Tutte [1947], and is quite easily seen to be equivalent to the min-max theorem, which was stated by Berge [1958]. So Theorem 5.2 is often called the *Tutte-Berge Formula*.

Just as we need a more general concept than in the bipartite case to get
a min-max formula, we also need a new algorithmic idea. It is the idea of
shrinking odd circuits. Let C be an odd circuit in G. We define $G' = G \times C$,
the subgraph obtained from G by shrinking C, as follows. $G \times C$ has node-set
$(V\backslash V(C)) \cup \{C\}$ and edge-set $E\backslash\gamma(V(C))$; for each edge e of G' and end v
of e in G, v is an end of e in G' if $v \notin V(C)$, and otherwise C is an end
of e in G'. (Notice that this is related to the concept of node identification
used in Section 3.5, in that we could obtain essentially the same graph by a
sequence of node identifications. But here we have a special convention for
naming the new node of G'.) We call $G \times C$ the graph obtained from G by
shrinking C. We illustrate in Figure 5.5 the result of shrinking an odd circuit.
A fundamental observation about odd-circuit shrinking is the following: If we

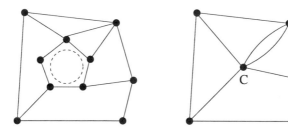

Figure 5.5. Shrinking an odd circuit

have a matching in $G \times C$, we can extend it to a matching in G in an obvious
way. See the example in Figure 5.6. We state this formally, as follows.

Proposition 5.4 *Let C be an odd circuit of G, let $G' = G \times C$, and let M' be
a matching of G'. Then there is a matching M of G such that $M \subseteq M' \cup E(C)$
and the number of M-exposed nodes of G is the same as the number of M'-
exposed nodes of G'.*

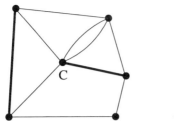

Figure 5.6. Extending a matching

Proof: Choose a node $w \in V(C)$ as follows. If C is covered by $e \in M'$, then choose w to be the node in $V(C)$ that is an end of e, and otherwise, choose w arbitrarily. Now deleting w from C results in a subgraph having a perfect matching M''. Take $M = M' \cup M''$. It is easy to check that M has the required properties.   ∎

Proposition 5.4 gives the inequality

$$\nu(G) \geq \nu(G \times C) + (|V(C)| - 1)/2, \qquad (5.2)$$

or equivalently, $\mathrm{def}(G) \leq \mathrm{def}(G \times C)$. It would be nice to know that this holds with equality; this is equivalent to the existence of a maximum matching using $(|V(C)| - 1)/2$ edges from $E(C)$. If this were true, finding a maximum matching in G could be reduced to finding a maximum matching in $G \times C$. Let us call an odd circuit C of G *tight* if (5.2) holds with equality. Not every odd circuit is tight; see the example of Figure 5.7. In fact, the algorithm we

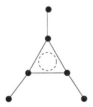

Figure 5.7. An odd circuit that is not tight

will describe will shrink odd circuits that are not necessarily tight, since it is difficult to identify tight ones. But we can use the idea of tight circuits to give a nonconstructive proof of the Tutte-Berge Formula, and this is what we do in the rest of the section. (We point out that the algorithm of the next section will give a proof of the formula that is independent of this one.) Let us call a node v of G *inessential* if there is a maximum matching of G not covering v; otherwise, we call v *essential*. Suppose that A satisfies (5.1) with equality. Choose $v \in A$ and consider $G' = G \backslash v$. Then $G' \backslash (A \backslash \{v\})$ has the same odd components as $G\backslash A$, and so $\nu(G') < \nu(G)$. In other words

> If A satisfies (5.1) with equality, then every $v \in A$ is essential.   (5.3)

Lemma 5.5 *Let $G = (V, E)$ be a graph and let $vw \in E$. If v, w are both inessential, then there is a tight odd circuit C using vw. Moreover, C is an inessential node of $G \times C$.*

Proof: There are maximum matchings M_1 and M_2 of G not covering v and w, respectively. Note that M_1 covers w and M_2 covers v; otherwise we could add vw to one of them and get a bigger matching. Consider the subgraph $H = (V, M_1 \triangle M_2)$. The component containing v consists of a path P beginning at v, since v is M_1-exposed. If P ends at an M_1-exposed node, it is M_1-augmenting, a contradiction. If P ends at an M_2-exposed node other than w, then attaching vw to it, we get an M_2-augmenting path, a contradiction. Therefore, P ends at w and so it forms with vw an odd circuit C. There is a maximum matching (M_1, for example) that contains $(|V(C)| - 1)/2$ edges from $E(C)$, so C is tight. Finally, $M_1 \backslash E(C)$ is a maximum matching of $G \times C$ not covering C, proving the last statement.   ∎

Proof of the Tutte-Berge Formula (Theorem 5.2): We already observed that the left-hand side is at most the right, so we need only show that there is a matching M and a set A such that the number of M-exposed nodes is exactly $oc(G\backslash A) - |A|$. The proof is by induction on the number of edges. The result is certainly true for graphs having no edges, so choose an edge vw.

Suppose that $\nu(G\backslash v) = \nu(G) - 1$. (The case for w is similar.) Then applying induction to $G\backslash v$, we have that there exists a set A' of nodes and a matching M' of $G\backslash v$ such that there are exactly $oc((G\backslash v)\backslash A') - |A'|$ M'-exposed nodes. Therefore, there is a matching M of G for which the number of M-exposed nodes is exactly $oc(G\backslash(A' \cup \{v\})) - |A' \cup \{v\}|$, and the result is proved for G.

Now we can suppose that both v and w are inessential. By Lemma 5.5, G has a tight circuit C. Applying induction to $G' = G \times C$, we get a matching M' and a set A' of nodes such that the number of M'-exposed nodes is exactly $oc(G'\backslash A') - |A'|$. Notice that the node C of G' cannot be in A', by (5.3), since it is an inessential node of G', by Lemma 5.5. Now we know from Proposition 5.4 that we can extend M' to a matching M of G having the same number of exposed nodes. But when we delete A' from G, we get the same number of odd components, because if one contains C, that node will now be replaced by $|V(C)|$ nodes and the component of $G\backslash A'$ containing them will still be odd. Thus we get a matching M and a set A' of nodes such that the number of M-exposed nodes is exactly $oc(G\backslash A') - |A'|$, as required.   ∎

Exercises

5.1. Let x be the integral feasible flow corresponding to the matching M of the bipartite graph G. Show that there is a one-to-one correspondence between

x-augmenting paths and M-augmenting paths. What is the connection between x-incrementing paths and M-alternating paths?

5.2. Let M be a matching of G and let p be the cardinality of a maximum matching. Using the method of proof of the Augmenting Path Theorem of Matchings, show that there are at least $p-|M|$ node-disjoint M-augmenting paths.

5.3. Use the result of Exercise 5.2 to show that, if M is a matching of cardinality 4000 in a graph having a matching of size 5000, then there is an M-augmenting path of length at most 9.

5.4. Suppose that $p > 0$ is the cardinality of a maximum matching of G, and let M be a matching of cardinality at most $p - \sqrt{p}$. Use the result of Exercise 5.2 to show that there is an M-augmenting path having at most $\sqrt{p}$ edges from M.

5.5. Let G be a graph and k a positive integer, $k \leq |V|/2$. Construct a graph G' having the property that G' has a perfect matching if and only if G has a matching of size k. Use this construction to prove the Tutte-Berge Formula from Tutte's Matching Theorem.

5.6. Let B denote the set of inessential nodes of $G = (V, E)$, C denote the set of nodes not in B but adjacent to at least one element of B, and D denote $V \setminus (B \cup C)$. ($\{B, C, D\}$ is called the *Gallai-Edmonds Partition* of G.) Prove that

(a) C is a minimizer in the Tutte-Berge formula;

(b) For every maximum matching M and every node $v \in C$, there is an edge $vw \in M$ with $w \in B$;

(c) Every maximum matching contains a perfect matching of $G[D]$.

5.7. Use Tutte's Matching Theorem to prove Petersen's Theorem, namely, if every node of G has degree 3 and $G \setminus e$ is connected for every edge e, then G has a perfect matching. Hint: First show that if $C \subseteq V$, $|C|$ odd, then $|\delta(C)| \geq 3$.

5.8. Show that Lemma 5.5 applied to a bipartite graph is equivalent to Kőnig's Theorem.

5.9. Suppose that G is connected and has the property that every node v is inessential. Use Lemma 5.5 to show that $\nu(G) = (|V| - 1)/2$.

5.2   MAXIMUM MATCHING

In this section we describe an algorithm, originally due to Edmonds [1965], to solve the following problem.

> *Maximum Matching Problem*
> Given a graph G, find a maximum matching of G.

The main techniques of the algorithm have been introduced in Section 5.1. We actually direct most of our efforts to showing how to find a perfect matching, if possible. Later we show how this method can be used to find a maximum matching.

Alternating Trees

We define a basic structure maintained by matching algorithms. Suppose we have a matching M of G and a fixed M-exposed node r of G. We build up iteratively sets A, B of nodes, such that each node in A is the other end of an odd-length M-alternating path beginning at r, and each node in B is the other end of an even-length M-alternating path beginning at r. A motivation for wanting such a set is that, if we find an edge vw such that $v \in B$ and $w \notin A \cup B$ is M-exposed, then the M-alternating path P from r to v together with vw gives an M-augmenting path. Such sets A, B could be built up, beginning with $A = \emptyset$, $B = \{r\}$, by the following rule:

If $vw \in E$, $v \in B$, $w \notin A \cup B$, $wz \in M$, then add w to A, z to B. (5.4)

The set $A \cup B$ and the edges used in its construction have the structure indicated in Figure 5.8. Namely, they form a tree T with "root" r having the

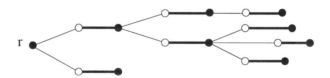

Figure 5.8. An alternating tree

following properties:

(a) Every node of T other than r is covered by an edge of $M \cap E(T)$;

(b) For every node v of T, the path in T from v to r is M-alternating.

Such a tree is called an M-*alternating tree*. Given an M-alternating tree T, we denote by $A(T)$ and $B(T)$ the sets of nodes at odd and even (respectively) distance from the root. We may refer to the elements of $A(T)$ as the *odd* nodes of T and to the elements of $B(T)$ as the *even* nodes of T. (In Figure 5.8 the odd nodes are white and the even nodes are black.) Notice that $|B(T)| = |A(T)| + 1$, since nodes other than r come in matched pairs, one in $A(T)$ and one in $B(T)$.

Let us state as subroutines written in the language of these trees, two of the techniques that we have already introduced. The first subroutine is an encoding of the rule (5.4), and the second is a statement of the augmentation step.

Use vw to extend T

Input: A matching M' of a graph G', an M'-alternating tree T, and an edge vw of G' such that $v \in B(T)$, $w \notin V(T)$ and w is M'-covered.

Action: Let wz be the edge in M' covering w. (Note that z is not a node of T.) Replace T by the tree having edge-set $E(T) \cup \{vw, wz\}$.

Use vw to augment M'

Input: A matching M' of a graph G', an M'-alternating tree T of G' with root r, and an edge vw of G' such that $v \in B(T)$, $w \notin V(T)$ and w is M'-exposed.

Action: Let P be the path obtained by attaching vw to the path from r to v in T. Replace M' by $M' \triangle E(P)$.

Here is a simple condition under which we can conclude that G has no perfect matching. We call an M-alternating tree T in a graph G *frustrated* if every edge of G having one end in $B(T)$ has the other end in $A(T)$.

Proposition 5.6 *Suppose that G has a matching M and an M-alternating tree T that is frustrated. Then G has no perfect matching.*

Proof: Clearly, every element of $B(T)$ is a single-node odd component of $G \backslash A(T)$. Since $|A(T)| < |B(T)|$, therefore G has no perfect matching. ∎

The Bipartite Case

The matching algorithm will need some other ideas, involving the handling of odd circuits. But since odd circuits do not arise when the graph is bipartite, we can now state an algorithm for perfect matching in a bipartite graph.

Perfect Matching Algorithm for Bipartite Graphs

Set $M = \emptyset$;
Choose an M-exposed node r and put $T = (\{r\}, \emptyset)$;
While there exists $vw \in E$ with $v \in B(T)$, $w \notin V(T)$
    If w is M-exposed
        Use vw to augment M;
        If there is no M-exposed node in G
            Return the perfect matching M and stop;
        Else
            Replace T by $(\{r\}, \emptyset)$, where r is M-exposed;
    Else
        Use vw to extend T;
Stop; G has no perfect matching.

The correctness of the algorithm is based on the following observation.

Proposition 5.7 *Suppose that G is bipartite, that M is a matching of G, and that T is an M-alternating tree such that no edge of G joins a node in $B(T)$ to a node not in $V(T)$. Then T is frustrated, and hence G has no perfect matching.*

Proof: We must show that every edge having an end in $B(T)$ has an end in $A(T)$. From the hypothesis, the only possible exception would be an edge joining two nodes in $B(T)$. But this edge, together with the paths joining them to the root of T, would form a closed path of odd length, which is clearly impossible in a bipartite graph. Hence T is frustrated, and so by Proposition 5.6, G has no perfect matching. ∎

The Blossom Algorithm for Perfect Matching

The basic ideas for a bipartite matching algorithm go back at least to Kőnig, and algorithms were well known by the decade of the 1950s. For the general problem, Tutte's Matching Theorem appeared in the 1940s, and approaches based on augmenting paths were suggested in the 1950s. However, the problem of finding an efficient algorithm remained open until the 1960s, when it was solved in the landmark paper of Edmonds [1965]. A main new idea that he used is the shrinking of odd circuits, which we have already introduced; now we will show the interesting way in which this concept and that of alternating trees come together in an algorithm.

Suppose that we have a graph G' obtained from G by a sequence of odd-circuit shrinkings. We call G' a *derived graph* of G. Nodes of G' are of two types: those that are nodes of G, which we call *original nodes* of G', and those that are not nodes of G, which we call *pseudonodes*. Given a node v of G', there corresponds a set $S(v)$ of nodes of G: If $v \in V$, that subset is just $\{v\}$, but if $v = C$ is a pseudonode, it is defined recursively to be the union of the sets $S(w)$ for all the nodes $w \in V(C)$. Clearly, the cardinality of this set will always be odd, since it is the sum of an odd number of odd numbers. Moreover, the collection of sets $S(v)$ for all nodes v of G' forms a partition of V. There is a simple extension of Proposition 5.6, which sometimes allows us to recognize, when we are working with one of its derived graphs, that G has no perfect matching.

Proposition 5.8 *Let G' be a derived graph of G, let M' be a matching of G', and let T be an M'-alternating tree of G' such that no element of $A(T)$ is a pseudonode. If T is frustrated, then G has no perfect matching.*

Proof: When we delete $A(T)$ from G, we get a component with node-set $S(v)$ for each $v \in B(T)$. Therefore, $\mathrm{oc}(G \backslash A(T)) > |A(T)|$, so G has no perfect matching. ∎

So the algorithm works with derived graphs of G, since if we find a perfect matching in a derived graph, we know there is a perfect matching in G, and if we find a certain kind of frustrated tree in a derived graph, we can conclude that G has no perfect matching. But how do we decide which circuits to shrink? The answer can be discovered by considering what happens if we try to solve the problem with only the tree extension and augmentation steps. For example, in Figure 5.9 we see an M-alternating tree in a graph G to which neither of the steps applies, yet it is clear that there does exist a perfect matching. Of course the tree is not frustrated, because of the existence of the

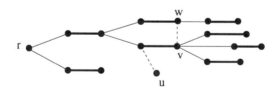

Figure 5.9. Augmenting path not found by tree extension

edge vw. In general, if there exists an edge vw with $v, w \in B(T)$, then this edge together with the path in T from v to w forms an odd circuit, which we can shrink to form a graph G'. (An odd circuit encountered in this way was called a "blossom" by Edmonds, and his algorithm is often called the "Blossom Algorithm".) Then something very nice happens: The matching and the alternating tree easily yield similar structures in G'. Consider, for example, the graph of Figure 5.10 obtained when we shrink the odd circuit in Figure 5.9. The matching edges that remain form a matching in G', and the tree edges that remain form a tree that is alternating with respect to this matching. Notice also that in G', an augmenting path becomes detectable by the straightforward approach we used before. Here is a formal description of

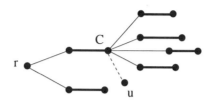

Figure 5.10. Tree structure after shrinking

the shrinking subroutine, including the updates of the matching and tree.

Use vw to shrink and update M' and T

Input: A matching M' of a graph G', an M'-alternating tree T, and an edge vw of G' such that $v, w \in B(T)$.

Action: Let C be the circuit formed by vw together with the path in T from v to w. Replace G' by $G' \times C$, M' by $M' \backslash E(C)$, and T by the tree (in G') having edge-set $E(T) \backslash E(C)$.

The next result confirms the intuition suggested by the examples. Its proof is Exercise 5.11.

Proposition 5.9   *After application of the shrinking subroutine, M' is a matching of G', T is an M'-alternating tree of G', and $C \in B(T)$.*   ∎

We are now ready to state the algorithm. In addition to the subroutines we have mentioned, it uses an algorithm to extend a matching of a derived graph of G to a matching of G. Of course, this algorithm repeatedly uses the proof of Proposition 5.4.

Blossom Algorithm for Perfect Matching

Input graph G and matching M of G;
Set $M' = M$, $G' = G$;
Choose an M'-exposed node r of G' and put $T = (\{r\}, \emptyset)$;
While there exists $vw \in E'$ with $v \in B(T)$, $w \notin A(T)$
    Case: $w \notin V(T)$, w is M'-exposed
        Use vw to augment M';
        Extend M' to a matching M of G;
        Replace M' by M, G' by G;
        If there is no M'-exposed node in G'
            Return the perfect matching M' and stop;
        Else
            Replace T by $(\{r\}, \emptyset)$, where r is M'-exposed;
    Case: $w \notin V(T)$, w is M'-covered
        Use vw to extend T;
    Case: $w \in B(T)$
        Use vw to shrink and update M' and T;
Return G', M', T and stop, G has no perfect matching.

Theorem 5.10 *The Blossom Algorithm terminates after $O(n)$ augmentations, $O(n^2)$ shrinking steps, and $O(n^2)$ tree-extension steps. Moreover, it determines correctly whether G has a perfect matching.*

Proof: It is easy to see that M' remains a matching throughout. Therefore, since each augmentation decreases the number of exposed nodes, there will be $O(n)$ augmentation steps. Between augmentations, each shrinking step decreases the number of nodes of G' while not changing the number of nodes not in T, and each tree-extension step decreases the number of nodes not in T while not changing the number of nodes of G'. Hence, the number of shrinking and tree-extension steps between augmentations is $O(n)$, and the total number of these steps is $O(n^2)$.

Finally, since each G' is a derived graph of G, if the algorithm halts with the conclusion that G has no perfect matching, then it has found a tree with the properties of Proposition 5.8, and so G has no perfect matching. ∎

The Blossom Algorithm for Maximum Matching

We have yet to explain how to solve the maximum matching problem. In fact, we can use the algorithm for perfect matching. Of course, when we apply that algorithm to G, if it terminates with a perfect matching, then that matching is maximum. If it terminates with a matching and a frustrated tree T in some derived graph, then we can extend the current matching to a matching of G, but it need not be maximum, because there may exist augmenting paths in other parts of G. Instead, we delete $V(T)$ from G'. If any exposed node remains, then we apply the Blossom Algorithm to the new G' beginning with the matching consisting of the edges of M' that were not deleted. (Notice that the new G' has no pseudonodes.) We continue this process until there are no exposed nodes left. Suppose that we now restore to G' all of the deleted parts including all of the matching edges. Let $T_1, \ldots, T_k$ be the frustrated trees we generated. Notice that the exposed nodes are exactly the roots of the T_i, so there are exactly k of them. Therefore, M' can be extended to a matching M of G having just k exposed nodes. Now let A be the union of the sets $A(T_i)$ for all i. If we delete A from G, each of the nodes of $B(T_i)$ for every i will give an odd component, and so we will have $\mathrm{oc}(G \backslash A) \geq |A| + k$. (In fact, this will turn out to hold with equality.) It follows that M is a maximum matching of G. (Notice that this argument provides the promised algorithmic proof of the Tutte-Berge Formula.)

Implementation of the Blossom Algorithm

Of course, the main new issue that comes up when we consider how to implement the matching algorithm, is how to represent the current derived graph. One natural way is to keep an explicit representation. That is, each time we shrink a circuit C, we would construct a representation for $G' \times C$ from the (known) representation of G'. This approach can be quite expensive in the worst case, but it also has some advantages. We will discuss it in more detail when we consider implementation of algorithms for minimum-weight perfect matching in Section 5.3. Another approach is to use a representation for G and to represent G' implicitly by keeping a representation of the partition of V determined by the sets $S(v)$ for all $v \in V(G')$. This is the approach that we will discuss here. The representation of G that we need will allow in constant time, given an edge e, to find its ends in G, as well as to find, for each $v \in V$, the edges in $\delta(v)$ in time proportional to their number.

When we shrink, the effect on the partition we are keeping is to merge some of its blocks. The operation that, given an edge $vw \in E$, finds the ends of that edge in the derived graph just requires identifying the sets $R(v)$ and $R(w)$, where $R(v)$ is the member of the partition containing v. Notice, then, that two of the basic operations required in the algorithm are the "find" and "merge" operations that we needed in Section 2.1 to implement Kruskal's Algorithm for the minimum spanning tree problem. This is an extremely well studied problem; see Tarjan [1983] for more details. We saw a method in Section 2.1 that can do $O(m)$ finds and $O(n)$ merges in time $O(m \log n)$. (In fact, a second method that accomplishes this was the subject of Exercise 2.15. That method turns out to be more important, in that it is the basis for more sophisticated methods which have better complexity. See Tarjan [1983] for more details.) We will evaluate the complexity of the matching algorithm assuming this subroutine is used. It will be apparent that a better way to do the finds and merges would lead to a better running time.

The algorithm constructs $O(n)$ trees. Let us analyze the time spent on one of them. We keep a list L of edges to be checked. Checking an edge means seeing whether it can be used to augment, shrink, or extend the tree. When the tree T is initialized to consist only of r, L is initialized to consist of $\delta(r)$. Now suppose that we choose an edge $e = uv$ from L. We delete e from L and compute $R(u)$ and $R(v)$. If they are equal, we know that e is not an edge of G'. (It has been "shrunken away.") So we do nothing. If neither of the corresponding nodes of G' is in $B(T)$, again, we do nothing. (Actually, one can show that this case will never occur.) If one of the nodes is not in G' (because of the deletion of a tree), again, we do nothing. If one of the nodes is in $B(T)$ and the other is in $V(G')\backslash V(T)$ and is M'-exposed, then we perform an augmentation. If one is in $B(T)$ and the other, w, is in $V(G')\backslash V(T)$ and is covered by an edge wz of M', we add e and wz to T; we also add all the edges of $\delta(z)$ (z is a node of G) to L. Finally, if both nodes are in $B(T)$ and are different, we trace the circuit, redefine T, and merge the appropriate sets, and for every node z of $A(T)$ on the circuit (they are all nodes of G), add each edge in $\delta(z)$ to L. It is a simple matter to verify that every edge that should be checked is checked, and no edge is checked more than once. The whole process requires, therefore, $O(m)$ finds. The number of merges is clearly $O(n)$, since each one decreases the number of sets in the partition. So the amount of work to find an augmentation is $O(m \log n)$. We have skipped over a few details that were treated for other problems in Chapter 2, such as how to maintain the tree, and how to recognize which nodes of G' are in $A(T)$ and in $B(T)$. It is easy to verify that none of these affects the above bound.

There is one more nontrivial thing to do. After an augmentation we need to go back to G and extend the matching to G. Of course, the partition now becomes trivial, so it is the updating of the matching we need to worry about. In fact, this can be done in time $O(n)$. One way is to maintain an explicit record of each shrinking, the names of the nodes and the edges of the circuit and the name of the new pseudonode, that is, enough information that we can easily apply the routine implicit in the proof of Proposition 5.4. As a consequence of this discussion we have the following theorem.

Theorem 5.11 *The Blossom Algorithm can be implemented to run in time* $O(nm \log n)$. ∎

There are implementations of the maximum matching algorithm with faster theoretical running times. They use two additional main ideas. One, which we have already mentioned, is that better ways to handle the find and merge operations enable replacing the $m \log n$ factor by something smaller (eventually by m). See Tarjan [1983] for more details. The other idea is to find at each step an augmenting path of least length. The original observations that led to this improvement are due to Hopcroft and Karp [1973]. They proved some general results, and showed how they could be applied to give an $O(\sqrt{n}m)$ algorithm for the bipartite case. Later, others extended this result to general graphs; see, for example, Micali and Vazirani [1980]. We shall restrict our description to the bipartite case. It is considerably simpler, and the running time obtained by Hopcroft and Karp remains essentially the best known.

It is easiest to give the description in the context of the maximum flow formulation of bipartite matching. If one finds at each step a shortest augmenting path, then it follows from Lemma 3.12 that the length of paths used for augmentation never decreases. Now one can conclude from this fact and Exercise 5.4 that, where p is the cardinality of a maximum matching, the first $p - \lfloor \sqrt{p} \rfloor$ augmentations are all on augmenting paths having at most $\sqrt{p}$ matching edges. Since there remain at most $\sqrt{p}$ augmentations to do, there are at most $2\sqrt{p}$ different lengths of augmenting paths used during the whole process. This suggests the problem of finding and performing all the augmentations on paths of a given length efficiently. Hopcroft and Karp showed how this can be done in time $O(m)$, and thus derived an algorithm having running time $O(\sqrt{p}m) = O(\sqrt{n}m)$. See Exercise 5.17.

Exercises

5.10. Derive from the bipartite matching algorithm a proof of König's Theorem.

5.11. Prove Proposition 5.9.

5.12. Show that the Blossom Algorithm can shrink a circuit that is not tight. Can this happen during the construction of the last alternating tree?

5.13. An *edge cover* of a graph $G = (V, E)$ (having no isolated nodes) is a subset D of edges such that every node is incident to at least one edge in D. Show how a minimum cardinality edge cover can be determined from a maximum matching. Hence prove that the minimum cardinality of an edge cover is $|V| - \nu(G)$.

5.14. Find a maximum matching and a minimizing set A in the Tutte-Berge Formula for the graph of Figure 5.11.

5.15. Let $T_1, \ldots, T_k$ be the trees at termination of the maximum matching algorithm, let $B = \cup(B(T_i);\ i = 1, \ldots, k)$, and let $B' = \cup(S(v) :\ v \in B)$. Prove that the elements of B' are precisely the inessential nodes of G. (Hint: To prove that nodes not in B' are essential, use (5.1).)

5.16. Find the Gallai-Edmonds Partition for the graph of Figure 5.11.

5.17. To show that the Hopcroft-Karp Algorithm to which we referred at the end of the section has the claimed running time, we need to show the following. If the shortest M-augmenting path has length k, then one can in time $O(m)$ find and perform augmentations on paths of length k, until the length of a shortest augmenting path has length larger than k. Consider the subgraph G' consisting of the nodes and edges of all shortest augmenting

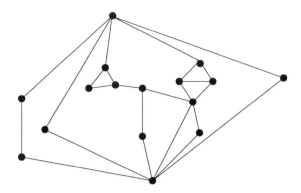

Figure 5.11. Maximum matching exercise

paths. Show how G' can be initially constructed in time $O(m)$. From results in Section 3.2, we know that G' gets smaller during the process of performing augmentations on paths of length k. Show, in the present situation, that an augmentation removes all of the nodes of the augmenting path from G'. We want to carry out the process while considering each edge of G' at most twice. Show how this can be done, by looking for augmenting paths in a "depth-first" manner.

5.18. (W. Anderson) *Slither* is a two-person game played on a graph $G = (V, E)$. The players, called *First* and *Second*, play alternately, with First playing first. At each step the player whose turn it is chooses a previously unchosen edge. The only rule is that at every step the set of chosen edges forms a simple path. The loser is the first player unable to make a legal play at his or her turn. Prove that, if G has a perfect matching, then First can force a win.

5.19. Prove that if First's first play in a game of slither is an edge uv such that u is an inessential node of G, then Second can force a win.

5.20. Let $\{B, C, D\}$ be the Gallai-Edmonds Partition of G. Prove that, if $D \neq \emptyset$, then First can force a win at slither.

5.21. Suppose that, in a game of slither the nodes of the path of played edges so far are all in C, where $\{B, C, D\}$ is the Gallai-Edmonds partition of G, and suppose the next player now chooses an edge having an end not in C. Prove that the other player can force a win.

5.3   MINIMUM-WEIGHT PERFECT MATCHINGS

In this section we describe an algorithm for finding a minimum-weight perfect matching of G. For this we need both the matching techniques of the last section and an optimality characterization that arises from linear programming. We begin with an integer-linear-programming formulation of the minimum weight perfect matching problem.

$$\text{Minimize } \sum(c_e x_e \: : \: e \in E) \tag{5.5}$$
$$\text{subject to}$$
$$x(\delta(v)) = 1, \text{ for all } v \in V \tag{5.6}$$
$$x_e \in \{0, 1\}, \text{ for all } e \in E. \tag{5.7}$$

It is easy to see that x is a feasible solution of this problem if and only if x is the characteristic vector of a perfect matching of G, and that x is an optimal solution if and only if x is the characteristic vector of a minimum-weight perfect matching of G. Of course, we get a lower bound on the minimum weight of a perfect matching by considering the linear-programming problem obtained by relaxing the integrality constraints (5.7).

$$\text{Minimize } \sum(c_e x_e \: : \: e \in E) \tag{5.8}$$
$$\text{subject to}$$
$$x(\delta(v)) = 1, \text{ for all } v \in V$$
$$x_e \geq 0, \text{ for all } e \in E.$$

We would not expect the resulting bound to be equal to the desired optimum. In the case in which G is bipartite, however, (5.8) does give the right value.

Minimum-Weight Perfect Matching in Bipartite Graphs

The fact that, for bipartite graphs, the linear-programming problem (5.8) has the same optimal value as the integer-linear-programming problem (5.5) is equivalent to a classical theorem of Birkhoff [1946].

Theorem 5.12 *(Birkhoff) Let G be a bipartite graph, and let $c \in \mathbf{R}^E$. Then G has a perfect matching if and only if (5.8) has a feasible solution. Moreover, if G has a perfect matching, then the minimum weight of a perfect matching is equal to the optimal value of (5.8).*

One can prove Birkhoff's Theorem from results of Chapter 4, or directly. (See Exercises 5.24 and 5.25.) We are going to describe an algorithm that will provide another proof, and which is a starting point for the main algorithm of the section. It is a primal-dual algorithm, like the one we used for the minimum-cost flow problem in Section 4.3. The dual linear-programming problem of (5.8) is

$$\text{Maximize } \sum(y_v \: : \: v \in V) \tag{5.9}$$
$$\text{subject to}$$
$$y_u + y_v \leq c_e, \text{ for all } e = uv \in E.$$

Given a vector $y \in \mathbf{R}^V$ and an edge $e = uv$, we denote by $\bar{c}_e = \bar{c}_e(y)$ the difference $c_e - (y_u + y_v)$. Thus y is feasible to (5.9) if and only if $\bar{c}_e \geq 0$ for all $e \in E$. In this case, we denote by $E_=$ (or $E_=(y)$) the set $\{e \in E : \bar{c}_e = 0\}$; its elements are the "equality edges" with respect to y. Then the complementary slackness conditions for the dual pair of problems (5.8) and (5.9) can be written:

$$x_e > 0 \text{ implies } \bar{c}_e = 0, \text{ for all } e \in E. \tag{5.10}$$

If x is the characteristic vector of a perfect matching M of G, these conditions are equivalent to

$$M \subseteq E_=. \tag{5.11}$$

Given a feasible solution y to (5.9), we can use the algorithm of Section 5.2 to search for a perfect matching having this property. If we succeed, we have a perfect matching whose characteristic vector is optimal to (5.8), as required. Otherwise, the algorithm will deliver a matching M of $G_= = (V, E_=)$ and an M-alternating tree such that nodes of $B(T)$ are joined by equality edges only to nodes in $A(T)$. See the example in Figure 5.12, where a perfect matching exists, but with respect to the current y (indicated at the nodes), there is none contained in $E_=$.

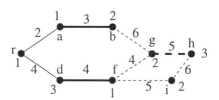

Figure 5.12. A frustrated tree in $G_=$

In that case, there is a natural way to change y, keeping in mind that we would like edges of M and of T to remain in $E_=$, and that we would like $\bar{c}_e$ to decrease for edges e joining nodes in $B(T)$ to nodes not in $A(T)$. Namely, we increase y_v by $\varepsilon > 0$ for all $v \in B(T)$, and we decrease y_v by ε for $v \in A(T)$. This has the desired effect; we will choose ε as large as possible so that feasibility in (5.9) is not lost, and as a result (provided that G has a perfect matching at all), some edge joining a node $u \in B(T)$ to a node $w \notin A(T)$ will enter $E_=$. Since G is bipartite, we will have $w \notin V(T)$, leading to an augmentation or a tree-extension step. In the example, we will choose $\varepsilon = 1$ because of the edge fg, and then that edge enters $E_=$, giving a tree-extension step. We get the following algorithm, due to Kuhn [1955] and Munkres [1957]. It is sometimes called the "Hungarian Algorithm."

Minimum-Weight Perfect Matching Algorithm for Bipartite Graphs

Let y be a feasible solution to (5.9), M a matching of $G_=$;
Set $T = (\{r\}, \emptyset)$, where r is an M-exposed node of G;
Loop
    While there exists $vw \in E_=$ with $v \in B(T)$, $w \notin V(T)$
        If w is M-exposed
            Use vw to augment M;
            If there is no M-exposed node in G
                Return the perfect matching M and stop;
            Else
                Replace T by $(\{r\}, \emptyset)$, where r is M-exposed;
        Else
            Use vw to extend T;
    If every $vw \in E$ with $v \in B(T)$ has $w \in A(T)$
        Stop, G has no perfect matching;
    Else
        Let $\varepsilon = \min\{\bar{c}_{vw} : v \in B(T), w \notin V(T)\}$;
        Replace y_v by $y_v + \varepsilon$ for $v \in B(T)$, $y_v - \varepsilon$ for $v \in A(T)$;

The correctness of the algorithm follows from the discussion that preceded it. Its running time is higher than for the unweighted case, because, in the worst case, we may have to do a dual change for every tree-extension step, which means $O(n^2)$ dual changes in all. So the bottleneck is the work involved in doing a dual change, in particular, in computing ε. A straightforward method requires looking at each edge, so we get an overall running time of $O(n^2 m)$. This can be improved, for example, to $O(n^3)$ using the method introduced first for the implementation of Prim's Algorithm for minimum spanning trees in Section 2.1, or to $O(nS(n,m))$ using methods of Chapter 4.

Minimum-Weight Perfect Matching in General

Of course, the result of Birkhoff's Theorem fails in general. Consider the example of Figure 5.13. There is a feasible solution of (5.8) of value 10.5, obtained by putting $x_e = 1/2$ for each edge of the two triangles. It is not too difficult to check that the minimum weight of a perfect matching is 14.

So in general, we need a better linear-programming bound than is provided by (5.8). The idea, due to Edmonds [1965a], is related to the one we used earlier to give a bound for the cardinality of a maximum matching. We call a cut D of G *odd* if it is of the form $\delta(S)$ where S is an odd-cardinality set of nodes. If D is an odd cut and M is a perfect matching, then M must contain at least one edge from D. It follows that, if x is the characteristic vector of a

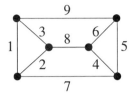

Figure 5.13. Linear programming bound

perfect matching, then for every odd cut D of G,

$$x(D) \geq 1. \tag{5.12}$$

This is called a *blossom inequality*. By adding these inequalities to the problem (5.8), we get a stronger linear-programming bound. (For example, if we add the inequality (5.12) to (5.8), for D consisting of all the horizontal edges in Figure 5.13, it is no longer possible to get a solution of value 10.5; in fact, the new optimal value for the resulting problem is 14, which is also the minimum weight of a perfect matching.) Let $\mathcal{C}$ denote the set of all odd cuts of G that are not of the form $\delta(v)$ for some node v. Then we are led to consider the linear-programming problem

$$\text{Minimize } \sum(c_e x_e : e \in E) \tag{5.13}$$

$$\text{subject to}$$

$$x(\delta(v)) = 1, \text{ for all } v \in V \tag{5.14}$$

$$x(D) \geq 1, \text{ for all } D \in \mathcal{C} \tag{5.15}$$

$$x_e \geq 0, \text{ for all } e \in E. \tag{5.16}$$

As we have indicated, (5.13) provides a better approximation to the minimum weight of a perfect matching, but a much stronger statement can be made. Its optimal value *is* the minimum weight of a perfect matching. This is the fundamental theorem of Edmonds [1965a]. It will be interpreted in Chapter 6 as the "Matching Polytope Theorem."

Theorem 5.13 *Let G be a graph, and let $c \in \mathbf{R}^E$. Then G has a perfect matching if and only if (5.13) has a feasible solution. Moreover, if G has a perfect matching, then the minimum weight of a perfect matching is equal to the optimal value of (5.13).*

The algorithm that we will describe will construct a perfect matching M whose characteristic vector x^* is an optimal solution to (5.13), and so M is a minimum-weight perfect matching. This will provide a proof of Theorem 5.13. The way in which we will know that x^* is optimal, is that we will have also a feasible solution to the dual linear-programming problem of (5.13) that satisfies the complementary slackness conditions with x^*. The dual problem is

$$\text{Maximize} \sum(y_v : v \in V) + \sum(Y_D : D \in \mathcal{C}) \tag{5.17}$$

subject to

$$y_v + y_w + \sum(Y_D : e \in D \in \mathcal{C}) \le c_e, \text{ for all } e = vw \in E \tag{5.18}$$

$$Y_D \ge 0, \text{ for all } D \in \mathcal{C}. \tag{5.19}$$

Given a vector (y, Y) as in (5.17) and an edge e, we denote by $\bar{c}_e = \bar{c}_e(y, Y)$ the difference between the right- and left-hand sides of (5.18). (We call $\bar{c}_e$ the *reduced cost* of the edge e.) Thus (y, Y) is feasible to (5.17) if and only if $Y_D \ge 0$ for all $D \in \mathcal{C}$ and $\bar{c}_e \ge 0$ for all $e \in E$. We use this notation to write the complementary slackness conditions for the dual pair of problems (5.13) and (5.17). They are

$$x_e > 0 \text{ implies } \bar{c}_e = 0, \text{ for all } e \in E \tag{5.20}$$

$$Y_D > 0 \text{ implies } x(D) = 1, \text{ for all } D \in \mathcal{C}. \tag{5.21}$$

If x is the characteristic vector of a perfect matching M of G, these conditions are equivalent to

$$e \in M \text{ implies } \bar{c}_e = 0, \text{ for all } e \in E \tag{5.22}$$

$$Y_D > 0 \text{ implies } |M \cap D| = 1, \text{ for all } D \in \mathcal{C}. \tag{5.23}$$

It is not obvious how an algorithm will work with the dual variables Y, but the answer is suggested by the matching algorithm of Section 5.2. We will be working with derived graphs G' of G, and such graphs have the property that

every odd cut of G' is an odd cut of G.

It follows from this, in particular, that every cut of the form $\delta_{G'}(v)$ for a pseudonode v of G' is an odd cut of G. These are the only odd cuts D of G' for which we will assign a positive value to Y_D. (Note, however, that such a cut of G' need not have this property in G—it is of the form $\delta(S)$ where S is an odd subset of G which becomes a pseudonode of G' after (repeated) odd-circuit shrinkings.) It follows that we can handle Y_D by replacing it by y_v, with the additional proviso that $y_v \ge 0$.

We take the same approach as in the bipartite case, trying to find a perfect matching in $G_=$ using tree-extension and augmentation steps. When we get stuck, we change y in the same way, except that the existence of edges joining two nodes in $B(T)$ will limit the size of ε. In particular, there may be an equality edge joining two such nodes. Then we shrink the circuit C, but there is now one small problem: How do we take into account the variables y_v for $v \in V(C)$, when those nodes are no longer in the graph? The answer is that we update c as well. Namely, we replace c_{vw} by $c_{vw} - y_v$ for each edge vw with

$v \in V(C)$ and $w \notin V(C)$. Notice that by this transformation, and the setting of $y_C = 0$ for the new pseudonode, the reduced costs $\bar{c}_e$ are the same for edges of $G \times C$ as they were for those edges in G. We will use c' to denote these updated weights, so when we speak of a 'derived graph' G' of G, the weight vector c' is understood to come with G', or we may refer to the *derived pair* (G', c'). The observation about the invariance of the reduced costs, however, means that we can avoid $\bar{c}'$ in favor of $\bar{c}$. The subroutine *Use vw to shrink and update M', T, and c'* is just the shrinking routine we used in the unweighted case with this additional updating of c' and setting $y_C = 0$.

An example of shrinking and updating weights is shown in Figure 5.14. On the left is the graph of Figure 5.13 with a feasible dual solution y (shown at the nodes) for which the edges of the left triangle are all equality edges. On the right we see the result of shrinking this circuit and updating the weights. Notice that an optimal perfect matching of this smaller graph extends to an optimal perfect matching of the original. (However, just as the algorithm in the unweighted case can shrink a circuit that is not tight, the one for the weighted case can shrink a circuit that does not have this property.)

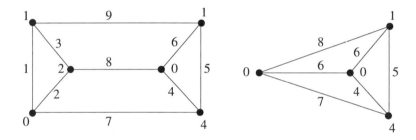

Figure 5.14. Shrinking and updating weights

The essential justification of the stopping condition of the algorithm is that, when we have solved the problem on a derived graph, then we have solved the problem on the original graph. For this to be correct, we have to be very specific about what we mean by "solved the problem." The proof of this result is immediate from the definitions.

Proposition 5.14 *Let G', c' be obtained from G, c by shrinking the odd circuit C of equality edges with respect to the dual-feasible solution y. Let M' be a perfect matching of G'' and, with respect to G', c', let (y', Y') be a feasible solution to (5.17) such that $M', (y', Y')$ satisfy conditions (5.22),(5.23) and such that $y'_C \geq 0$. Let M be the perfect matching of G obtained by extending M' with edges from $E(C)$. Let (y, Y) be defined as follows. For $v \in V \backslash V(C)$, $y_v = y'_v$. (For $v \in V(C)$, y_v is already defined.) For $D \in \mathcal{C}$, we put $Y_D = Y'_D$ if $Y'_D > 0$; we put $Y_D = y'_C$ if $D = \delta'(C)$; otherwise, we put $Y_D = 0$. Then with respect to G, c, (y, Y) is a feasible solution to (5.17) and $M, (y, Y)$ satisfy the conditions (5.22), (5.23).* ∎

Now let us describe the dual variable change. It is the same one used in the bipartite case, but with different rules for the choice of ε. First, we need to consider edges $e = uv$ with $u, v \in B(T)$ when choosing ε, so we will need $\varepsilon \le \bar{c}_e/2$ for such edges. Second, we need to ensure that y_v remains nonnegative if v is a pseudonode, so we need $\varepsilon \le y_v$ for such nodes. Since it is the nodes in $A(T)$ whose y-values are decreased by the change, these are the ones whose y-values affect the choice of ε. To illustrate, a slight change in the example of Figure 5.12 leads to the one in Figure 5.15. Here we see that a choice of $\varepsilon = 1$ will no longer be allowable, because of the edge bf. So we take $\varepsilon = 1/2$, and after the dual change, that edge will be an equality edge.

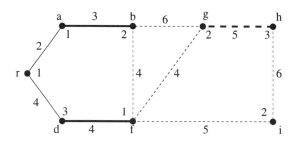

Figure 5.15. A dual change where G is nonbipartite

Change y

Input: A derived pair (G', c'), a feasible solution y of (5.17) for this pair, a matching M' of G' consisting of equality edges, and an M'-alternating tree T consisting of equality edges in G'.

Action: Let $\varepsilon_1 = \min(\bar{c}_e : e$ joins in G' a node in $B(T)$ to a node not in $V(T))$;
$\varepsilon_2 = \min(\bar{c}_e/2 : e$ joins in G' two nodes in $B(T))$;
$\varepsilon_3 = \min(y_v : v \in A(T), v$ is a pseudonode of $G')$.
$\varepsilon = \min(\varepsilon_1, \varepsilon_2, \varepsilon_3)$.
Replace y_v by
$y_v + \varepsilon$, if $v \in B(T)$;
$y_v - \varepsilon$, if $v \in A(T)$;
y_v, otherwise.

The last ingredient of the algorithm that we need is a way to handle the expansion of pseudonodes. Note that we no longer have the luxury of expanding them all after finding an augmentation. The reason is that expanding the pseudonode v when y_v is positive would mean giving a value to a variable Y_D, where D is a cut of the current derived graph that is not of the form $\delta_{G'}(u)$

for some pseudonode u. Therefore, we can expand a pseudonode v *only if* $y_v = 0$. Moreover, in some sense we *need* to do such expansions, because a dual variable change may not result in any equality edge that could be used to augment, extend, or shrink (because the choice of ε was determined by some odd pseudonode). However, in this case, unlike the unweighted case, we are still in the process of constructing a tree, and we do not want to lose the progress that has been made. So the expanding step should, as well as updating M' and c', also update T. The example in Figure 5.16 suggests what to do. Suppose that the odd node v of the M-alternating tree T is a pseudonode and expansion of the corresponding circuit C with the updating of M leaves us with the graph on the right. There is a natural way to update

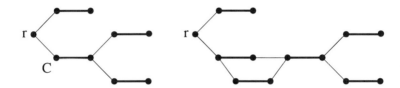

Figure 5.16. Expanding an odd pseudonode

T after the expansion, by keeping a subset of the edges of C. In Figure 5.17 we show the new alternating tree after the pseudonode expansion illustrated in Figure 5.16.

Figure 5.17. Tree update after pseudonode expansion

Expand odd pseudonode v and update M', T, and c'

Input: A matching M' consisting of equality edges of a derived graph G', an M'-alternating tree T consisting of equality edges, and an odd pseudonode v of G' such that $y_v = 0$.

Action: Let f, g be the edges of T incident with v, let C be the circuit that was shrunk to form v, let u, w be the ends of f, g in $V(C)$, and let P be the even-length path in C joining u to w. Replace G' by the graph obtained by expanding C. Replace M' by the matching obtained by extending M' to a matching of G'. Replace T by the tree having edge-set $E(T) \cup E(P)$. For each edge st with $s \in V(C)$ and $t \notin V(C)$, replace c'_{st} by $c'_{st} + y_s$.

We leave to the Exercises the proof of the next result, verifying that this routine does the right thing.

Proposition 5.15 *After the application of the expand subroutine, M' is a matching contained in $E_=$, and T is an M'-alternating tree whose edges are all contained in $E_=$.* ∎

We can now state an algorithm for finding a minimum-weight perfect matching.

Blossom Algorithm for Minimum-Weight Perfect Matching

Let y be a feasible solution to (5.9), M' a matching of $G_=$, $G' = G$;
Set $T = (\{r\}, \emptyset)$, where r is an M'-exposed node of G';
Loop
    Case: There exists $e \in E_=$ whose ends in G' are $v \in B(T)$
             and an M'-exposed node $w \notin V(T)$
        Use vw to augment M';
        If there is no M'-exposed node in G'
            Extend M' to perfect matching M of G and stop;
        Else
            Replace T by $(\{r\}, \emptyset)$, where r is M'-exposed;
    Case: There exists $e \in E_=$ whose ends in G' are $v \in B(T)$
             and an M'-covered node $w \notin V(T)$
        Use vw to extend T;
    Case: There exists $e \in E_=$ whose ends in G' are $v, w \in B(T)$
        Use vw to shrink and update M', T, and c';
    Case: There is a pseudonode $v \in A(T)$ with $y_v = 0$
        Expand v and update M', T, and c';
    Case: None of the above
        If every $e \in E$ incident in G' with $v \in B(T)$ has its
             other end in $A(T)$ and $A(T)$ contains no pseudonode
        Stop, G has no perfect matching;
        Else
            Change y.

Theorem 5.16 *The Blossom Algorithm terminates after performing $O(n)$ augmentation steps, and $O(n^2)$ tree-extension, shrinking, expanding, and dual change steps. Moreover, it returns a minimum-weight perfect matching, or determines correctly that G has no perfect matching.*

Proof: There are clearly $O(n)$ augmentation steps, so it suffices to show that there are $O(n)$ of the other kinds of steps while a single tree is being constructed. Each dual change step leads to one of the others, so we can restrict ourselves to extension, shrinking, and expanding. Note that every shrinking creates a new even node of T and every expansion is of an odd pseudonode of T. Consider the sum of $|S(v)|$ over odd pseudonodes v of T, minus the sum of $|S(v)|$ over pseudonodes of G' not in T. (Recall the definition of $S(v)$ given on page 138.) An expansion decreases this quantity, and a shrinking or tree-extension does not increase it. Therefore, there are $O(n)$ expansions. Similarly, each shrinking step increases the sum of $|S(v)|$ over even pseudonodes of T, and no other step decreases it, so the number of shrinkings is $O(n)$.

Finally, each tree-extension increases the number of edges of T by 2, whereas the total decrease of this number due to shrinkings is at most n, and expansions do not decrease it. So the number of tree-extension steps is also O(n), as required.

If the algorithm stops with the conclusion that there is no perfect matching, this is correct, for the same reason as for the unweighted algorithm. If it stops with the claim that a minimum-weight matching has been found, then before expansion, it had found a perfect matching of a derived graph, and a dual solution y satisfying the optimality conditions (5.22), (5.23) (with $Y = 0$) and with the additional property that $y_v \geq 0$ for every pseudonode v. Now Proposition 5.14, applied repeatedly, gives us a perfect matching M of G and a feasible solution (y, Y) of (5.17) such that conditions (5.22) and (5.23) are satisfied. Therefore, M is an optimal perfect matching and (y, Y) is an optimal solution to (5.17). ∎

Notice that we now have a proof of Theorem 5.13, except for the part that says that there is no perfect matching if and only if (5.13) has no feasible solution. This part is left as an exercise. There are some other useful consequences of the algorithm, which we point out now.

Optimal Dual Solutions of Matching Problems

The matching algorithm can be used to prove several important properties of optimal dual solutions. The first such property is an immediate consequence of the algorithm. Namely, if we consider the family $\mathcal{S}$ of sets S such that $Y_{\delta(S)} > 0$, we can choose each such S to be $S(v)$ for a pseudonode v of a derived graph G' of G, and moreover, the final derived graph G'' is a derived graph of G'. It follows that, for any two sets of this family, either they are disjoint, or one is contained in the other. Such a family of sets is called *nested*, and in this case, (y, Y) is also said to be *nested*. So the algorithm proves the following result.

Theorem 5.17 *If the dual linear-programming problem (5.17) has an optimal solution, then it has one that is nested.* ∎

Since we have proved that the linear-programming problem (5.13) has an optimal solution that is integral (if it has an optimal solution at all), we might wonder whether the same might be true of the dual linear-programming problem (5.17). We can see one reason why this might not be the case—if we have ugly numbers for the weights, then the right-hand side of (5.17) would be ugly, and we could not hope for such a property. So we should expect to have such a solution, in general, only when the weights are all integers. But even this is too much to ask; consider the example of the complete graph on four nodes in which all weights are 1. Here the only optimal dual solution has $y_v = 1/2$ for all nodes v. (The odd cuts do not help, since all of them are of the form $\delta(v)$.) However, we can prove the existence of an integral solution with a slightly stronger hypothesis. This result was stated and proved algorithmically by Barahona and Cunningham [1989]. However, it is a consequence of a result of Seymour [1981], which we will see in Section 5.4.

Theorem 5.18 *If $c \in \mathbf{Z}^E$, $c(E(C))$ is an even integer for every circuit C of G, and (5.17) has an optimal solution, then it has an optimal solution that is nested and integral.*

There is an easy corollary which tells us that the four-node example above is the worst that can happen with integral weights. (We say a vector p is "half-integral" if $2p$ is integral.)

Corollary 5.19 *If $c \in \mathbf{Z}^E$ and (5.17) has an optimal solution, then it has an optimal solution that is nested and half-integral.*

Proof: Obviously $2c$ has the property that its sum on every circuit is an even integer. Therefore, by Theorem 5.18, (5.17) has an integral nested optimal solution (y, Y) when c is replaced by $2c$. Now $\frac{1}{2}(y, Y)$ is the required solution. ∎

Proof of Theorem 5.18: We show that the weighted matching algorithm keeps both the property that y is integral and that the sum of the weights on the edges of any circuit of any derived graph is an even integer. We use the fact that, if y and c' are integral, then a path of equality edges from u to v has weight congruent to $y_u + y_v$ (mod 2). (To see this, just add up the equations $y_a + y_b = c'_e$ for path edges e having ends a, b in G'.)

First, suppose that y is integral, but that we shrink an odd circuit C and create for the first time a circuit C' with odd weight. (Say that this shrinking occurs in the derived pair (G, c) and the new derived pair is (G', c') where $G' = G \times C$.) Then C' must go through the pseudonode C of G'. Let u, v be the nodes of C that are ends in G of the edges of C' incident with the pseudonode C of G'. Then $c'(E(C')) = c(E(C')) - y_u - y_v$. Take a path P from u to v in C having as few edges as possible. Then P together with C' forms a circuit C'' in G. Now (congruences are all mod 2)

$$c(E(C'')) = c(E(C')) + c(E(P)) \equiv c(E(C')) + y_u + y_v \equiv c'(E(C')) \equiv 1,$$

a contradiction.

Now suppose that y is integral, but expansion of a pseudonode C creates for the first time a circuit C' having odd weight. Suppose that the new derived pair is (G, c), and that the old one is (G', c'), where $G' = G \times C$. Then C' must have nodes in both $V(C)$ and $V \setminus V(C)$. Let P be a simple path formed by a subsequence of C' joining nodes $u, v \in V(C)$, but using only edges of G'. Let P' be a path in C joining u to v and having as few edges as possible. Then $E(P)$ is the edge-set of a circuit in G', so $c'(E(P))$ is even, and so $c(E(P)) \equiv y_u + y_v$. Also, since the edges of P' are all equality edges,

$c(E(P')) \equiv y_u + y_v$. This means that we could replace P by P' in C' and the weight would still be odd. (The new path will be closed, but might no longer be a circuit.) Repeating this argument, we eventually get a closed path in $G[V(C)]$ with odd weight. But this implies that $G[V(C)]$ has a circuit of odd weight. This is a contradiction, since this circuit would already have existed when C was shrunk.

Now suppose that a dual change is about to make y nonintegral for the first time. Then there are nodes $u, v \in B(T)$ joined by an edge e such that $\bar{c}_e$ is odd. The path in T from u to v consists of equality edges, and so its weight is congruent to $y_u + y_v$. It follows that the circuit formed by e together with this path has weight congruent to

$$y_u + y_v + c_e \equiv \bar{c}_e \equiv 1.$$

This contradicts the fact that circuit weights remain even as long as y remains integral. Therefore, y remains integral throughout, and so the optimal solution (y, Y) available at termination is integral too.   ∎

The Triangle Inequality

Our last result on properties of optimal dual solutions applies to a class of weight functions that frequently arise in practice. If $G = (V, E)$ is a complete graph and $c \in \mathbf{R}^E$, we say that c satisfies the *triangle inequality*, if $c_{uv} + c_{vw} \geq c_{uw}$ for all $u, v, w \in V$. Of course, weights that arise from a distance function, such as the Euclidean distances discussed in Chapter 1, satisfy the triangle inequality. In case weights are nonnegative and satisfy the triangle inequality, we can show that there is an optimal dual solution (y, Y) for which $y \geq 0$. This result will be applied to postman problems and Euclidean matching problems in the next two sections.

Theorem 5.20 *Let $G = (V, E)$ be a complete graph having an even number of nodes and let $c \in \mathbf{R}^E$ such that $c \geq 0$ and c satisfies the triangle inequality. If the perfect matching algorithm is started with $(y, Y) = 0$, then the optimal dual solution at termination of the algorithm will satisfy $y \geq 0$.*

Proof: We show that y remains nonnegative throughout the execution of the algorithm. If y_v is decreased by a dual variable change, then v is an odd node of the current tree T. Let e, f be the two edges of T to which v is incident, let u be the other end of e in G, let w be the other end of f in G, and let $g = uw$. Extend the current dual solution to a feasible solution (y, Y) of (5.17), let $\mathcal{C}_e$ denote the family of odd cuts D for which $Y_D > 0$ and $e \in D$, and similarly for $\mathcal{C}_f$ and $\mathcal{C}_g$. Notice that $\mathcal{C}_g$ is the disjoint union of $\mathcal{C}_e$ and $\mathcal{C}_f$.

We need to show that the amount ε of the dual change is at most y_v. Since $\bar{c}_e = \bar{c}_f = 0$, we have

$$2\varepsilon \leq \bar{c}_g \quad = \quad c_{uw} - y_u - y_w - Y(\mathcal{C}_g)$$

$$
\begin{aligned}
&\leq & c_{uv} + c_{vw} - y_u - y_w - Y(\mathcal{C}_e) - Y(\mathcal{C}_f) \\
&= & \bar{c}_e + y_v + \bar{c}_f + y_v \\
&= & 2y_v,
\end{aligned}
$$

as required.   ■

Implementation of the Weighted Matching Algorithm

Recall that we described an implementation of the unweighted matching algorithm that worked on the original graph G and represented the current derived graph G' implicitly. Namely, it kept a disjoint set structure to represent the sets $S(v)$ corresponding to nodes v of G'. It needed only to do two operations on this structure, finds and merges. The "splits" (dividing some $S(v)$ into disjoint sets) that would correspond to pseudonode expansions were unnecessary, because the structure was simply reinitialized to a set of singletons after an augmentation. In the weighted algorithm it would be necessary to handle splits, so an implementation representing the current derived graph in this way would be more complicated (but it can be done). We describe instead an implementation of the weighted algorithm that represents derived graphs explicitly. Because of this choice, the implementation is actually quite simple to describe.

So suppose that we have at hand a representation of G'. (We keep for each edge its ends, and for each node the list of its incident edges, which we may refer to as its "incidence list.") The main work in the algorithm (aside from the updating of the representation, which we describe later) is the dual variable change. This is because we may have to do such an operation before being able to do any of the other steps, and its outcome tells us which of the steps we can do. We can perform a dual change in the following quite simple way. For each node in $B(T)$ we run through its incidence list, checking for each edge e whether its other end is outside T or is another node of $B(T)$. In each of those two cases, we compute $\bar{c}_e$ or $\frac{1}{2}\bar{c}_e$, and we take the minimum of all of these numbers. We also find the minimum of y_v for all pseudonodes $v \in A(T)$. After this, changing y is easy. So a dual change takes time $O(m)$. The other basic steps, tree-extension, shrinking, expanding, and augmenting, are no more complex than dual changes and the bound on their number is no larger, so we easily get a bound of $O(n^2 m)$ for all the work of the algorithm, except for the updating of the graph representation.

We keep a representation of G as well as the representation of the current G'. Of course, we keep for each pseudonode formed (and not yet expanded), the circuit C that was shrunk to form it. We are going to show that these structures can be updated in time $O(m)$ after each shrinking or pseudonode expansion. In fact, the structures other than the graph representation are trivial to update, so we concentrate on the graph representation. The update after the shrinking of an odd circuit C is quite easy. We run through the incidence lists for all the nodes in $V(C)$. For each edge encountered, we either add it to the incidence list for C and appropriately update its edge-end information or (if its other end is in $V(C)$ also) put it, with its edge-end information, on C's "inactive" list. The expansion of a pseudonode C is more complex. We go through C's inactive list, updating the incidence lists, and restoring the edge-end information; this takes care of edges that have both ends in $V(C)$ in the new G'. Then we have to deal with the edges having one end in $V(C)$, and this is harder. We find the subsets $S(v)$ of V for all $v \in V(C)$. This requires tracing $O(n)$ pointers in total. Now we use the representation of G. We run through the incidence list of C, find which $S(v)$ one of its ends is in, and update the relevant incidence list and edge-end information. In both cases we do a constant amount of work per edge considered, so the time required is certainly $O(m)$. Since there are $O(n^2)$ shrinkings and expansions, the time required for maintaining the representation of G' throughout the algorithm is $O(n^2m)$. Therefore, we have the following result.

Theorem 5.21 *The Blossom Algorithm for finding a minimum-weight perfect matching can be implemented to run in time $O(n^2m)$.* ∎

It is easy to spot aspects of the above implementation that are susceptible to improvement. The obvious ones are the dual changes, and the graph representations. We have already mentioned that ideas like those used for the graph representation in the unweighted algorithm can be extended to the present situation, but not easily. Similarly, the ideas that we have used in the bipartite case to give an improved bound for performing a dual change can be applied, and again, complications arise. An implementation of a weighted matching algorithm having a better running time can be found in Gabow [1990]. In practice, the approach we have described seems to work quite well; see Applegate and Cook [1993]. We also point out that the running time we have derived looks quite bad ($O(n^4)$) for dense graphs. But this is misleading, since there are techniques that allow doing most of the work on a sparse subgraph. These are described below.

Price and Repair

When solving large instances of the minimum-weight perfect matching problem, it is impractical to work with the edge set of any graph G that is not extremely sparse. The storage requirements are one obvious reason, but equally important is the fact that the practical running time of the Blossom Algorithm is greatly influenced by the number of edges in G. (This is certainly true for the implementation that we described above, but it also applies to more sophisticated implementations.) A standard procedure in combinatorial optimization for dealing with such a situation is to use a two-stage approach to handle large dense graphs. We first choose a sparse subgraph G^* of G and find the optimal perfect matching contained in it. (A typical choice for G^* is to take the k edges of least weight meeting each node v, for some small value of k, for example, $k = 5$.) Next, we compute the reduced costs of the edges in G that have not been included in G^*. (We call this the *pricing* step.) If any edge e has negative reduced cost (that is, the dual solution does not satisfy the inequality (5.18) for e), then we must *repair* the matching by adding e to G^* and recomputing the optimal matching and dual solution. After repairing all negative reduced cost edges, we repeat the pricing step, and again make repairs. We continue these price-repair rounds until we have no edges having negative reduced cost. At this point, by setting $x_e = 0$ for all edges not in G^*, the complementary slackness conditions hold for the entire graph G, and thus we have an optimal perfect matching. This process is a simple version of the general column generation approach that we outlined in Section 3.6. It was first used in this context by Derigs and Metz [1991].

It is an easy matter to carry out the pricing step efficiently, by taking advantage of the nested structure of the final derived graph. We leave this as Exercise 5.39.

Repairing the matching is a bit more difficult. We could, of course, just find all of the negative edges, add them to G', and recompute the optimal perfect matching from scratch. This would be rather time consuming, however, since we would be repeating much of our earlier work. Instead, we would like to modify the final derived graph G' so that we can add the new edge $uv = e$ and continue the Blossom Algorithm. We describe a procedure, due to Ball and Derigs [1983], that accomplishes this goal. (An alternative approach is described in Cunningham and Marsh [1978].)

The procedure uses the fact that if either u or v happens to be a node in G', then the modification is easy. Indeed, if u is a node of G', then we can decrease the value of y'_u until $\bar{c}_e = 0$. To maintain the complementary slackness conditions, we need to set $x_h = 0$ for the matching edge h that meets node u. We can now continue the Blossom Algorithm, starting from the exposed node u.

The difficulty arises when both u and v are contained in pseudonodes of G'. In this case, we cannot simple alter the value of y_u or y_v since this would disturb the structure of G'; for example, one of the shrunk odd circuits will no longer consist of equality edges. To handle this, we first make dual changes to make u a node of G', expanding all pseudonodes containing u. Ball and Derigs accomplish this by allowing the Blossom Algorithm to do the work. Their idea is to add a new node p and an edge f joining p and u (giving f the reduced cost 0) and apply the Blossom Algorithm from the exposed node p. Of course, we will find no augmenting path, but during the search we will perform dual changes that allow us to expand pseudonodes. We stop the process as soon as u becomes a node of the derived graph. Then, deleting p and f, we proceed as above.

This two-stage approach to matching problems makes it possible to solve large dense instances in reasonable amounts of computer time. For example,

Cook and Rohe [1997] report on the solution of a Euclidean matching instance having over 5,000,000 nodes.

Maximum-Weight Matchings

Consider the following problem.

Maximum-Weight Matching Problem

Given: A graph $G = (V, E)$ and $c \in \mathbf{R}^E$.

Objective: To find a matching M of G such that $c(M)$ is maximum.

Notice that the maximum matching problem of Section 5.2 is a direct special case of this problem—one just takes $c_e = 1$ for all $e \in E$. We will explain briefly how dropping the requirement that the matching be perfect changes the problem. (The change to maximization is only cosmetic; it arises from the fact that, if we ask for a minimum-weight matching, then the empty matching will be optimal unless there are negative weights, so it seems more natural to maximize.) There are several aspects to cover: how to solve this problem by reducing it to the problem solved already, how to modify the algorithm to handle it directly, and how to obtain a linear-programming description for it.

Let us begin by showing a reduction of the maximum-weight matching problem to the maximum-weight (or minimum-weight) perfect matching problem. Make a copy G^* of G, give all of the edges of the copy the same weight as the corresponding edge in G, and then join each node of G to its copy in G^* by an edge of weight zero. Let Q be the set consisting of the latter edges and let $\hat{G}$ denote the resulting graph. This construction is illustrated in Figure 5.18. Moreover, given a matching M of G, we can choose its copy M^* in G^* and

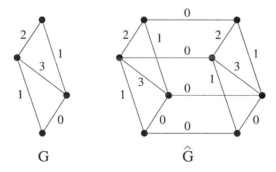

Figure 5.18. Construction of $\hat{G}$

then complete $M \cup M^*$ to a perfect matching of $\hat{G}$ with edges from Q. Thus to every matching of G there corresponds a perfect matching of $\hat{G}$ of exactly twice the weight. Now given a perfect matching $\hat{M}$ of $\hat{G}$, we can delete the nodes covered by edges of $\hat{M} \cap Q$, and we are left with matchings M_1 of G and M_2 of G^*. Clearly, if $\hat{M}$ is a maximum-weight perfect matching of $\hat{G}$, then M_1 must be a maximum-weight matching of G. Hence, we can solve the maximum-weight matching problem in G by solving a maximum-weight perfect matching problem in $\hat{G}$.

Now we give a linear-programming problem whose optimal value is the maximum weight of a matching of G. Let S be a subset of V of odd cardinality. Since any matching of G cannot intersect $\gamma(S)$ in more than $(|S|-1)/2$ edges, the characteristic vector x of any matching of G satisfies $x(\gamma(S)) \leq (|S|-1)/2$. This observation leads to the following linear-programming problem, whose optimal value is clearly an upper bound on the maximum weight of a matching.

$$\text{Maximize } \sum(c_e x_e : e \in E) \qquad (5.24)$$
$$\text{subject to}$$
$$x(\delta(v)) \leq 1, \text{ for all } v \in V$$
$$x(\gamma(S)) \leq (|S| - 1)/2, \text{ for all } S \subseteq V, |S| \geq 3, |S| \text{ odd}$$
$$x_e \geq 0, \text{ for all } e \in E.$$

In fact, it is again true that the bound is exact.

Theorem 5.22 *The maximum weight of a matching of G is equal to the optimal value of (5.24).*

This result can be proved by applying Theorem 5.13 to $\hat{G}$ constructed above. (See Exercise 5.34.) It can also be proved by modifying the weighted perfect matching algorithm to solve the current problem directly. We sketch such a modification now. We denote by $\mathcal{O}$ the set

$$\{S \subseteq V : |S| \text{ odd and at least } 3\}.$$

The dual linear-programming problem to (5.24) is

$$\text{Minimize } \sum(y_v : v \in V) + \sum(\tfrac{1}{2}(|S| - 1)z_S : S \in \mathcal{O}) \qquad (5.25)$$
$$\text{subject to}$$
$$y_u + y_v + \sum(z_S : e \in \gamma(S), S \in \mathcal{O}) \geq c_e \text{ for all } e \in E$$
$$y_v \geq 0 \text{ for all } v \in V$$
$$z_S \geq 0 \text{ for all } S \in \mathcal{O}.$$

The crucial difference from (5.17) is the nonnegativity requirement on y_v. Given a feasible solution (y, z) of (5.25), we denote by $E_=$ the set of edges e for which the corresponding constraint of (5.25) holds with equality. By writing down the complementary slackness conditions for the dual pair of problems (5.24),(5.25) and considering the case when a feasible solution x of (5.24) is the characteristic vector of a matching M of G, we get the following optimality conditions:

$$M \subseteq E_= \tag{5.26}$$

$$\text{for all } S \in \mathcal{O}, \ z_S > 0 \text{ implies } |M \cap \gamma(S)| = \tfrac{1}{2}(|S| - 1) \tag{5.27}$$

$$\text{for all } v \in V, \ y_v > 0 \text{ implies } v \text{ is } M\text{-covered.} \tag{5.28}$$

The changes that are required in the algorithm relate to the nonnegativity requirement on y and the associated optimality condition (5.28). Of course, the latter condition says that we do not need to worry about an exposed original node v if $y_v = 0$. But what about an exposed pseudonode S? We do not need to worry about it if we can expand it and update the matching so that the element $v \in S$ ultimately left exposed has $y_v = 0$. Now the following observation is useful: An exposed pseudonode can be expanded so that an arbitrary element is left exposed. The proof is just the same as the proof of Proposition 5.4. Therefore, we do not need to worry about an exposed pseudonode precisely if it contains at least one node v such that $y_v = 0$. Therefore, the algorithm chooses the node r from which to search for an augmentation to be either an M'-exposed original node v of G' with $y_v > 0$ or an M'-exposed pseudonode S of G' such that $y_v > 0$ for every $v \in S$. Let us call such a node of G' *bad*. If no bad node exists, of course, it means that we are done, except for extending M' to G.

Another way in which the algorithm is different is that we must ensure that y remains nonnegative after a dual variable change. The last change is the trickiest. We need to allow a different kind of augmentation, which occurs when there is a node $v \in B(T)$ that is not bad. Then we use the even-length M'-alternating path from v to r to change M'. Notice that this change does not increase the cardinality of M', but it does decrease the number of bad nodes, which is what matters.

Exercises

5.22. Show that (5.8) has a feasible solution if and only if there exists a set of edges and odd circuits such that every node of G is a node of exactly one of the odd circuits or is incident to exactly one of the edges, but not both.

5.23. Show that the converse of Theorem 5.12 is false. That is, show that there exists a nonbipartite graph with the property that, for every $c \in \mathbf{R}^E$, the optimal value of (5.8) is equal to the minimum weight of a perfect matching of G.

5.24. Prove Theorem 5.12 by transforming the minimum-weight perfect matching problem in a bipartite graph to a minimum-cost flow problem, and applying Theorem 4.5.

5.25. Prove Theorem 5.12 as follows. Let x be an optimal solution of (5.8) that has a minimum number of nonintegral components. If x is not integral, show that there exists a circuit, for each of whose edges e, x_e is not integral. Use this circuit to find another optimal solution having fewer nonintegral components.

5.26. Prove Proposition 5.15.

5.27. Find a minimum-weight perfect matching and an optimal solution of the dual problem for the example of Figure 5.19.

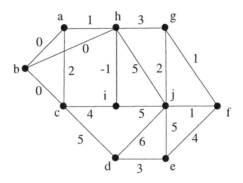

Figure 5.19. Optimal matching example

5.28. Use the matching algorithm to prove that G has a perfect matching if and only if (5.13) has a feasible solution. (Hint: To prove "if," show that if there is no perfect matching, then the dual problem is unbounded.)

5.29. Prove that the optimal value of the following problem is the minimum weight of a perfect matching of G.

$$\text{Minimize } \sum(c_e x_e : e \in E) \tag{5.29}$$
$$\text{subject to}$$
$$x(\delta(v)) = 1 \text{ for all } v \in V$$
$$x(\gamma(S)) \le (|S| - 1)/2 \text{ for all } S \subseteq V, |S| \ge 3, |S| \text{ odd}$$
$$x_e \ge 0 \text{ for all } e \in E.$$

5.30. Let $c_e = 1$ for all $e \in E$, and let $\{B, C, D\}$ be the Gallai-Edmonds Partition of G. Define y_v to be 0 if $v \in B$, 1 if $v \in C$, and $1/2$ if $v \in D$. Define z_S to be 1 if S is a component of $G[B]$ and 0 otherwise. Show that (y, z) is an optimal solution of (5.24). Derive properties of maximum matchings by applying complementary slackness.

5.31. Let $G = (V, E)$ be a graph, let $c \in \mathbf{R}^E$, and let k be a positive integer. Show how the problem of finding a minimum-weight matching having cardinality k can be reduced to a minimum-weight perfect matching problem.

5.32. Let (z, Z) be the variables of the dual linear-programming problem to (5.29), where $z \in \mathbf{R}^V$, $Z \in \mathbf{R}^{\mathcal{O}}$. Let (y^*, Y^*) be an optimal solution to (5.17). Write each positive Y_D^* as $Y_{\delta(S)}^*$ for some $S \in \mathcal{O}$ such that $D = \delta(S)$. Let $Z_S^* = -2Y_{\delta(S)}^*$ and let $z_v^* = y_v^* + \sum (Y_{\delta(S)}^* : v \in S)$. Prove that (z^*, Z^*) is an optimal solution to the dual of (5.29).

5.33. Prove that if c is integral, and the dual of (5.29) has an optimal solution, then it has an optimal solution (z^*, Z^*) such that Z^* is integral and z^* is half-integral.

5.34. Prove Theorem 5.22 by applying Theorem 5.13 to the graph $\hat{G}$ constructed from G in the reduction of the maximum-weight matching problem to the maximum-weight perfect matching problem. (Hint: Let x be an optimal solution to (5.24). Define a vector $\hat{x} \in \mathbf{R}^{E(\hat{G})}$ as follows. If $e \in E$, define $\hat{x}_e$ to be x_e. If e^* is the edge of G^* corresponding to $e \in E$, define $\hat{x}_{e^*}$ to be x_e. If $e(v)$ is the edge in Q that joins $v \in V$ to its copy in G^*, define $\hat{x}_{e(v)}$ to be $1 - x(\delta(v))$. Now prove that $\hat{x}$ is a feasible solution of (5.13) defined on $\hat{G}$.)

5.35. Let M be a perfect matching of G, and define the "cost" of an M-alternating circuit C of G to be $c(E(C) \backslash M) - c(E(C) \cap M)$. Prove that M is of minimum weight with respect to c if and only if there is no M-alternating circuit of negative cost. Hint: Consider symmetric differences.

5.36. Given a graph $G = (V, E)$, $c \in \mathbf{R}^E$, and a perfect matching M of G, we form a digraph $G' = (V, E')$ with arc-cost vector c', as follows. There is an arc $vw \in E'$ for every path v, vu, u, uw, w of G such that $uw \in M$; the arc has cost $c_{vu} - c_{uw}$. Show that

(a) If M is not a minimum-weight perfect matching of G, then G' has a dicircuit of negative cost.

(b) If G' has a dicircuit of negative cost and G is bipartite, then G has a perfect matching of smaller weight than M.

(c) Statement (b) is not true for general graphs.

5.37. (J. Geelen) Show that one can compute a minimum-weight edge cover by computing a maximum-weight matching. Hint: You may want to consider the edge weights $c_{uv}' = w_u + w_v - c_{uv}$, where w_v is $\min(c_e : e \in \delta(v))$.

5.38. Let r, s be nodes of a graph $G = (V, E)$ and let $c \in \mathbf{R}^E$ with $c \geq 0$. We want to find a minimum-weight simple (r, s)-path having an odd number of edges. Make $\hat{G}$ as in the construction that reduces the maximum-weight matching problem to the maximum-weight perfect matching problem, and delete nodes s^* and r^*. Explain how a minimum-weight perfect matching of the new graph solves this shortest odd path problem.

5.39. Let $G = (V, E)$ be a graph and $c \in \mathbf{R}^E$. Suppose we apply the Blossom Algorithm to compute a minimum-weight perfect matching in G. Let G' be the final derived graph that is produced and let (y, Y) be the corresponding dual solution. Describe a method for computing the reduced costs (relative to (y, Y)) corresponding to each pair of nodes $u, v \in V$.

5.4 T-JOINS AND POSTMAN PROBLEMS

Postman Problems

A postman wishes to deliver mail along all the edges of a graph $G = (V, E)$ and return to his starting point; that is, he wants to find a closed path, traversing every edge at least once. Such a path is called a *postman tour* of G. It may be that it is necessary to traverse some edges while not delivering mail, just to get to other edges where mail has not yet been delivered. There is a cost c_e associated with each such traversal of edge e. (We will assume that $c \geq 0$; we also assume that G is connected.) The *postman problem* is the problem of finding a postman tour of minimum cost, or equivalently, a postman tour that minimizes the cost of the "extra" edge-traversals. Of course, if there exists such a tour with no extra edge traversals, then that tour is an optimal postman tour. A closed edge-simple path P such that $E(P) = E$ is called an *Euler tour* of G. So if G has an Euler tour, then it is an optimal postman tour. There is a well-known necessary and sufficient condition for a graph to have an Euler tour. See Exercise 5.40 for a proof, as well as an algorithm to construct a tour when one exists. The *degree* of a node v of a (loopless) graph G is just the number of edges of G incident with v (assuming that G has no loops). We denote the degree of node v by $\deg_G(v)$.

Theorem 5.23 *A connected graph G has an Euler tour if and only if every node of G has even degree.* ∎

This theorem gives an easy way to recognize when there is an optimal postman tour of a special kind, but it also can be used more generally. Suppose that we have a postman tour of G. Let x_e be the number of extra traversals of edge e for each $e \in E$. Construct the graph G^x by making $1 + x_e$ copies of edge e for each e. (In Figure 5.20 we show G, x, and G^x for the path P with node-sequence $a, b, c, d, e, f, c, g, d, e, f, b, c, g, a$.) Then it is easy to see

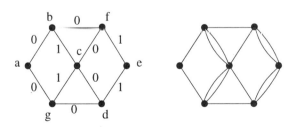

Figure 5.20. G, x, and G^x

that G^x has an Euler tour. Conversely, suppose that $x \in \mathbf{Z}^E$ is nonnegative, and has the property that G^x has an Euler tour. Then there is a postman tour of G that uses edge e x_e extra times. In view of these observations and Theorem 5.23, the postman problem is equivalent to the problem:

$$\text{Minimize } \sum(c_e x_e : e \in E) \tag{5.30}$$
$$\text{subject to}$$
$$x(\delta(v)) \equiv |\delta(v)| \pmod 2, \text{ for all } v \in V$$
$$x_e \geq 0, \text{ for all } e \in E$$
$$x_e \text{ integer, for all } e \in E.$$

From this description of the problem, we can immediately see that there is an optimal solution for which x is $\{0, 1\}$-valued. The reason is that if some $x_e \geq 2$, we can simply lower x_e by 2, giving another solution with cost no worse than the given one. Therefore, we can look for a set J of edges (the set whose characteristic vector is x). We call a set J a *postman set* of G if for every $v \in V$, v is incident with an odd number of edges from J if and only if v has odd degree in G. Then our problem can be stated as

---

Postman Problem

Given: A graph $G = (V, E)$ and $c \in \mathbf{R}^E$ such that $c \geq 0$.

Objective: To find a postman set J such that $c(J)$ is minimum.

We will show that this problem belongs to a class of problems that can be solved with matching techniques. In fact, although the method we will use works in the wider context of "T-joins," it was introduced by Edmonds [1965] specifically to solve the postman problem.

Before going on, let us point out that the postman problem can be stated also for digraphs. It is again true that the problem can be reduced to the problem of finding x of minimum cost so that the resulting digraph G^x has a (directed) Euler tour. However, in this case the resulting problem is a bit easier, and can be solved by network flow methods. See Exercises 4.4 and 4.6.

T-Joins

Let $G = (V, E)$ be a graph, and let $T \subseteq V$ such that $|T|$ is even. A *T-join* of G is a set J of edges such that

$$|J \cap \delta(v)| \equiv |T \cap \{v\}| \pmod 2, \text{ for all } v \in V.$$

In other words, J is a T-join if and only if the odd-degree nodes of the subgraph (V, J) are exactly the elements of T. We are interested in the following problem.

Optimal T-Join Problem

Given: A graph $G = (V, E)$, a set $T \subseteq V$ such that $|T|$ is even, and a cost vector $c \in \mathbf{R}^E$.

Objective: Find a T-join J of G such that $c(J)$ is minimum.

Here are some examples.

Postman sets. Let $T = \{v \in V : |\delta(v)| \text{ is odd}\}$. Then the T-joins are precisely the postman sets. We have shown that (assuming that $c \geq 0$), finding an optimal T-join solves the postman problem.

Even set. Let $T = \emptyset$. Then a T-join is exactly an *even set*, that is, a set $A \subseteq E$ such that every node of (V, A) has even degree. A set is even if and only if it can be decomposed into edge-sets of edge-disjoint circuits. (See Exercise 5.41.) Of course, the optimal T-join problem is trivial if the costs are nonnegative, because the subgraph having no edges will be optimal. More generally, because of the decomposability property mentioned above, the empty set will be optimal if and only if G has no negative-cost circuit. It follows that we can find a negative-cost circuit or determine that none exists by solving an optimal T-join problem.

(r, s)-paths. Let $r, s \in V$, and let $T = \{r, s\}$. It is easy to see that every T-join J contains the edge-set of an (r, s)-path. (Proof: If not, the component of the subgraph (V, J) that contains r has only one node of odd degree.) Therefore, the minimal T-joins are precisely the edge-sets of the simple (r, s)-paths. Now suppose that we have a T-join that is not minimal. Then deleting the edge-set of a simple (r, s)-path, we are left with a nonempty even set. Therefore, if every circuit has nonnegative cost, then solving the optimal T-join problem will find a least-cost simple (r, s)-path. (If negative-cost circuits are allowed, then the least-cost (r, s)-path problem is difficult, just as was the case for digraphs.) It may seem that we are discussing a problem that we already know how to solve. In fact, the shortest path problem in an undirected graph with nonnegative costs can be solved by ordinary shortest path techniques, but when negative costs are allowed, that solution method fails. See Exercise 2.37. We will be able to fill the gap with the methods of this chapter, that is, to solve the undirected shortest path problem in the case where negative costs are allowed, but negative cost circuits are not.

The following observation about T-joins will often be useful.

Proposition 5.24 *Let J' be a T'-join of G. Then J is a T-join of G if and only if $J \triangle J'$ is a $(T \triangle T')$-join of G.*

Proof: It will be enough to prove the "only if" part. The other part can be deduced by applying this one with J replaced by $J \bigtriangleup J'$ and T replaced by $T \bigtriangleup T'$. So suppose that J is a T-join and J' is a T'-join. Let $v \in V$. Then $|(J \bigtriangleup J') \cap \delta(v)|$ is even if and only if $|J \cap \delta(v)| \equiv |J' \cap \delta(v)| \pmod 2$, which is true if and only if v is an element of neither or both of T and T', that is, if and only if $v \notin T \bigtriangleup T'$. ∎

Solving the Optimal T-Join Problem

We will show how to use matching to find an optimal T-join when the edge-costs are nonnegative. This will give a solution to the postman problem. In addition, we will show that it is easy to transform a problem with negative costs into one with nonnegative costs, so this will allow the solution of the optimal T-join problem in general.

When $c \geq 0$, there is always an optimal T-join that is minimal. The minimal T-joins can be characterized, as follows.

Proposition 5.25 *Every minimal T-join is the union of the edge-sets of $|T|/2$ edge-disjoint simple paths, which join the nodes in T in pairs.*

Proof: It is easy to see that any set that is the union of such paths is a T-join, so it will be enough to show that any T-join J contains such a set of edge-disjoint paths. Let $u \in T$, and let H be the component of (V, J) that contains u. If there is no node other than u in $T \cap V(H)$, then H has only one node of odd degree, which is impossible. Therefore, a node $v \neq u$ is also a node of both T and H. It follows that there is a simple (u, v)-path P such that $E(P) \subseteq J$. Now $J \backslash E(P)$ is a T'-join, where $T' = T \backslash \{u, v\}$, by Proposition 5.24, and we can repeat the argument. ∎

Now suppose that an optimal T-join J is expressed as a union of edge-sets of paths as in Proposition 5.25, and P is one of these paths, with P joining nodes $u, v \in T$. Suppose that there is a (u, v)-path P' in G that has smaller cost than P. Notice that by Proposition 5.24, $J \bigtriangleup E(P) \bigtriangleup E(P')$ is a T-join. What is its cost? Because $E(P) \subseteq J$, it is

$$c(J \backslash E(P)) + c(E(P')) - 2c((J \backslash E(P)) \cap E(P'))$$
$$\leq c(J) - c(E(P)) + c(E(P')) < c(J),$$

a contradiction. Therefore, P is a minimum-cost (u, v)-path, and we have proved the following result.

Proposition 5.26 *Suppose that $c \geq 0$. Then there is an optimal T-join that is the union of $|T|/2$ edge-disjoint shortest paths joining the nodes of T in pairs.* ∎

For any pair u, v of nodes in T, let $d(u, v)$ be the cost of a least-cost (u, v)-path in G. Let $|T| = 2k$. It follows from Proposition 5.26 that (if $c \geq 0$) the minimum cost of a T-join is

$$\text{Minimize } \sum (d(u_i, v_i) : \; i = 1, \ldots, k) \tag{5.31}$$

$$\text{subject to}$$

$$u_1 v_1, \ldots, u_k v_k \text{ is a pairing of the elements of } T.$$

So, given the values $d(u, v)$, the only thing to decide is the best way to pair the nodes in T. But this is a matching problem! We form a complete graph $\hat{G} = (T, \hat{E})$, give edge uv weight $d(u, v)$, and find a minimum-weight perfect matching of $\hat{G}$. This determines a pairing of the nodes of T and therefore a set of paths joining those nodes in the corresponding pairs. These paths may not be edge-disjoint, because of the fact that we allow $c_e = 0$, but their symmetric difference will be a T-join of minimum cost, which is what we want. We illustrate this technique on the example of Figure 5.21, where the elements of T are indicated in black. In Figure 5.22, the complete graph $\hat{G}$ with weights

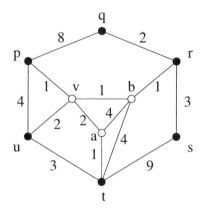

Figure 5.21. T-join example

$d(u, v)$ and a d-optimal perfect matching M are indicated. Taking the symmetric difference of the paths of G corresponding to the matching edges of $\hat{G}$, we get the optimal T-join indicated in Figure 5.23.

Negative Costs

The method described above for finding a minimum-cost T-join works only if the costs are required to be nonnegative. Here we show that if negative costs do occur, it is possible to transform the optimal T-join problem into one having nonnegative costs. This gives us, for example, a polynomial-time algorithm to find a least-cost path in an undirected graph having some negative edge costs,

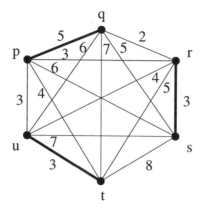

Figure 5.22. $\hat{G}$, d, and M

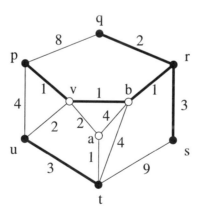

Figure 5.23. Optimal T-join

but no negative-cost circuit. That problem is not solvable by the techniques of Chapter 2.

Given $c \in \mathbf{R}^E$, let $N = \{e \in E : c_e < 0\}$. Let T' be the set of nodes of G that have odd degree in the subgraph (V, N). Then N is a T'-join, and it follows from Proposition 5.24 that J is a T-join if and only if $J \triangle N$ is a $(T \triangle T')$-join. Now

$$
\begin{aligned}
c(J) &= c(J \backslash N) + c(J \cap N) \\
&= c(J \backslash N) - c(N \backslash J) + c(N \backslash J) + c(J \cap N) \\
&= |c|(J \triangle N) + c(N).
\end{aligned}
$$

(Here $|c|$ is the vector defined by $|c|_e = |c_e|$.) Since $c(N)$ is a constant that does not depend on J, it follows that J is an optimal T-join with respect to cost vector c if and only if $J \triangle N$ is an optimal $(T \triangle T')$-join with respect to cost vector $|c|$. Since the latter vector is nonnegative, we have reduced the optimal T-join problem for general cost vectors to the special case where the cost vector is nonnegative.

Let us summarize the method of solving the optimal T-join problem.

Optimal T-Join Algorithm

Step 1. Identify the set N of edges having negative cost, and the set T' of nodes incident with an odd number of edges from N. Replace c by $|c|$ and T by $T \triangle T'$.

Step 2. Find a least-cost (u, v)-path P_{uv} with respect to cost vector c for each pair u, v of nodes from T. Let $d(u, v)$ be the cost of P_{uv}.

Step 3. Form a complete graph $\hat{G} = (T, \hat{E})$ with uv having weight $d(u, v)$ for each $uv \in \hat{E}$. Find a minimum-weight perfect matching M in $\hat{G}$.

Step 4. Let J be the symmetric difference of the edge-sets of paths P_{uv} for $uv \in M$.

Step 5. Replace J by $J \triangle N$.

Of course, if $c \geq 0$, Steps 2 through 4 will do the job. The above method for solving the optimal T-join problem requires the solution, at worst, of an all-pairs shortest path problem and the solution of a minimum-weight perfect matching problem on a complete graph of at most n nodes. So its running time is $O(n^4)$. We illustrate the algorithm on the example at the left in Figure 5.24. Again, the T nodes are indicated in black, so this is a problem of finding a least-cost (r, s)-path. On the right, we indicate the new costs and the new T. In Figure 5.25 we show, first, the optimal T-join for the graph at the right of the previous figure, and then the optimal T-join of the original problem.

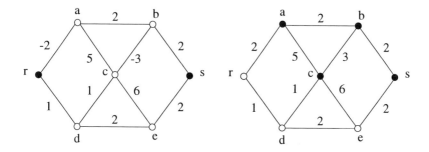

Figure 5.24. G, T, c and $G, T \triangle T', |c|$

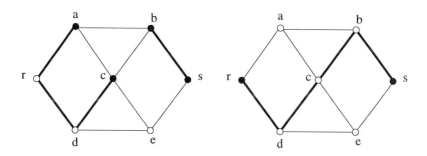

Figure 5.25. $|c|$-optimal $(T \triangle T')$-join and c-optimal T-join

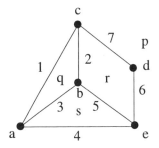 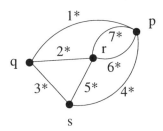

Figure 5.26. A plane embedding and its dual

Maximum Cuts and T-Joins

The *maximum cut problem* is, given an undirected graph $G = (V, E)$ and $c \in \mathbf{R}^E$, to find a cut $D = \delta(A)$ of G such that $c(D)$ is maximum. This is a well-known hard problem—hard in the same sense that the traveling salesman problem is hard. It remains hard when $c_e = 1$ for every edge e, that is, in the case when a maximum cardinality cut is desired. One special case when it is efficiently solvable is related to network flow theory. This is the case when there is at most one edge e such that $c_e > 0$; see Exercise 5.43. Here we show that a second important special case, when G is a planar graph, is efficiently solvable by reduction to an optimal T-join problem. This result is due to Hadlock [1975].

For a plane embedding of a graph $G = (V, E)$, let F denote the set of faces of the embedding. We define the *planar dual* G^* as follows. The nodeset of G^* is F. For each $f, g \in F$ and $e \in E$ such that e is an edge of the boundary of both f and g, we make an edge $e^* = fg$ of G^*. (See Figure 5.26.) Because there is a one-to-one correspondence between the edges of G and G^*, we will consider E to be the edgeset of G^* also. The following result is derived from standard theorems in graph theory. See West [1996], for example.

Proposition 5.27 *A set $D \subseteq E$ is a cut of G if and only if D is an even set of G^*.* ∎

It follows from Proposition 5.27 that, to solve the maximum cut problem on a planar graph, we can solve an optimal T-join problem (with $T = \emptyset$) on its planar dual.

Application to Two-layer Embeddings

We describe, a bit informally, an application of the planar maximum cut problem to the "two-layer embedding of electrical networks." The ideas here are due to Chen, Kajitani, and Chan [1983] and Pinter [1984]. A connected graph $G = (V, E)$ is drawn in the plane, but there are edges crossing, as in Figure 5.27. We will assume that no three edges cross at the same point. We

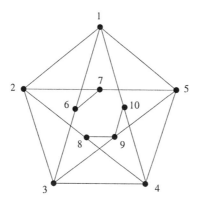

Figure 5.27. A nonplanar graph drawn in the plane

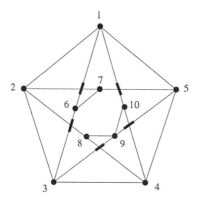

Figure 5.28. A feasible solution

want to "embed" the graph to avoid such crossings, by moving parts of the graph to a second level. Every node will be on one level or the other, but some edges will have to change levels, perhaps more than once, and we want to minimize the total number of such changes, called "vias."

There is a simple way to construct a feasible solution, as follows. Initially, put all of the graph on level 1. For each crossing of two edges e, f, change the level of f, say, just before the crossing and just after it, introducing two vias. In this way we get a feasible solution using two vias for every crossing in the original drawing. For the problem in Figure 5.27, we show in Figure 5.28 such a feasible solution. Here the parts of edges on level 2 are drawn with thick lines. (All the nodes are on level 1.) So there is a via at every point where a line changes width, a total of 10 vias. The original drawing of G defines a planar graph G' as follows. Each edge e of $G = (V, E)$ is divided by the crossings into *segments*; $k + 1$ segments if e is involved in k crossings. We define V' to be $V \cup C$, where C is the set of crossings. There is an edge of G' for each segment, and that edge joins the two nodes of G' at its ends.

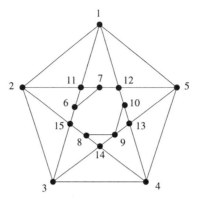

Figure 5.29. The planar graph G'

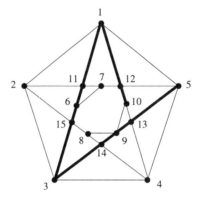

Figure 5.30. G' with the initial choice of vias

Clearly, G' is a planar graph. For the problem of Figure 5.27, G' is shown in Figure 5.29. It is clear that in a solution it is unnecessary for an edge e to change levels more than once between crossings, so we need at most one via in any segment. Conversely, it is easy to see that we can construct a feasible solution from the knowledge of which segments have vias. Therefore, the problem is to choose the set of segments on which vias occur, that is, to choose a certain subset of edges of G'. The set of edges of G' corresponding to segments having vias in the initial solution are shown as thick edges in Figure 5.30.

Now, given the initial feasible solution, an arbitrary second solution differs from it by: (a) changing the level of certain elements of V; (b) changing the "over-under" relationships for certain elements of C. For example, consider the second feasible solution illustrated in Figure 5.31. It has eight vias, and differs from the solution in Figure 5.28 by "switching" of the nodes $6, 9, 10, 13$. So finding an optimal solution can be stated as finding the best subset of nodes of G'; these are the nodes of G whose level must be changed and the crossings

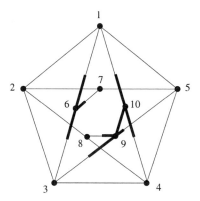

Figure 5.31. A second feasible solution

where the levels of the corresponding parts of the edges must be switched. Consider an edge uv of G'. It is easy to see that:

(a) If both or neither of u, v are switched, then uv has a via in the second solution if and only if it has a via in the first solution;

(b) If exactly one of u, v is switched, then uv has a via in the second solution if and only if it does not have a via in the first solution.

Therefore, if P is the subset of edges of G' having vias in the initial solution and Q is the subset having vias in the second solution, then

$$Q = P \triangle \delta'(A)$$

where A is the set of switched nodes. It follows that $|Q| = |P| + |\delta'(A)\backslash P| - |P\backslash\delta'(A)|$, and so, if we define c_e to be 1 if $e \in P$ and -1 if $e \notin P$, then

$$|Q| = |P| - c(\delta'(A)).$$

Therefore, if we find a maximum-weight cut $\delta'(A)$ of G' with respect to c, then switching the initial solution at the elements of A gives a solution with a minimum number of vias. Since G' is a planar graph, we can find a set A by solving an optimal even set problem (T-join problem with $T = \emptyset$) on the planar dual of G'. For the example of Figure 5.27, we need to solve the maximum cut problem on G' with weights c_e, where $c_e = 1$ for each edge that is thick in Figure 5.30 and $c_e = -1$ for each edge that is thin in Figure 5.30. It turns out that a maximum cut with respect to these weights is $\delta'(\{6, 9, 10, 13\})$, so the solution already indicated in Figure 5.31 is optimal.

LP Problems Related to T-Joins

A set $S \subseteq V$ is T-*odd* if $|S \cap T|$ is odd. In this case $\delta(S)$ is called a T-*cut*. Let J be a T-join and S a T-odd set of nodes. Consider the subgraph

$(S, J \cap \gamma(S))$. It has an even number of nodes of odd degree, but in (V, J), S contains an odd number of nodes of odd degree (namely, those of $T \cap S$). Therefore, we must have

$$|J \cap \delta(S)| \text{ is an odd integer.}$$

It follows that $|J \cap \delta(S)| \geq 1$, and hence that the characteristic vector x of any T-join of G satisfies

$$x(\delta(S)) \geq 1. \tag{5.32}$$

Therefore, the following linear-programming problem provides a lower bound for the cost of any T-join.

$$\text{Minimize } \sum(c_e x_e \,:\, e \in E) \tag{5.33}$$
$$\text{subject to}$$
$$x(D) \geq 1, \text{ for all } T\text{-cuts } D$$
$$x_e \geq 0, \text{ for all } e \in E.$$

Let us observe that, although this lower bound is valid even if we allow negative costs, it does not give much information in that case, because then (5.33) will usually be unbounded. On the other hand, Edmonds and Johnson [1973] proved that, if $c \geq 0$, this lower bound is exact.

Theorem 5.28 *If $G = (V, E)$, $T \subseteq V$ with $|T|$ even, and $c \in \mathbf{R}^E$ with $c \geq 0$, then the minimum cost of a T-join of G is equal to the optimal value of (5.33).*

Edmonds and Johnson gave a direct algorithm (that is, they did not use the reduction to matching) for finding an optimal T-join. Their algorithm uses the optimality conditions for (5.33), and constructs, as well as an optimal T-join, a feasible solution to the dual problem

$$\text{Maximize} \sum(Z_D \,:\, D \text{ a } T\text{-cut}) \tag{5.34}$$
$$\text{subject to}$$
$$\sum(Z_D \,:\, e \in D, \, D \text{ a } T\text{-cut}) \leq c_e, \text{ for all } e \in E \tag{5.35}$$
$$Z_D \geq 0, \text{ for all } T\text{-cuts } D \tag{5.36}$$

having the same value. This proves Theorem 5.28. We are going to give a different proof based on the method we have described for finding an optimal T-join.

Proof of Theorem 5.28. Let J^* be a T-join of G of minimum c-cost. Since we have already observed that $c(J^*)$ is at least the optimal value of (5.33), it will be enough to show that there exists a feasible solution of the dual problem (5.34) having objective value $c(J^*)$. We prove this in two steps.

First, we show that it is true in the special case when $T = V$, and then we reduce the general case to this one by a trick.

Suppose that $T = V$, and let $\hat{G}$ and d denote the complete graph and edge weights constructed in the solution of the optimal T-join problem. Then $c(J^*)$ is equal to the minimum d-weight of a perfect matching of $\hat{G}$. By Theorem 5.13 this is equal to the maximum value of the corresponding dual problem, namely, (5.17) with G replaced by $\hat{G}$ and c replaced by d. Moreover, since d arises from lengths of least-cost paths and $c \geq 0$, it follows that $d \geq 0$ and d satisfies the triangle inequality. Therefore, we can choose such an optimal solution (y, Y) so that $y \geq 0$, by Theorem 5.20. Now if D is a T-cut of the form $\delta(v)$ for some node v, we put $Z_D = Y_D + y_v$, and if D is a T-cut not of this form, we just take $Z_D = Y_D$. Then $\sum(Z_D : D \text{ a } T\text{-cut}) = c(J^*)$, and for each edge $e = uv$ of G

$$\sum(Z_D : D \text{ a } T\text{-cut with } e \in D) \leq d(u, v).$$

Since $d(u, v) \leq c_{uv}$ for all $e = uv \in E$ and $Z \geq 0$, Z is a feasible solution of (5.34) having objective value equal to $c(J^*)$, as required.

Now suppose that $T \neq V$. We use a reduction due to A. Frank. We make a copy v' of each node $v \in V \backslash T$, and make a new graph G' having node set $V' = V \cup \{v' : v \in V \backslash T\}$, and edge set $E' = E \cup \{vv' : v \in V \backslash T\}$. We define cost c'_e for $e \in E'$ by $c'_e = c_e$ if $e \in E$ and $c'_e = 0$ otherwise. Now we declare $T' = V'$ and consider the problem of finding a T'-join of G' having minimum c'-cost. It is easy to see that every such set is of the form $J^* \cup (E' \backslash E)$, where J^* is a T-join of G of minimum c-cost. Therefore, by the first part of the proof, there is a feasible solution Z of (5.34) (where G, c, T are replaced by G', c', T') having objective value $c(J^*)$. Since $c'_{vv'} = 0$, if some T'-cut D contains vv', then $Z_D = 0$. It follows that each T'-cut D with $Z_D > 0$ is of the form $\delta'(S)$ where S contains an even number of nodes from $V' \backslash T$, and therefore an odd number from T; that is, it is actually a T-cut of G. Therefore, Z gives a feasible solution of (5.34) (with respect to G, c, T) having objective value $c(J^*)$, as required. ∎

Suppose that G and c have the property that $c \in \mathbf{Z}^E$ and the cost with respect to c of any circuit of G is an even integer. Then it is easy to see that $\hat{G}$ and d also have this property. Therefore, by Theorem 5.18, there is an optimal dual solution for the matching problem on $\hat{G}$ with weight vector d that is integral. Now, it is clear that the proof will deliver an integral Z. We therefore have a proof of the following result of Seymour [1981].

Theorem 5.29 *Let $G = (V, E)$ be a graph and $c \in \mathbf{Z}^E$. Suppose that every circuit of G has even c-cost. Then (5.34) has an optimal solution that is integral.* ∎

Of course, one can conclude that if the costs are integral, then the T-join dual problem (5.34) has an optimal solution that is half-integral. The exercises describe some other aspects of this theorem.

Finally, we describe a linear-programming problem whose optimal value is the minimum cost of a T-join, even without the assumption that $c \geq 0$. Let J be a T-join and let S be a subset of V. Then

$$|J \cap \delta(S)| \equiv |S \cap T| \pmod{2}.$$

Therefore, if $F \subseteq \delta(S)$ with $|F| + |S \cap T|$ odd, then $J \cap \delta(S) \neq F$. It follows that

$$|J \cap (\delta(S) \backslash F)| + |F \backslash J| \geq 1.$$

Therefore, the characteristic vector x^* of J satisfies

$$x^*(\delta(S) \backslash F) - x^*(F) \geq 1 - |F|.$$

So we get a lower bound for the optimal cost of a T-join from the following linear-programming problem.

$$\text{Minimize } \sum(c_e x_e : e \in E) \qquad (5.37)$$

$$\text{subject to}$$

$$x(\delta(S) \backslash F) - x(F) \geq 1 - |F| \text{ for all } S \subseteq V \text{ and } F \subseteq \delta(S)$$

$$\text{such that } |F| + |S \cap T| \text{ is odd}$$

$$0 \leq x_e \leq 1 \text{ for all } e \in E.$$

In fact, this lower bound is exact.

Theorem 5.30 *If $G = (V, E)$, $T \subseteq V$ with $|T|$ even, and $c \in \mathbf{R}^E$, then the minimum cost of a T-join of G is equal to the optimal value of (5.37).*

Proof: The proof uses the trick for handling negative weights, together with Theorem 5.28. We have already observed that the optimal value of (5.37) provides a lower bound on the minimum weight of a T-join, so it is enough to show a T-join J^* whose characteristic vector satisfies complementary slackness with a feasible solution of the dual of (5.37). This dual problem, obtained by first writing the constraints $x_e \leq 1$ as $-x_e \geq -1$, is:

$$\text{Maximize } \sum(1 - |F|)Y_{S,F} - \sum w_e \qquad (5.38)$$

$$\text{subject to}$$

$$\sum(Y_{S,F} : e \in \delta(S) \backslash F) - \sum(Y_{S,F} : e \in F) - w_e \leq c_e, \text{ for all } e \in E;$$

$$Y_{S,F} \geq 0, \ w_e \geq 0.$$

Here it is understood that $Y_{S,F}$ is defined only when $S \subseteq V$, $F \subseteq \delta(S)$, and $|F| + |S \cap T|$ is odd (and, of course, w_e is defined for $e \in E$). Let $N = \{e \in E : c_e < 0\}$, let $T' = \{v \in V : |N \cap \delta(v)| \text{ is odd}\}$, and let J' be a $(T \triangle T')$-join of G having minimum cost with respect to $|c|$. Then by

Theorem 5.28 and the Duality Theorem, there is a feasible solution Z' of (5.34) (with c replaced by $|c|$ and T replaced by $(T \triangle T')$) satisfying complementary slackness with the characteristic vector x' of J'. Let J^* denote $J' \triangle N$ and let x^* be its characteristic vector. Since J^* is a T-join, x^* is a feasible solution of (5.37). Let $S \subseteq V$ such that $\delta(S)$ is a $(T \triangle T')$-cut. Then $|S \cap (T \triangle T')|$ is odd, so $|S \cap T| + |N \cap \delta(S)|$ is odd. Now define F to be $N \cap \delta(S)$ and define $Y^*_{S,F}$ to be $Z'_{\delta(S)}$; for all other (S, F), define $Y^*_{S,F}$ to be 0. For $e \in N$, we define w^*_e to be $-c_e - \sum(Z'_D : e \in D)$, and for $e \notin N$, we define w^*_e to be 0. We claim that x^* and (Y^*, w^*) are feasible solutions to (5.37) and (5.38) that satisfy complementary slackness. Exercise 5.50 asks you to verify this claim.

∎

Exercises

5.40. Suppose that every node of G has even degree and that G is connected. Let P be an edge-simple closed path in G, and suppose that $E(P) \neq E$. Show that there exists an edge-simple closed path P' of G having no edge in common with P and having at least one node in common with P. Use this idea to prove Theorem 5.23 and to give an algorithm that constructs an Euler tour of G.

5.41. Prove that if every node of G has even degree, then its edge-set is the union of edge-disjoint circuits.

5.42. Let J be a T-join, and define the weight w_e of edge e to be c_e if $e \in J$ and to be $-c_e$ if $e \notin J$. Show that J is of minimum cost with respect to c if and only if G has no circuit of positive weight with respect to w.

5.43. Consider the maximum cut problem for (G, c) where at most one edge f has $c_f > 0$. Show how the solvability of this problem follows from the solvability of the minimum (r, s)-cut problem of Chapter 3.

5.44. Find an optimal dual solution for the optimal T-join problem of Figure 5.21.

5.45. Let G be a graph having edge-connectivity at least 2, and let $c \in \mathbf{R}^E$ with $c \geq 0$. Prove that G has a postman set of cost at most $\frac{1}{3}c(E)$. (Hint: Use Theorem 5.28.)

5.46. Suppose that G is complete, $|V|$ is even, and c satisfies the triangle inequality. Prove that there is an optimal V-join that is a perfect matching. Use this fact and Theorem 5.28 to prove that the optimal perfect matching problem on G has an optimal dual solution (y, Y) such that $y \geq 0$.

5.47. Prove that in a bipartite graph the minimum cardinality of a T-join is equal to the maximum number of edge-disjoint T-cuts.

5.48. Use the result of Exercise 5.47 to give a proof of Seymour's Theorem 5.29. (Hint: Replace each edge of G by a path of length c_e.)

5.49. Prove Theorem 5.18 from Theorem 5.29. (Hint: Add a large even number to each edge-weight.)

5.50. Verify the claim at the end of the proof of Theorem 5.30.

5.51. Prove that the minimum weight of an even set of G is the optimal value of

$$\text{Minimize } \sum(c_e x_e : e \in E)$$
$$\text{subject to}$$
$$x(\delta(S)\backslash F) - x(F) \geq 1 - |F|, \text{ for all } S \subseteq V \text{ and } F \subseteq \delta(S)$$
$$\text{such that } |F| \text{ is odd}$$
$$0 \leq x_e \leq 1, \text{ for all } e \in E.$$

5.5   GENERAL MATCHING PROBLEMS

In this section we describe a number of optimization problems more general than matching problems, that can be solved by similar methods. We give linear-programming descriptions as well as results on polynomial-time solvability. These results, for the most part, first appeared without proof in Edmonds and Johnson [1970]. Complete proofs of the polyhedral results were given, for example, by Aráoz, Cunningham, Edmonds, and Green-Krótki [1983]. Edmonds and Johnson proved these results by generalizing the matching algorithm. However, we instead emphasize techniques that take advantage of the algorithmic and linear-programming results established in earlier sections.

Let $G = (V, E)$ be a graph, and let $b \in \mathbf{Z}^V$. A *b-factor* of G is a set $M \subseteq E$ such that $|M \cap \delta(v)| = b_v$ for each $v \in V$. So a 1-factor is just a perfect matching. Another interesting choice for b is 2—a 2-factor is a collection of circuits covering all of the nodes exactly once. The minimum-weight 2-factor problem is a useful relaxation of the traveling salesman problem. Note that the problem of deciding whether a bipartite graph has a b-factor is solvable by network flow methods—see Exercise 3.18. In addition, the problem of finding a b-factor of minimum weight in a bipartite graph can be solved as a minimum-cost flow problem.

As a further generalization, suppose that we are given, as well as b, a vector $u \in (\mathbf{Z} \cup \{\infty\})^E$. A *u-capacitated perfect b-matching* (or *(b, u)-matching* for short) is a vector $x \in \mathbf{Z}^E$ such that

$$x(\delta(v)) = b_v, \text{ for all } v \in V$$
$$0 \leq x_e \leq u_e, \text{ for all } e \in E.$$

Of course, when each u_e is 1, a (b, u)-matching is just the characteristic vector of a b-factor. At the other extreme, when each $u_e = \infty$, a (b, u)-matching is called a *perfect b-matching*. Notice that, when G is a bipartite graph, the problem of finding a perfect b-matching of minimum weight is what we have called the transportation problem in Chapter 4.

Further extensions were also studied by Edmonds and Johnson [1970], including problems on "bidirected graphs." We refer the reader to the papers of Aráoz, Cunningham, Edmonds, and Green-Krótki [1983] and Schrijver [1983a] for discussions.

Linear Programs for General Matching Problems

Here we describe generalizations to b-factors and (b, u)-matchings of the linear-programming description of the perfect matching problem, given in Theorem 5.13. Let us consider first the case of b-factors. Since a b-factor determines a spanning subgraph of G having degree b_v at node v for each $v \in V$, it is easy to see that there can be no b-factor unless $b(V)$ is even. It follows that the set $T = \{v \in V : b_v \text{ is odd}\}$ has even cardinality. Moreover, any b-factor is a T-join for this choice of T. Hence, the linear-programming description (5.37) for the optimal T-join problem gives a set of linear constraints that are satisfied by the characteristic vector of any b-factor. (Notice that the condition that $|T \cap S| + |F|$ is odd is equivalent to the condition that $b(S) + |F|$ is odd.) Of course, such a vector also satisfies the constraints $x(\delta(v)) = b_v$ for all $v \in V$. Therefore, the following linear-programming problem gives a lower bound for the value of a minimum-weight b-factor:

$$\text{Minimize } \sum(c_e x_e : e \in E) \tag{5.39}$$

$$\text{subject to}$$

$$x(\delta(v)) = b_v, \text{ for all } v \in V$$

$$x(\delta(S) \setminus F) - x(F) \geq 1 - |F|, \text{ for all } S \subseteq V \text{ and } F \subseteq \delta(S)$$

$$\text{such that } |F| + b(S) \text{ is odd}$$

$$0 \leq x_e \leq 1, \text{ for all } e \in E.$$

A result announced in Edmonds [1965a] states that this bound is exact.

Theorem 5.31 *Let $G = (V, E)$, $b \in \mathbf{Z}^V$, and $c \in \mathbf{R}^E$. Then G has a b-factor if and only if (5.39) has a feasible solution. Moreover, if G has a b-factor, then the minimum weight with respect to c of a b-factor of G is equal to the optimal value of (5.39).*

The proof of this result follows fairly easily from the similar theorem on T-joins. However, our proof also uses a bit of polyhedral theory, so it is postponed to Section 6.1.

Now we present a linear-programming description of the problem of finding a (b, u)-matching of minimum weight. Let $S \subseteq V$ and $F \subseteq \delta(S)$ such that $b(S) + u(F)$ is odd. (Note that ∞ is not odd!) For any (b, u)-matching x^*,

$$x^*(\delta(S)) \equiv b(S) \pmod{2}.$$

It follows that we cannot have $x^*(\delta(S)) = x^*(F)$, and so $x^*(\delta(S)\setminus F) > 0$ or $x^*(F) < u(F)$. Therefore, x^* satisfies

$$x^*(\delta(S)\setminus F) - x^*(F) \geq 1 - u(F).$$

This leads to the following linear-programming estimate for the minimum weight of a (b, u)-matching.

$$\text{Minimize } \sum(c_e x_e : e \in E) \tag{5.40}$$
$$\text{subject to}$$
$$x(\delta(v)) = b_v, \text{ for all } v \in V$$
$$x(\delta(S \setminus F)) - x(F) \geq 1 - u(F), \text{ for all } S \subseteq V \text{ and } F \subseteq \delta(S)$$
$$\text{such that } u(F) + b(S) \text{ is odd}$$
$$0 \leq x_e \leq u_e, \text{ for all } e \in E.$$

The following result is due to Edmonds and Johnson [1970].

Theorem 5.32 *Let $G = (V, E)$, $b \in \mathbf{Z}^V$, $u \in (\mathbf{Z} \cup \{\infty\})^E$, and $c \in \mathbf{R}^E$. Then G has a (b, u)-matching if and only if (5.40) has a feasible solution. Moreover, if G has a (b, u)-matching, then the minimum weight with respect to c of a (b, u)-matching of G is equal to the optimal value of (5.40).*

Proof: It is clear that any (b, u)-matching is a feasible solution of (5.40) having the same objective value, and it follows that the optimal value of (5.40) is at most the minimum weight of a (b, u)-matching. Now suppose that x^* is a feasible solution of (5.40); it will be enough to show that there exists a (b, u)-matching having weight no larger than that of x^*.

First, suppose that u_e is finite for all $e \in E$. Define a graph $G' = (V, E')$ by replacing each edge e by u_e copies, each having the same ends as e, and define $c' \in \mathbf{R}^{E'}$ by putting $c'_{e'} = c_e$ for each copy e' of e. It is easy to see that the optimal value of a (b, u)-matching of G is equal to the optimal value of a b-factor of G'. The reason is that, given a (b, u)-matching x of G, we can make a b-factor of G' by choosing exactly x_e copies of e from G' for all $e \in E$, and given a b-factor of G', we can make a (b, u)-matching x of G by taking x_e to be the number of copies of e that are in the b-factor; moreover, this correspondence preserves weight. Therefore, by Theorem 5.31, the optimal value of a (b, u)-matching of G is equal to the optimal value of (5.39) with G replaced by G'. So it is enough to show how to construct from x^* a feasible solution of (5.39) having the same weight. To do this, we simply define $\hat{x}_{e'}$ to be x_e^*/u_e for every copy e' of $e \in E$. It is easy to see that $\hat{x}$ has the same weight as x^*, that $\hat{x}(\delta'(v)) = b_v$ for all v, and that $0 \leq \hat{x} \leq 1$. (Here δ' is δ for the graph G'.)

Now consider the inequality

$$\hat{x}(\delta'(S)\backslash F') - \hat{x}(F') \geq 1 - |F'|,$$

where $F' \subseteq \delta'(S)$ and $b(S)+|F'|$ is odd. Suppose first that F' consists precisely of the copies of the edges of some subset F of $\delta(S)$. Then $b(S) + u(F)$ is odd, and the left-hand side of the above inequality is just

$$x^*(\delta(S)\backslash F) - x^*(F) \geq 1 - u(F) = 1 - |F'|,$$

as required. Now suppose that there exists some $e \in \delta(S)$ such that one copy, e', of e is in F' and another, e'', is not in F'. Then

$$\hat{x}(\delta'(S)\backslash F') - \hat{x}(F') + |F'| - 1 \geq \hat{x}_{e''} + 1 - \hat{x}_{e'} - 1 = 0,$$

as required. This completes the proof of the theorem for the case when every u_e is finite.

Now notice that, no matter what u is, the feasible region of (5.40) is bounded, since $x_e \leq b_v$ for every edge $e = uv$. Therefore, we can replace each infinite u_e by a sufficiently large integer so that the feasible solutions of (5.40) do not change, and the (b,u)-matchings do not change. Therefore, the theorem follows from the case when u is finite, which we have already proved. ∎

Algorithms for General Matching Problems

We will show that the minimum-weight (b,u)-matching problem can be solved in strongly polynomial time, that is, the polynomial-time bound will not depend on the sizes of the numbers in b, u, or the weight function c. This immediately implies that the b-factor problem and the perfect b-matching problem each can be solved in strongly polynomial time. As we will see, however, the (b,u)-matching algorithm actually involves a reduction to the perfect b-matching problem.

Let $G = (V, E)$ be a graph, $b \in \mathbf{Z}^V$, $u \in (\mathbf{Z} \cup \{\infty\})^E$, and $c \in \mathbf{R}^E$. If $u_e = \infty$ for some edge $e = rv$, then we can replace u_e by the minimum of b_r and b_v without altering the set of (b,u)-matchings of G. Therefore, we may assume that each of the capacities is finite.

We define a new graph $G' = (V', E')$ by replacing each edge of G by a path of length three: for each edge $e = rv$ we introduce two new nodes $p_{e,r}$ and $p_{e,v}$ and three edges $rp_{e,r}$, $p_{e,r}p_{e,v}$, and $p_{e,v}v$. Define $b' \in \mathbf{Z}^{V'}$ by letting $b'_v = b_v$ if $v \in V$ and $b'_{p_{e,v}} = u_e$ for each new node $p_{e,v}$. Define $c' \in \mathbf{R}^{E'}$ by $c'_{rp_{e,r}} = c'_{p_{e,v}v} = c_{rv}$ and $c'_{p_{e,r}p_{e,v}} = 0$ for all $e = rv \in E$. (See Figure 5.32.) It is easy to see that the minimum-weight (b,u)-matching problem in G can be solved by computing a minimum-weight perfect b'-matching in G', using the

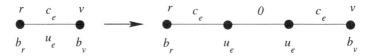

Figure 5.32. Reduction to perfect b-matching

new weight function c'. (Exercise 5.52.) Therefore, we turn our attention to the perfect b-matching problem.

It will be both interesting and convenient to describe an algorithm for the closely related *maximum-weight b-matching problem*, rather than working directly with perfect b-matchings. A vector $x \in \mathbf{Z}^E$ is a *b-matching* if x is nonnegative and $x(\delta(v)) \le b_v$ for all $v \in V$. Given a vector $w \in R^E$, the maximum-weight b-matching problem is to find a b-matching x such that $w^T x$ is maximum.

An algorithm for maximum-weight b-matching can be used to solve minimum-weight perfect b-matching problems, by setting $w_e = -c_e + \Delta$ for a sufficiently large Δ (but not *too* large, since we want the new numbers to have size that is polynomial in the size of the original data). We leave this as Exercise 5.53.

The maximum-weight b-matching problem can be solved via a direct primal-dual augmenting path algorithm, similar to the Blossom Algorithm for finding maximum-weight matchings. This was announced in Edmonds [1965a], and a full description of the method was given by Pulleyblank [1973]. The result is not a polynomial-time algorithm, however, since the number of augmentations is bounded in terms of the b_v values. This is similar to the situation we had for the minimum-cost flow problem. Indeed, the idea of successive scaling that we described in Section 4.4 can be applied to b-matchings to obtain an algorithm that is polynomial in the size of the graph G and $\log(b_v)$ for each node v, as was shown by Cunningham and Marsh (see Marsh [1979]). In the case of minimum-cost flows, the dependence of the running-time bound on the sizes of the input numbers was removed by the Scale-and-Shrink Algorithm, to obtain a strongly polynomial algorithm for the problem. This technique has not been directly applied to b-matchings, but it has been used indirectly by Anstee [1987] to solve b-matching problems in strongly polynomial time by using a strongly polynomial minimum-cost flow algorithm as a subroutine. The algorithm we present also follows this line and is due to J. Edmonds, as described by Gerards [1995].

For a vector $x \in \mathbf{R}^I$, when I is some index set, we let

$$\| x \|_\infty = \text{maximum } (|x_i| : i \in I)$$

and

$$\| x \|_1 = \sum (|x_i| : i \in I).$$

These values are the L_∞ and L_1 norms of x, respectively. The b-matching algorithm makes use of the following "proximity" result, due to Edmonds (see Gerards [1995]).

Theorem 5.33 *(b-Matching Proximity Theorem) Let $G = (V, E)$, let $b, b' \in \mathbf{Z}^V$ be nonnegative vectors, and let $w \in \mathbf{R}^E$. Then for any maximum-weight b-matching x there exists a maximum-weight b'-matching x' satisfying*

$$\| x - x' \|_\infty \leq 2 \| b - b' \|_1 .$$

Before proving this theorem, we need to present a useful decomposition lemma. For vectors $x, y \in \mathbf{R}^I$, we say that y is *conformal* to x if $x_i \geq 0$ implies $y_i \geq 0$, and $x_i \leq 0$ implies $y_i \leq 0$, for each $i \in I$.

Lemma 5.34 *Let $G = (V, E)$ be a graph and let $x \in \mathbf{Z}^E$. Then x is the sum of vectors $y \in \mathbf{Z}^E$ with y conformal to x, $(y(\delta(v)) : v \in V)$ conformal to $(x(\delta(v)) : v \in V)$, and $\| y \|_\infty \leq 2$.*

Proof: We will prove the result by induction on $\| x \|_1$. For any path P in G we define $q^P \in \mathbf{Z}^E$ by letting

$$q_e^P = \text{the number of times } P \text{ traverses } e, \text{ if } x_e \geq 0,$$
$$q_e^P = \text{minus the number of times } P \text{ traverses } e, \text{ if } x_e < 0.$$

Suppose that $x(\delta(v_0)) > 0$ for some node v_0. Let $P = (v_0, e_1, v_1, \ldots, e_t, v_t)$ be a longest path satisfying

$$x_{e_i} > 0 \text{ if } i \text{ is odd and } x_{e_i} < 0 \text{ if } i \text{ is even;} \tag{5.41}$$

$$\text{each edge } e \text{ is traversed at most } |x_e| \text{ times;} \tag{5.42}$$

$$v_i \neq v_j \text{ if } i \equiv j \pmod 2 \text{ and } i < j \leq t - 1. \tag{5.43}$$

(Notice that the third condition ensures that no edge is traversed more than twice, and thus $\| q^P \|_\infty \leq 2$.)

Suppose that the reason P cannot be extended to a longer path is that $v_t = v_i$ for some $i < t$ with $i \equiv t \pmod 2$. Consider the closed path $C = (v_i, e_{i+1}, v_{i+1}, \ldots, e_t, v_t)$. Let $y = q^C$. Then y and $x - y$ are conformal to x. Moreover, $y(\delta(v)) = 0$ for all $v \in V$ and $(x - y)(\delta(v)) = x(\delta(v))$ for all $v \in V$. By induction, we can decompose $x - y$ as described in the lemma. Adding y gives a decomposition of x as required.

So we may assume that $v_t \neq v_i$ for each $i < t$ with $i \equiv t \pmod 2$. Let $y = q^P$. Then y is conformal to x. If t is even, $y(\delta(v_t)) < 0$. Also, since P cannot be extended to a longer path we know that each edge $e = v_t v$ with $x_e > 0$ is traversed by P exactly x_e times. Hence, $x(\delta(v_t)) < 0$. Also $y(\delta(v_0)) > 0$, so we can apply induction to $x - y$, then add y to obtain the required decomposition of x. A similar argument handles the case when t is odd.

The case when $x(\delta(v_0)) < 0$ for some node v_0 can be handled in a similar manner, by requiring that the path P satisfy

$$x_{e_i} < 0 \text{ if } i \text{ is odd and } x_{e_i} > 0 \text{ if } i \text{ is even} \tag{5.44}$$

instead of the condition (5.41) above.

Finally, if $x(\delta(v)) = 0$ for all $v \in V$, then we can construct a path P as above and find a closed path C that allows us to apply induction to $x - q^C$. ∎

Proof of the b-Matching Proximity Theorem: By considering incremental changes to b, we may assume that $\| b - b' \|_1 = 1$. Let x be a maximum-weight b-matching and let x' be a maximum-weight b'-matching. If x' happens to be a b-matching and x happens to be a b'-matching, then the result follows by taking x as the required b'-matching. (To see this, notice that $w^T x \geq w^T x'$, since x is a maximum-weight b-matching and x' is some b-matching. Therefore, since x' is a maximum-weight b'-matching, x must also be a maximum-weight b'-matching.) If this is not the case, then we will use Lemma 5.34 to produce a similar situation.

Suppose that $b'_r = b_r - 1$, for the unique node r for which b'_r and b_r differ. (The case $b'_r = b_r + 1$ is similar.) Clearly x' is a b-matching, so if x is not a b'-matching, then we must have $x(\delta(r)) = b_r$. Now consider the decomposition that is obtained by applying Lemma 5.34 to $x - x'$. For some vector y in the decomposition, we must have $y(\delta(r)) \geq 1$. It follows that $x - y$ is a b'-matching and $x' + y$ is a b-matching. (The nonnegativity of $x - y$ and $x' + y$ is implied by the decomposition's conformality with $x - x'$; the degree constraints on $x - y$ and $x' + y$ are implied by the decomposition's conformality with $((x - x')(\delta(v)) : v \in V)$.) Since $x' + y$ is a b-matching and since x is a maximum-weight b-matching, we have $w^T x \geq w^T (x' + y)$, and hence $w^T (x - y) \geq w^T x'$. Since x' is a maximum-weight b'-matching, it follows that $x'' = x - y$ is a maximum-weight b'-matching with $\| x'' - x \|_\infty = \| y \|_\infty \leq 2$. ∎

We will use the b-Matching Proximity Theorem to piece together two special-case b-matching algorithms into a method for the general problem.

The first special case is when b_v is even for each $v \in V$. To handle this case, we build a bipartite graph H as follows. For each $v \in V$ we create a new node v' and for each edge rv we create the pair of edges $r'v$ and rv'. Define $\bar{b}_v = \bar{b}_{v'} = \frac{1}{2}b_v$ for each $v \in V$, and define $\bar{w}_{r'v} = \bar{w}_{rv'} = w_{rv}$ for each $rv \in E$.

We can find a maximum-weight $\bar{b}$-matching $\bar{x}$ of H in strongly polynomial time by converting the problem to a minimum-cost flow problem, using the reductions given in Section 4.1. Defining $x_{rv} = \bar{x}_{r'v} + \bar{x}_{rv'}$ for each edge $rv \in E$, we obtain a maximum-weight b-matching of G. Indeed, if there is a b-matching of G of greater weight than that of x, then there is a half-integer solution to the $\bar{b}$-matching problem in H of greater weight than that of $\bar{x}$. This contradicts the fact that we have an optimal solution to the corresponding minimum-cost flow problem.

The second special case is when all of the b_v values are bounded by some polynomial function of the size of the graph G. In this case, a maximum-weight b-matching can be computed in strongly polynomial time by reducing the problem in a natural way to a maximum-weight matching problem. The technique is to create a graph G_b by splitting each node of G into b_v copies, and replacing each edge rv by $b_r b_v$ edges connecting the b_r copies of r to the b_v copies of v, where each of the $b_r b_v$ new edges receives the weight w_{rv}. More formally, $G_b = (V_b, E_b)$ where

$$V_b = \{q_{v,i} : v \in V, 1 \le i \le b_v\},$$
$$E_v = \{q_{r,j} q_{v,i} : rv \in E, 1 \le j \le b_r, 1 \le i \le b_v\}.$$

This construction was given by Tutte [1954]. It is easy to see that a maximum-weight matching in G_b corresponds to a maximum-weight b-matching in G. We leave this as Exercise 5.54.

With these two pieces, we can prove the following result.

Theorem 5.35 *For any graph $G = (V, E)$, $b \in \mathbf{Z}^V$, and $w \in \mathbf{R}^E$, a maximum-weight b-matching can be found in strongly polynomial time.*

Proof: Let $b'_v = 2\lfloor b_v \rfloor$ for all $v \in V$. Since b' is even, by the special-case algorithm we described above, we can find a maximum-weight b'-matching x' in G in strongly polynomial time. By the b-Matching Proximity Theorem, there exists a maximum-weight b-matching x with

$$\| x - x' \|_\infty \le 2 \| b - b' \|_1 \le 2|V|.$$

We will use the existence of x to guide us in completing the construction of a maximum-weight b-matching.

First, define the b-matching z by letting

$$z_e = \text{maximum } (0, x'_e - 2|V|)$$

for all $e \in E$. Then $x \ge z$ and

$$(x - z)_e \le \| x - x' \|_\infty + \| x' - z \|_\infty \le 4|V|$$

for all $e \in E$. So $x - z$ is a b''-matching, where b'' is defined by letting

$$b''_v = \text{minimum } (4|V|\deg_G(v), b_v - z(\delta(v)))$$

for all $v \in V$. Now since the b''_v values are bounded by a polynomial in the size of G, we can form the graph $G_{b''}$ and solve the resulting maximum-weight matching problem in strongly polynomial time. Let x'' denote the maximum-weight b''-matching that we obtain with this method.

Since x'' is a maximum-weight b''-matching and $x - z$ is some b''-matching, we know that $w^T x'' \geq w^T (x - z)$, and therefore $w^T (x'' + z) \geq w^T x$. So $z + x''$ is a maximum-weight b-matching. ∎

From a practical point of view, if we have an application that involves the solution of large (b, u)-matching problems, we may not want to follow the algorithm exactly as we have described it. Suppose we are faced, for example, with a 2-factor problem or some other b-factor problem with small b_v values. Then it is quite efficient to solve the problem by a direct primal-dual augmenting path algorithm, along the lines of the Blossom Algorithm for minimum-weight perfect matchings. An outline of how to carry out such a direct algorithm is contained in a report by Edmonds, Johnson, and Lockhart [1969] (see also Miller and Pekney [1995]). On the other hand, if we reduce the b-factor problem to a perfect b-matching problem by the edge splitting method and solve it with a perfect b-matching algorithm, then the size of the instances that can be handled in a limited amount of time goes down considerably—this is particularly true for dense graphs, like those that arise in geometric applications. Reducing the problem further to a perfect matching problem on the graph G_b puts an even greater restriction on the size of the instances that can be handled. These remarks need to be qualified, however, by the fact that it is somewhat more difficult to write an efficient computer code for b-factors than it is for matchings, and by the fact that for small instances (up to several thousand nodes) the reduction to matching is quite acceptable.

If we are faced with a problem instance containing large b_v values, then a direct augmenting path algorithm will not be a viable alternative. In such a case, we may want to carry out the minimum-cost flow approximation and then apply a direct algorithm to solve the remaining b''-matching problem.

Exercises

5.52. Let $G = (V, E)$ be a graph, $b \in \mathbf{Z}^V$, $u \in \mathbf{Z}^E$, and $c \in \mathbf{R}^E$. Let G' be the graph obtained by replacing each edge by a path of length three, as indicated in Figure 5.32. Show that a minimum-weight (b, u)-matching in G can be found by solving a minimum-weight perfect b'-matching problem in G'.

5.53. Let $G = (V, E)$ be a graph, $b \in \mathbf{Z}^V$, and $c \in \mathbf{R}^E$. Let

$$\Delta = CB + C$$

where $C = \| c \|_\infty$ and $B = \| b \|_1$. Show that the minimum-weight perfect b-matching problem can be solved by computing a maximum-weight b-matching in G, with weight function $w \in \mathbf{R}^E$ defined as $w_e = -c_e + \Delta$ for all $e \in E$.

5.54. Let $G = (V, E)$ be a graph, $b \in \mathbf{Z}^V$, and $w \in \mathbf{R}^E$. Show that the maximum-weight b-matching problem in G can be solved by computing a maximum-weight matching in G_b.

5.55. (Tutte) Let $G = (V, E)$ be a graph. Define a 2-*node cover* as a nonnegative vector $y \in \mathbf{Z}^V$ such that $y_r + y_v \geq 2$ for each $rv \in E$. Define the *size* of a vector to be the sum of its components. Show that the maximum size of a 2-matching is equal to the minimum size of a 2-node cover.

5.56. Let $G = (V, E)$ be a graph, $b \in \mathbf{Z}^V$, $u \in \mathbf{Z}^E$, and $w \in \mathbf{R}^E$. A u-*capacitated b-matching* is a nonnegative vector $x \in \mathbf{Z}^E$ such that $x_e \leq u_e$ for all $e \in E$ and $x(\delta(v)) \leq b_v$ for all $v \in V$. Show that the maximum-weight u-capacitated b-matching problem can be solved in strongly polynomial time by a reduction to the maximum-weight b-matching problem.

5.57. Let $G = (V, E)$ be a graph, $a' \in \mathbf{Z}^V$, $a'' \in \mathbf{Z}^V$, and $w \in \mathbf{R}^E$. A $[a', a'']$-*matching* is a nonnegative vector $x \in \mathbf{Z}^E$ such that $a'_v \leq x(\delta(v)) \leq a''_v$ for all $v \in V$. Show the maximum-weight $[a', a'']$-matching problem can be solved in strongly polynomial time by a reduction to a u-capacitated b-matching problem. (Hint: Build a new graph by creating a new node v' for each $v \in V$, and new edges vv' for $v \in V$ and $r'v'$ for all pairs of nodes $r, v \in V$. Let $b_v = b_{v'} = a''_v$ for $v \in V$, and let $u_{vv'} = a''_v - a'_v$ for $v \in V$, and $u_e = \infty$ for all other edges in the new graph.)

5.6 GEOMETRIC DUALITY AND THE GOEMANS-WILLIAMSON ALGORITHM

Recall the geometric matching problem we introduced in Chapter 1: We are given an even cardinality set of points V in the plane and we wish to find a perfect matching of V that minimizes the sum of the lengths of the edges. If the lengths are determined by the Euclidean distance between the points, then we have an instance of the *Euclidean matching problem*. This problem is a popular subject in computational geometry. Clearly, we can solve the problem by applying the minimum-weight perfect matching algorithm to the complete graph on the node-set V, taking the weight c_{uv} of the edge uv to be $d(u, v)$, where $d(u, v)$ is the Euclidean distance between u and v. Note that these weights satisfy the triangle inequality. Therefore, using the results of Theorem 5.17 and Theorem 5.20, there is a perfect matching M and a nested solution (y, Y) to (5.17) that satisfies the optimality conditions (5.20), (5.21), and has $y \geq 0$. The dual solution (y, Y) can be presented in a nice geometric way, giving a visual proof of the optimality of the matching. The idea is due to Jünger and Pulleyblank [1993], and we describe it below.

First, suppose that we construct, for each node v, a circular disk D_v centered at v, having radius $r_v \geq 0$. We call such a disk a *control zone*. If no pair of disks overlap, then we can see immediately that any solution to the perfect matching problem must have length at least as large as the sum of the radii of the disks. These disks will represent the variables y.

The objects corresponding to Y are a bit more complicated. We call a pair of compact sets (N, I) a *moat* if

$$I \subset N, \ |I \cap V| \text{ is odd, and } N \backslash \text{ interior}(I) \text{ contains no points in } V.$$

The *width* of a moat is the minimum distance between points in I and points in $\mathbf{R}^2 \backslash \text{ interior}(N)$. A simple example of moat is given in Figure 5.33. Notice

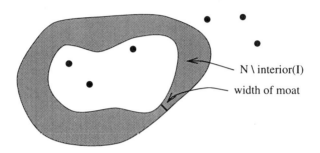

Figure 5.33. A moat

that in any perfect matching of V, at least one edge must intersect the moat in at least its width. Therefore, if we have a disjoint packing of control zones and moats, then the sum of the control zone radii and moat widths is a lower bound on the length of any perfect matching of V.

Now consider our dual solution (y, Y). Let $\mathcal{S} = \{S \in \mathcal{O} : Y_S > 0\}$. Define $r_v = y_v$ for all $v \in V$, and $w_S = Y_S$ for all $S \in \mathcal{S}$. It follows from the fact that (y, Y) is feasible to (5.17) that there is room between each pair of nodes (u, v) to pack control zones of radius r_u and r_v centered at u and v plus moats of widths w_S surrounding each set $S \in \mathcal{S}$ separating u and v. But it is not clear exactly how to construct these moats and control zones.

To do this, we begin by drawing the control zones of radius r_v for each $v \in V$. Next, we construct moats starting with minimal members of $\mathcal{S}$. We choose a minimal unprocessed $S \in \mathcal{S}$. For each $v \in S$, we first construct a disk B_v centered at v of radius $\rho_v = r_v + \sum (w_T : v \in T \subset S, \ T \in \mathcal{S})$. (If $\rho_v = 0$, we set $B_v = \{v\}$.) Now we construct a new disk B'_v of radius $\rho_v + w_S$, centered at v. The moat corresponding to S consists of $N = \cup_{v \in S}(B'_v)$ and $I = \cup_{v \in S}(B_v)$. See Figure 5.34. Note that this construction has the property that the sum of the separating moat widths plus control zone radii of any two nodes is no greater than the distance between them, which ensures that we never have overlap. Also,

$$\sum (d(u, v) : uv \in M) = \sum_{v \in V} r_v + \sum_{S \in \mathcal{S}} w_S,$$

as required.

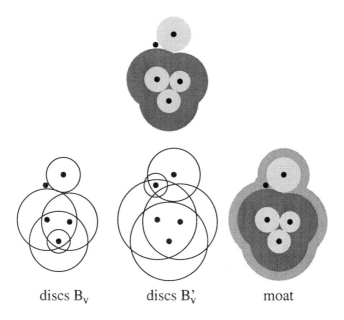

$$\text{discs } B_v \qquad \text{discs } B'_v \qquad \text{moat}$$

Figure 5.34. Moat construction

An example of how a packing of control zones and moats establishes the optimality of a matching is given in Figure 5.35. To verify the optimality of the matching we simply check the following conditions.

- Each edge of the matching lies completely within moats and control zones.

- The amount of the matching lying within each moat is exactly the width of the moat.

- The amount of the matching lying within each control zone is exactly the radius of the control zone.

These requirements constitute the geometric interpretation of the complementary slackness conditions for the linear-programming problems (5.13) and (5.17).

The Goemans-Williamson Algorithm

Moats and control zones provide a nice way to view a minimum-weight perfect matching algorithm proposed by Goemans and Williamson [1995]. This algorithm is one of a very general family of algorithms designed by Goemans and Williamson. The algorithm works on Euclidean matching problems, as well as any other instance specified by a complete graph and edge weights satisfying the triangle inequality. Unlike the Blossom Algorithm, the matching produced by the Goemans-Williamson algorithm is not necessarily optimal. However, the algorithm does have a good performance guarantee: The weight

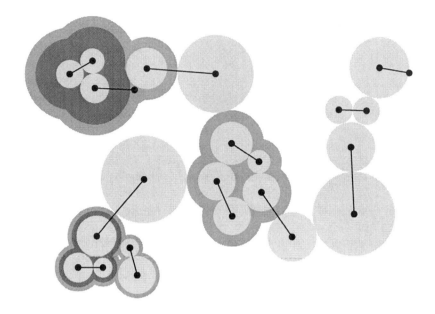

Figure 5.35. Optimal matching and packing

of the matching it finds is never more than twice the weight of an optimal matching. For Euclidean instances, the guarantee is provided by a disjoint packing of moats and control zones, with the sum of the control-zone radii and the moat widths being at least one-half the weight of the matching produced by the algorithm. In the general triangle-inequality case, the guarantee is given as a solution (y, Y) to the (dual) linear-programming problem (5.17), having objective value at least one-half of the weight of the matching.

We will describe the algorithm for the general case, and then interpret its steps in terms of moats and control zones.

Suppose we have a graph $G = (V, E)$ and nonnegative edge weights $c \in \mathbf{R}^E$ that satisfy the triangle inequality. The Goemans-Williamson algorithm will construct an *even spanning forest* of G, that is, a spanning forest for which all components have an even number of nodes. It is easy to see that there exists a perfect matching whose weight is no greater than the sum of the forest's edge lengths, as we outline below.

Define an edge of an even tree as being *even* or *odd* according as the number of nodes in the subtrees created by deleting the edge is even/odd. An edge of a tree is called a *leaf* if one of its ends has degree 1.

- Let F be an even spanning forest. If any tree T of F contains an even edge, then by deleting the edge we obtain an even spanning forest of length no greater than that of F.

- Suppose all edges in F are odd, and some tree T has ≥ 4 nodes. Let T' be the tree obtained from T by deleting all of T's leaves. Then some leaf r of T' is adjacent to a leaf w of T. There must be another leaf w' of T also adjacent to v, or else v would be of degree two in T and one of the adjacent edges would be even. Remove w and w' from T and join them to each other, creating a new two-node tree in the forest. By the triangle inequality, $c_{ww'} \leq c_{wv} + c_{w'v}$.

Combining these two steps, we can reduce any even forest to a perfect matching.

We now describe the matching algorithm. At each stage we have a forest (not even until the end) F, and a solution (y, Y) to (5.17). For any edge $e = vw$, we let $\bar{c}_e$ denote the reduced cost of e with respect to (y, Y), that is

$$\bar{c}_e = c_e - y_v - y_w - \sum (Y_D : e \in D, \ D \subseteq V, \ |D| > 1, \ |D| \text{ odd}).$$

Let $\mathcal{C}$ denote the node sets of the trees in F. For each set $C \in \mathcal{C}$, let

$$parity(C) \;=\; \begin{cases} 0 & \text{if } |C| \text{ is even} \\ 1 & \text{if } |C| \text{ is odd}. \end{cases}$$

Initially F contains no edges, $\mathcal{C} = \{\{v\} : v \in V\}$, and each component of y and Y is set to 0.

Goemans-Williamson Algorithm

Step 1. If there exist no set $C \in \mathcal{C}$ with $|C|$ odd, then remove any even edges from F and stop.

Step 2. Find an edge $e = vw$, with $v \in C_i \in \mathcal{C}$, $w \in C_j \in \mathcal{C}$, $C_i \neq C_j$, that minimizes $\varepsilon = \bar{c}_e / (parity(C_i) + parity(C_j))$. (Notice that at least one of $|C_i|$ and $|C_j|$ must be odd.)

Step 3. For each $C \in \mathcal{C}$ such that $C = \{v\}$ for some node $v \in V$, add ε to y_v, and for each $C \in \mathcal{C}$ such that $|C| > 1$ and $|C|$ is odd, add ε to Y_C.

Step 4. Add edge e to F, and update $\mathcal{C}$ by removing the sets C_i and C_j and adding the set $C_i \cup C_j$. Go to Step 1.

The algorithm terminates with an even forest F^* and a dual solution (y^*, Y^*). From the choice of ε in Step 2, it follows that (y^*, Y^*) is a feasible solution to (5.17), that is, all reduced costs are nonnegative. Therefore, a lower bound on the weight of any perfect matching is

$$LB = \sum (y_v^* : v \in V) + \sum (Y_D^* : D \subseteq V, \ |D| > 1, \ |D| \text{ odd}).$$

We claim that sum of the weights of the edges in F^* is at most twice LB.

Let us denote by $\mathcal{C}^k$ and ε^k the family of sets $\mathcal{C}$ and real number ε that are present when Step 2 is executed for the kth time. Now consider an edge $e = vw$ that is present in F^*. Let $v \in C_i \in \mathcal{C}^k$ and $w \in C_j \in \mathcal{C}^k$, and define

$$c_e^k = \begin{cases} 0 & \text{if } C_i = C_j \\ \varepsilon^k(parity(C_i) + parity(C_j)) & \text{otherwise.} \end{cases}$$

Then

$$c_e = c_e^1 + c_e^2 + \cdots + c_e^t$$

where t is the total number of iterations of the algorithm. (Notice that if edge e is added to F in iteration s, then $c_e^k = 0$ for all $k > s$.) We will show that for each $k = 1, \ldots, t$,

$$\sum (c_e^k : e \in E(F^*)) \le 2\varepsilon^k |\{C \in \mathcal{C}^k : |C| \text{ odd}\}|. \tag{5.45}$$

Since

$$LB = \sum (\varepsilon^k |\{C \in \mathcal{C}^k : |C| \text{ odd}\}| : k = 1, \ldots, t),$$

it follows that $c(E(F^*)) \le 2LB$, as we claimed above.

Let $k \in \{1, \ldots, t\}$ and consider the forest $\bar{F}^*$ obtained from F^* by shrinking each set $C \in \mathcal{C}^k$ into a single node. Label the nodes of $\bar{F}^*$ as odd/even depending on the parity of the corresponding sets in $\mathcal{C}^k$. Let V^{even} and V^{odd} be the even and odd nodes of $\bar{F}^*$. Since F^* contains no even edges, no leaf of F^* is in V^{even}.

Lemma 5.36 *Let F be a forest whose nodes are partitioned into $V^{even} \cup V^{odd}$ in such a way that no leaf of a tree of F is in V^{even}. Then*

$$\sum_{V^{odd}} \deg_F(v) \le 2|V^{odd}|.$$

Proof: For any forest F

$$\sum_{V(F)} \deg_F(v) \le 2|V(F)| = 2|V^{odd}| + 2|V^{even}|.$$

We have

$$\sum_{V^{odd}} \deg_F(v) = \sum_{V(F)} \deg_F(v) - \sum_{V^{even}} \deg_F(v) \le \sum_{V(F)} \deg_F(v) - 2|V^{even}|,$$

since every even node has degree at least two. Therefore

$$\sum_{V^{odd}} \deg(v) \le 2|V^{odd}| + 2|V^{even}| - 2|V^{even}| = 2|V^{odd}|,$$

as required.   ∎

Now since $\sum(c_e^k : e \in E(F^*)) = \varepsilon^k \sum(\deg_{\bar{F}^*}(v) : v \in V^{odd})$, the inequality (5.45) follows. This completes the proof of the claim.

For Euclidean instances, the lower bound LB can be interpreted as a packing of moats and control zones. In fact, the algorithm itself can be viewed as a process that grows moats and control zones in a greedy fashion. In Step 2, we grow a control zone around each of the sets $C \in \mathcal{C}$ that have $|C| = 1$, and we grow a moat around each of the sets $C \in \mathcal{C}$ that have $|C| > 1$ and $|C|$ odd. We grow the moats and control zones uniformly, until a pair of them touch one another. This touching pair determines the next edge that is added to F^*.

The algorithm is illustrated on a four node example in Figure 5.36. As a final step in this example, the even edge $(2,3)$ is deleted, resulting in the even forest F consisting of the edges $(1,2)$ and $(3,4)$.

Exercises

5.58. Suppose we are given a set V of points in the plane, and we define the distance between points u at location (x_u, y_u) and v at location (x_v, y_v) to be the maximum of $|x_u - x_v|$ and $|y_u - y_v|$. If we wish to obtain a lower bound for the solution by packing control zones and moats, what shape should these now be?

5.59. Define a control zone and moat lower bound for the Euclidean traveling salesman problem. Give an example where the optimal tour has greater length than the best bound that can be obtained in this way.

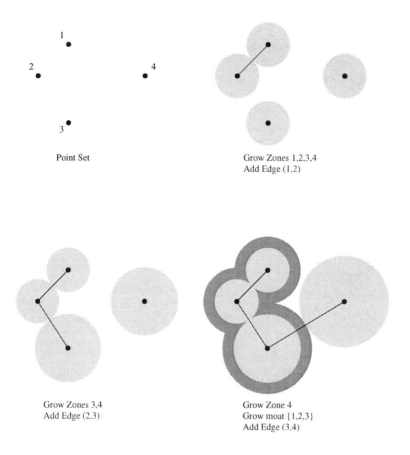

Figure 5.36. Example of the Goemans-Williamson Algorithm

CHAPTER 6

Integrality of Polyhedra

6.1 CONVEX HULLS

Consider the line segment L joining two points u and v in $\mathbf{R}^n$. For any given $w \in \mathbf{R}^n$, the mathematical programming problem $\max\{w^T x : x \in L\}$ is particularly easy to solve: Either u or v is an optimal solution. Indeed, suppose $w^T u \geq w^T v$ and let $x \in L$. By elementary geometry we have $x = \lambda u + (1 - \lambda)v$ for some scalar λ between 0 and 1. Thus

$$w^T x = \lambda w^T u + (1 - \lambda)w^T v \leq \lambda w^T u + (1 - \lambda)w^T u = w^T u$$

and so, u is an optimal solution.

The simple argument used above generalizes naturally to the case of several vectors: If

$$x = \lambda_1 v_1 + \cdots + \lambda_k v_k \tag{6.1}$$

for some vectors $v_1, \ldots, v_k$ and nonnegative scalars $\lambda_1, \ldots, \lambda_k$ such that $\lambda_1 + \cdots + \lambda_k = 1$, then

$$w^T x = \lambda_1 w^T v_1 + \cdots + \lambda_k w^T v_k \leq \max\{w^T v_i : i = 1, \ldots, k\}.$$

Equation (6.1) expresses x as a *convex combination* of the vectors $v_1, \ldots, v_k$. What distinguishes a convex combination from an ordinary linear combination are the requirements that the λ_i be nonnegative and sum to 1.

The *convex hull* of a finite set S (denoted by *conv.hull*(S)) is the set of all vectors that can be written as a convex combination of S. We have proven the following result.

Proposition 6.1 *Let $S \subseteq \mathbf{R}^n$ be a finite set and let $w \in \mathbf{R}^n$. Then*

$$\max\{w^T x : x \in S\} = \max\{w^T x : x \in conv.hull(S)\}.$$

∎

At first sight, this proposition may seem to be of little use if our objective is to maximize over S, since we have replaced a finite problem by an infinite one. As we shall see, however, the convex hull of S has nice geometric properties that can often be exploited.

To get a feeling for convex hulls, consider a set, S, of points in the plane. If $|S| = 2$, then conv.hull(S) is just the line segment joining the two points. If $|S| = 3$ and the points do not all lie on a line, then conv.hull(S) is the triangle having the three points as vertices. In general, conv.hull(S) is the unique (convex) polygon that contains S and has all of its vertices in S (see Figure 6.1). So conv.hull(S) can be described by listing, say in clockwise order,

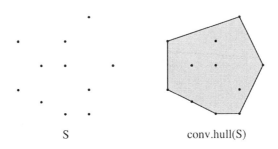

S   conv.hull(S)

Figure 6.1. Convex hull

those points in S that will be vertices. The problem of finding this listing is a well-studied topic in computational geometry (see Edelsbrunner [1987]).

Given a set $S = \{s_1, \ldots, s_k\} \subseteq \mathbf{R}^n$, we can easily certify that a vector $v \in \mathbf{R}^n$ is contained in conv.hull(S): We just specify scalars $\lambda_1, \ldots, \lambda_k$ such that

$$
\begin{array}{lll}
(i) & \sum_{i=1}^k \lambda_i = 1 & \qquad\qquad (6.2)\\[2mm]
(ii) & \sum_{i=1}^k \lambda_i s_i = v & \\[2mm]
(iii) & \lambda_i \geq 0, \text{ for all } i = 1, \ldots, k. &
\end{array}
$$

Notice that (6.2) is just a linear system in the variables $\lambda_1, \ldots, \lambda_k$. Testing membership in conv.hull(S) is equivalent to determining whether or not (6.2) has a solution $(\lambda_1, \ldots, \lambda_k)$. It follows that $v \notin$ conv.hull(S) can be certified with the help of Farkas' Lemma from the theory of linear inequalities.

Proposition 6.2 *Let $S \subseteq \mathbf{R}^n$ be a finite set and let $v \in \mathbf{R}^n \backslash conv.hull(S)$. Then there exists an inequality $w^T x \leq t$ that separates v from conv.hull(S), that is, $w^T s \leq t$ for all $s \in conv.hull(S)$ but $w^T v > t$.*

Proof: Writing S as $\{s_1, \ldots, s_k\}$, we know that (6.2) has no solution. Thus, by Farkas' Lemma there exist $z \in \mathbf{R}$ (corresponding to equation (i)) and $y \in \mathbf{R}^n$ (corresponding to equation (ii)) such that

$$y^T s_i + z \leq 0, \text{ for all } i = 1, \ldots, k$$
$$y^T v + z > 0.$$

Let $w = y$ and $t = -z$. Since $w^T s_i \leq t$ for all $i = 1, \ldots, k$, Proposition 6.1 implies that $w^T s \leq t$ for all $s \in conv.hull(S)$, which proves the result. ∎

Digression: Matching Polyhedra

A large part of the study of polyhedral methods in combinatorial optimization was motivated by a theorem of Edmonds on matchings in graphs. Throughout the chapter we will return to his celebrated result to illustrate the general concepts we develop. (This material is not required for the central topics of the chapter.) The perfect matching algorithm of Chapter 5 actually proves Edmonds' theorem, as we describe below.

For a graph $G = (V, E)$, let $\mathcal{PM}(G) \subseteq \mathbf{R}^E$ denote the set of characteristic vectors of its perfect matchings. Suppose $\mathcal{PM}(G)$ is nonempty. Then, given a weight vector $(-w_e : e \in E)$, the Blossom Algorithm for minimum-weight perfect matching will find a vector $\bar{x} \in \mathcal{PM}(G)$ that minimizes the inner product $w^T \bar{x}$, that is, a minimum-weight perfect matching in G. At the same time, the algorithm determines a dual solution that proves $\bar{x}$ is indeed an optimal solution. As we discussed in Chapter 5, the form of the dual solution actually proves that $\bar{x}$ minimizes $w^T x$ over all solutions to the system of inequalities

$$x(\delta(v)) = 1, \text{ for all } v \in V \qquad (6.3)$$
$$x(\delta(S)) \geq 1, \text{ for all } S \subseteq V, |S| \geq 3 \text{ and odd}$$
$$x_e \geq 0, \text{ for all } e \in E.$$

This is a stronger statement, since each vector in $\mathcal{PM}(G)$ is a solution of (6.3). It can be described neatly as follows.

Theorem 6.3 *(Perfect Matching Polytope Theorem) For any graph $G = (V, E)$, the convex hull of $\mathcal{PM}(G)$ is identical to the set of solutions of the linear system (6.3).*

Proof: Let $G = (V, E)$ be a graph and let P denote the set of solutions of (6.3). Since $\mathcal{PM}(G) \subseteq P$, from Proposition 6.1 we can conclude that conv.hull($\mathcal{PM}(G)$) $\subseteq P$. To prove the reverse inclusion, suppose $\bar{x} \notin$ conv.hull($\mathcal{PM}(G)$). We must show $\bar{x} \notin P$. By Proposition 6.2 we have an inequality $\sum (w_e x_e : e \in E) \leq t$ that separates $\bar{x}$ from conv.hull($\mathcal{PM}(G)$). Now, using the perfect matching algorithm with weight vector $(w_e : e \in E)$, the dual solution provided shows that $\sum (w_e x_e : e \in E) \leq t$ for all solutions of (6.3). So $\sum (w_e x_e : e \in E) \leq t$ also separates $\bar{x}$ from P, and hence $\bar{x} \notin P$.

∎

In a similar way, we can restate a number of the other linear-programming results from Chapter 5 in terms of convex hulls. For example, Theorem 5.30 on T-joins is equivalent to:

Theorem 6.4 *Let $G = (V, E)$ and $T \subseteq V$ with $|T|$ even, then the convex hull of the characteristic vectors of T-joins of G is the set of feasible solutions of*

$$\text{Minimize } \sum (c_e x_e : \ e \in E)$$

$$\text{subject to}$$

$$x(\delta(S) \backslash F) - x(F) \geq 1 - |F| \text{ for all } S \subseteq V \text{ and } F \subseteq \delta(S)$$

$$\text{such that } |F| + |S \cap T| \text{ is odd}$$

$$0 \leq x_e \leq 1 \text{ for all } e \in E.$$

∎

We will use this interpretation of the T-join theorem to give a proof of Theorem 5.31 on b-factors which we stated in Section 5.5. First, we restate the theorem as:

Theorem 6.5 *Let $G = (V, E)$ and $b \in \mathbf{Z}^V$. Then the convex hull P of characteristic vectors of b-factors of G is the set Q of feasible solutions of*

$$\text{Minimize } \sum (c_e x_e : \ e \in E)$$

$$\text{subject to}$$

$$x(\delta(v)) = b_v, \text{ for all } v \in V$$

$$x(\delta(S) \setminus F) - x(F) \geq 1 - |F|, \text{ for all } S \subseteq V \text{ and } F \subseteq \delta(S)$$

$$\text{such that } |F| + b(S) \text{ is odd}$$

$$0 \leq x_e \leq 1, \text{ for all } e \in E.$$

Proof: We have already pointed out in Section 5.5 that $P \subseteq Q$. Now suppose that $\hat{x} \in Q$. Then by Theorem 6.4, $\hat{x}$ is a convex combination $\sum_{i=1}^{k} \lambda_i x^i$ of characteristic vectors of T-joins, where $T = \{v : b_v \text{ is odd}\}$. Choose this expression so that $\alpha = \sum_{i=1}^{k} \lambda_i \sum (|x^i(\delta(v)) - b_v| : v \in V)$ is as small as possible. If $\alpha = 0$, then every x^i is the characteristic vector of a b-factor, so $\hat{x} \in P$, as required. Otherwise, we may assume that there is a node u such that $x^1(\delta(u)) > b_u$. Since $\hat{x}(\delta(u)) = b_u$, we may assume that $x^2(\delta(u)) < b_u$.

Suppose that x^i is the characteristic vector of E_i for $i = 1$ and 2. We construct a closed edge-simple path as follows:

- The first edge is an element of $(E_1 \backslash E_2) \cap \delta(u)$.
- If the path enters $v \in V$ on an edge in $E_2 \backslash E_1$, it leaves *if possible* on an edge in $E_1 \backslash E_2$.
- If the path enters $v \in V \backslash \{u\}$ on an edge in $E_1 \backslash E_2$, it leaves *if possible* on an edge in $E_2 \backslash E_1$.
- If the path enters u on an edge in $E_1 \backslash E_2$, it terminates.

It is easy to see that it is possible to construct this path, since $E_1 \triangle E_2$ is an even set of G and $|E_1 \cap \delta(u)| > |E_2 \cap \delta(u)|$. Let E' be the edge-set of the path. Now define y^{k+i} to be the characteristic vector of $E_i \triangle E'$. Since E' is an even set and E_i is a T-join, y^{k+i} is the characteristic vector of a T-join for $i = 1$ and 2. for $i = 1$ and 2. It is easy to see that $|y^{k+i}(\delta(u)) - b_u| < |x^i(\delta(u)) - b_u|$ for $i = 1$ and 2. It is also straightforward to check the following.

Claim. For all $v \in V \backslash \{u\}$, $|(y^{k+1} + y^{k+2})(\delta(v)) - b_v| \leq |(x^1 + x^2)(\delta(v)) - b_v|$.

Now define y^i to be x^i for $1 \leq i \leq k$, and define μ_i to be $\lambda_i - \varepsilon$ for $i = 1$ and 2, to be ε for $i = k + 1$ and $k + 2$, and to be λ_i for $3 \leq i \leq k$, where ε is sufficiently small and positive. Then $\sum_{i=1}^{k+2} \mu_i y^i$ is an expression for $\hat{x}$ as a convex combination of characteristic vectors of T-joins for which α is smaller, a contradiction. ∎

Exercises

6.1. Prove the claim made in the proof of Theorem 6.5.

6.2   POLYTOPES

A *polyhedron* is the solution set of a finite system of linear inequalities. The notion should be a familiar one since a linear-programming problem can be described as the problem of optimizing a linear function over a polyhedron. Considering the solution set as a geometric object allows us to gain considerable insight into linear inequalities, duality, and, what is important here, the use of linear programming in combinatorial optimization.

The polyhedron defined by the linear system

$$
\begin{aligned}
-1x_1 - 1x_2 &\leq 5 \\
-2x_1 + 1x_2 &\leq -1 \\
1x_1 + 3x_2 &\leq 18 \\
1x_1 + 0x_2 &\leq 6 \\
1x_1 - 2x_2 &\leq 2.
\end{aligned}
$$

is illustrated in Figure 6.2. Note that this polyhedron is bounded: it contains no infinite half-line. A polyhedron of this type is called a *polytope*. That is, a

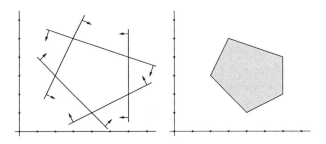

Figure 6.2. A polyhedron

polyhedron $P \subseteq \mathbf{R}^n$ is a polytope if there exist $l, u \in \mathbf{R}^n$ such that $l \leq x \leq u$ for all $x \in P$. As we shall see later, a polytope can be described "internally" as the convex hull of a finite set of vectors and, conversely, the convex hull of a finite set of vectors is always a polytope.

An inequality $w^T x \leq t$ is *valid* for a polyhedron P if $P \subseteq \{x : w^T x \leq t\}$. As usual, we call the solution set of an equation $w^T x = t$ a *hyperplane* if $w \neq 0$. With respect to a polyhedron P, it is called a *supporting hyperplane* if $w^T x \leq t$ is valid for P and $P \cap \{x : w^T x = t\} \neq \emptyset$. It should be evident that the intersection of a polyhedron with one of its supporting hyperplanes is a special type of subset (see Figure 6.3). In the plane, these subsets are the edges and corners of a polygon. In general, we call the intersection a *face* of the polyhedron. By convention, we also call the null set and the polyhedron itself faces, and we refer to the other faces as *proper* faces. A description of a face in terms of the polyhedron's defining system of inequalities can be obtained as follows.

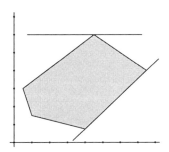

Figure 6.3. Supporting hyperplanes

Proposition 6.6 *A nonempty set $F \subseteq P = \{x : Ax \leq b\}$ is a face of P if and only if for some subsystem $A°x \leq b°$ of $Ax \leq b$ we have $F = \{x \in P : A°x = b°\}$.*

Proof: Suppose F is a face of P. Then there exists a valid inequality $w^T x \leq t$ such that $F = \{x \in P : w^T x = t\}$. Consider the linear-programming problem $\max\{w^T x : Ax \leq b\}$. The set of optimal solutions is precisely F. Now let y^* be an optimal solution to the dual problem $\min\{y^T b : y^T A = w^T, y \geq 0\}$ and let $A°x \leq b°$ be those inequalities $a_i^T x \leq b_i$ whose corresponding dual variable y_i^* is positive. By complementary slackness, we have $F = \{x : Ax \leq b, A°x = b°\}$ as required.

Conversely, if $F = \{x \in P : A°x = b°\}$ for some subsystem $A°x \leq b°$ of $Ax \leq b$, then add the inequalities $A°x \leq b°$ to obtain an inequality $w^T x \leq t$. Every $x \in F$ satisfies $w^T x = t$, and every $x \in P \setminus F$ satisfies $w^T x < t$, as required.   ∎

This result implies that a polyhedron has only finitely many distinct faces. In terms of linear programming, this says that there are only a finite number of possibilities for the optimal solution set to a linear-programming problem.

The minimal (by inclusion) faces of a polyhedron have a special structure.

Proposition 6.7 *Let F be a minimal nonempty face of $P = \{x : Ax \leq b\}$. Then $F = \{x : A°x = b°\}$ for some subsystem $A°x \leq b°$ of $Ax \leq b$. Moreover, the rank of the matrix $A°$ is equal to the rank of A.*   ∎

We leave the proof as an exercise.

A vector $v \in P$ is called a *vertex* if $\{v\}$ is a face of P. If v is a vertex, we will also refer to the set $\{v\}$ as a vertex. We call P *pointed* if it has at least one vertex. (An example of a polyhedron having no vertex is $\{(x_1, x_2) \in \mathbf{R}^2 : x_1 \geq 0\}$.) The above proposition implies that a vertex v is the unique solution to $A°x = b°$ for some subsystem $A°x \leq b°$ of $Ax \leq b$. Furthermore, since the rank of $A°$ is equal to the rank of A for any minimal face, P has a vertex if and only if A has full column rank. Therefore, we have the following result.

Proposition 6.8 *If a polyhedron P is pointed then every minimal nonempty face of P is a vertex.*   ∎

This will play a central role in the use of linear programming in combinatorial optimization. For example, if we want to show that a linear-programming problem always has an integral optimal solution as the objective function varies, it suffices to show that all vertices of the underlying polyhedron are integral.

The geometric intuition of a vertex is that it is a "corner" of the polyhedron. More precisely, we have the following result.

Proposition 6.9 *Let* $P = \{x : Ax \leq b\}$ *and* $v \in P$. *Then* v *is a vertex of* P *if and only if* v *cannot be written as a convex combination of vectors in* $P \setminus \{v\}$.

Proof: Suppose v is a vertex and let $A^\circ x \leq b^\circ$ be a subsystem of $Ax \leq b$ such that $\{v\} = \{x \in P : A^\circ x = b^\circ\}$. Now suppose v can be written as a convex combination $\lambda_1 u_1 + \cdots + \lambda_k u_k$ of vectors $u_1, \ldots, u_k \in P$, for some positive constants $\lambda_1, \ldots, \lambda_k$ with $\sum(\lambda_i : i = 1, \ldots, k) = 1$. Then we must have $A^\circ u_i = b^\circ$ for all $i = 1, \ldots, k$. But v is the unique solution to $A^\circ x = b^\circ$, so the conclusion follows.

Conversely, suppose v cannot be written as a convex combination of vectors in $P \setminus \{v\}$. Let $A^\circ x \leq b^\circ$ consist of those inequalities in $Ax \leq b$ which v satisfies with equality, and let $F = \{x : A^\circ x = b^\circ\}$. We must show that v is the unique member of F. So suppose there exists a vector $u \in F \setminus \{v\}$ and let $L = \{v + \lambda(u - v) : \lambda \in \mathbf{R}\}$ be the line through v and u. Clearly, $L \subseteq F$. Now since $a_i v < b_i$ for each inequality of $Ax \leq b$ not in $A^\circ x \leq b^\circ$, for a sufficiently small number $\varepsilon > 0$ we have $v + \varepsilon(u - v) \in P$ and $v - \varepsilon(u - v) \in P$. But $v = 1/2(v + \varepsilon(u - v)) + 1/2(v - \varepsilon(u - v))$, contrary to our assumption. ∎

Twisting this characterization a bit, we get the following description of a polytope.

Theorem 6.10 *A polytope is equal to the convex hull of its vertices.*

Proof: Let P be a (nonempty) polytope. Since P is bounded, P must be pointed. (See Exercise 6.2.) Let $v_1, \ldots, v_k$ be the vertices of P. Clearly, conv.hull$\{v_1, \ldots, v_k\} \subseteq P$. So suppose there exists

$$u \in P \setminus \text{conv.hull}(\{v_1, \ldots, v_k\}).$$

Then by Proposition 6.2, there exists an inequality $w^T x \leq t$ that separates u from conv.hull$\{v_1, \ldots, v_k\}$. Let $t^* = \max\{w^T x : x \in P\}$ and consider the face $F = \{x \in P : w^T x = t^*\}$. Since $u \in P$, we have $t^* > t$. So F contains no vertex of P, a contradiction. ∎

We can push this connection a little further, to obtain the equivalence that we promised at the beginning of the section.

Theorem 6.11 *A set* P *is a polytope if and only if there exists a finite set* V *such that* P *is the convex hull of* V.

Proof: One direction of this equivalence immediately follows from Theorem 6.10. For the other direction, let $V = \{v_1, \ldots, v_k\} \subseteq \mathbf{R}^n$ be a finite set and let $P = \text{conv.hull}(V)$. We must show that P is the solution set of some system of linear inequalities. The idea of the proof is to define a polytope consisting of the coefficients and right-hand-sides of all valid inequalities for P and apply Theorem 6.10 to select a finite subset of these valid inequalities.

To carry this out, consider the set $Q \subseteq \mathbf{R}^{n+1}$ defined as

$$\{\left(\begin{smallmatrix}\alpha\\\beta\end{smallmatrix}\right) : \alpha \in \mathbf{R}^n, \beta \in \mathbf{R}, \ -1 \le \alpha \le 1, \ -1 \le \beta \le 1, \ \alpha^T v \le \beta \ \text{ for all } \ v \in V\}.$$

Since V is a finite set, Q is a polytope. Thus, by Theorem 6.10, we know that Q is equal to the convex hull of its vertices $\left(\begin{smallmatrix}a_1\\b_1\end{smallmatrix}\right), \ldots, \left(\begin{smallmatrix}a_m\\b_m\end{smallmatrix}\right)$. We claim that the linear system

$$a_1^T x \le b_1, \quad a_2^T x \le b_2, \quad \cdots \quad , a_m^T x \le b_m \tag{6.4}$$

defines P.

We first show that P is contained in the solution set of (6.4). So, suppose $\bar{x} \in P$. Then $\bar{x} = \lambda_1 v_1 + \cdots + \lambda_k v_k$ for some nonnegative constants $\lambda_1, \ldots, \lambda_k$ such that $\lambda_1 + \cdots + \lambda_k = 1$. Thus, for each $i = 1, \ldots, m$ we have

$$a_i^T \bar{x} = a_i^T (\lambda_1 v_1 + \cdots + \lambda_k v_k) = \lambda_1 a_i^T v_1 + \cdots + \lambda_k a_i^T v_k \le \lambda_1 b_i + \cdots + \lambda_k b_i = b_i.$$

So $\bar{x}$ satisfies all inequalities in (6.4).

Going in the other direction, we show that the solution set of (6.4) is contained in P. This time let $\bar{x}$ be a solution to (6.4) and suppose $\bar{x} \notin P$. Then, by Proposition 6.2, there exists an inequality $w^T x \le t$ that separates $\bar{x}$ from P. By scaling $w^T x \le t$ by a positive constant if necessary, we may assume that $-1 \le w \le 1$ and $-1 \le t \le 1$. That is, we may assume that $\left(\begin{smallmatrix}w\\t\end{smallmatrix}\right) \in Q$. So we may write $\left(\begin{smallmatrix}w\\t\end{smallmatrix}\right) = \gamma_1 \left(\begin{smallmatrix}a_1\\b_1\end{smallmatrix}\right) + \cdots + \gamma_m \left(\begin{smallmatrix}a_m\\b_m\end{smallmatrix}\right)$ for some nonnegative constants $\gamma_1, \ldots, \gamma_m$ such that $\gamma_1 + \cdots + \gamma_m = 1$. Therefore

$$w^T \bar{x} = \gamma_1 a_1^T \bar{x} + \cdots + \gamma_m a_m^T \bar{x} \le \gamma_1 b_1 + \cdots + \gamma_m b_m = t.$$

But this is a contradiction, since $w^T \bar{x} > t$. So $\bar{x} \in P$, which completes the proof. ∎

For a generalization of Theorem 6.11 to unbounded polyhedra we refer the reader to Chvátal [1983] or Schrijver [1986].

The above theorem is the driving force behind polyhedral combinatorics. As we have seen in Chapters 2, 3, 4, and 5, an important ingredient in an algorithmic approach to a combinatorial problem is a min-max relationship to prove optimality of solutions. With Theorem 6.11, we have a general method for obtaining such min-max relations:

- Represent the combinatorial problem as an optimization problem over a finite set of vectors S, for example, by considering characteristic vectors.

- Find a linear description of conv.hull(S).

- Apply the Duality Theorem of Linear Programming and Proposition 6.1 to obtain a min-max relation for the combinatorial problem.

Of course, Theorem 6.11 only states that the second step is possible; it does not give us a good method for finding the description. The art of polyhedral combinatorics is in establishing conditions under which this step can be readily carried out. Also, note that the number of inequalities in a linear defining system may be far greater than the size of the original combinatorial problem. So the third step does not directly imply a polynomial-time algorithm for the optimization problem. It does, however, provide a basis for such an algorithm, as in the case of matchings in graphs. Moreover, in Chapter 7 we illustrate how even partial knowledge of a linear description of conv.hull(S) can be useful in an optimization routine.

Digression: Matching Polyhedra

Another way to describe Edmonds' Perfect Matching Polytope Theorem is to say that for a graph G the polyhedron $PM(G)$ defined by (6.3) has only integral vertices. Edmonds' algorithmic proof provides a general technique for establishing such a result. When faced with a problem of this type (that is, verifying that all vertices of a polytope are integral) it often pays to think through the matching algorithm to see if such a primal-dual approach can be adopted. Throughout this chapter, we will establish a number of other general techniques that are also useful to have available. Indeed, we have already covered enough material to present a purely "polyhedral" approach to Edmonds' theorem.

Let's begin with the special case of bipartite graphs. For this class, the matching polytope result is just a restatement of Theorem 5.12 from page 145. It is considerably simpler than Edmonds' theorem, since we may remove the blossom inequalities from the defining system.

Theorem 6.12 *(Birkhoff's Theorem) Let G be a bipartite graph. Then the convex hull of the perfect matchings of G is defined by*

$$x(\delta(v)) = 1, \; for \; all \; v \in V \tag{6.5}$$
$$x_e \geq 0, \; for \; all \; e \in E.$$

Proof: Let P be the polytope defined by (6.5). Clearly, each perfect matching of G is a member of P. We need only show that all vertices of P are integral. So suppose $\tilde{x}$ is a nonintegral vertex and let $\tilde{E} = \{e \in E : 0 < \tilde{x}_e < 1\}$ be the "fractional" edges. Since $\tilde{x}(\delta(v)) = 1$ for all nodes v, each node that meets an edge in $\tilde{E}$ meets at least two such edges. So $\tilde{E}$ contains an even circuit C. (The graph G, being bipartite, has no odd circuits.) Let $d \in \mathbf{R}^E$ be a vector that is 0 for all edges not in C, and alternately 1 and -1 for the edges of C. Then for $\varepsilon > 0$ and sufficiently small, both $\tilde{x} + \varepsilon d$ and $\tilde{x} - \varepsilon d$ are in P. But $\tilde{x} = \frac{1}{2}(\tilde{x} + \varepsilon d) + \frac{1}{2}(\tilde{x} - \varepsilon d)$, a contradiction. ∎

In general, removing the blossom inequalities from the definition of $PM(G)$ will introduce nonintegral vertices. Indeed, letting $FPM(G)$ (the *fractional matching polytope*) denote the polytope defined by (6.5), we have the following characterization.

Theorem 6.13 *(Fractional Matching Polytope Theorem)* *Let G be a graph and let $x \in FPM(G)$. Then x is a vertex of $FPM(G)$ if and only if $x_e \in \{0, \frac{1}{2}, 1\}$ for all $e \in E$ and the edges e for which $x_e = \frac{1}{2}$ form node disjoint odd circuits.*

Proof: To see the sufficiency of the condition, let $\tilde{x}$ satisfy the above properties and define $w_e = -1$ if $\tilde{x}_e = 0$ and $w_e = 0$ if $\tilde{x}_e > 0$. Then $\tilde{x}$ is the unique member of $FPM(G) \cap \{x : w^T x = 0\}$.

We will verify the necessity of the conditions in two steps. First we show that every vertex of $FPM(G)$ is half-integral, that is, $x_e \in \{0, \frac{1}{2}, 1\}$ for all $e \in E$. To this end, construct a bipartite graph G' from G by replacing each node $v \in V$ by two nodes v', v'' and replacing each edge $e = uv \in E$ by the two edges $e' = u'v'', e'' = u''v'$. Let $w \in \mathbf{R}^E$ be a vector of edge weights and extend w in the obvious way to a vector w' on the edges of G' (letting $w_{e'}$ and $w_{e''}$ be equal to w_e). Now by Birkhoff's Theorem, $\max\{(w')^T x' : x' \in FPM(G')\}$ has an integral optimal solution. Using the correspondence $x_e = \frac{1}{2}(x'_{e'} + x'_{e''})$, this gives a half-integral solution to $\max \{w^T x : x \in FPM(G)\}$, which is what we need.

To complete the proof, notice that if $\tilde{x} \in FPM(G)$ is half-integral, then the fractional edges must form node disjoint circuits. But, following the proof of Birkhoff's Theorem, we know if $\tilde{x}$ is a vertex then there can be no such circuit of even length. ∎

Building on this result, we give Schrijver's [1983] short inductive proof of Edmonds' theorem.

2nd Proof of the Perfect Matching Polytope Theorem: Let $\tilde{x}$ be a vertex of $PM(G)$. Then, by Proposition 6.7, there exists a set $\tilde{E} \subseteq E$ and a family $\mathcal{W}$ of odd subsets $S \subseteq V$, with $3 \leq |S| \leq |V| - 3$, such that $\tilde{x}$ is the unique solution to

$$x_e = 0, \text{ for all } e \in \tilde{E}$$
$$x(\delta(v)) = 1, \text{ for all } v \in V$$
$$x(\delta(S)) = 1, \text{ for all } S \in \mathcal{W}.$$

Case 1: $\mathcal{W} = \emptyset$. Then $\tilde{x}$ is a vertex of the fractional matching polytope $FPM(G)$. So by the above theorem, the fractional edges of $\tilde{x}$ form disjoint odd circuits. But since $\tilde{x}$ satisfies the blossom inequalities, no such circuit can exist. (The blossom inequality corresponding to the node set of the circuit "cuts off" such a fractional solution.) So in this case, $\tilde{x}$ is integral.

Case 2: $\mathcal{W} \neq \emptyset$. Choose $S \in \mathcal{W}$ and let G^1 and G^2 be the graphs obtained from G by shrinking S and $V \setminus S$ to pseudonodes, respectively. Let $\tilde{x}^1$ and $\tilde{x}^2$ be the restrictions of $\tilde{x}$ to the edges of G^1 and G^2.

It is easy to check that $\tilde{x}^1$ and $\tilde{x}^2$ satisfy the inequalities (6.3) with respect to the graphs G^1 and G^2. So, working by induction (both G^1 and G^2 have fewer nodes than G), we may assume that $\tilde{x}^1$ and $\tilde{x}^2$ are contained in the convex hull of perfect matchings of G^1 and G^2, respectively. This means that we can write $\tilde{x}^1$ and $\tilde{x}^2$ as convex combinations of perfect matchings of G^1 and G^2. Now these combinations can be glued together to write $\tilde{x}$ as a convex combination of perfect matchings of G. (This involves a technical argument that takes some effort to see.) But, since $\tilde{x}$ is a vertex, this implies that $\tilde{x}$ is itself a perfect matching of G, which completes the proof. ∎

Exercises

6.2. Let P be a bounded polyhedron. Show that P is pointed. (This is used in the proof of Theorem 6.10, so you should not use this result in your proof.)

6.3. Let P be a polyhedron and suppose that $P \subseteq \{x : x \geq 0\}$. Show that P is pointed.

6.4. Prove Proposition 6.7.

6.5. A *matchable set* of a graph G is the set of ends of some matching of G. Let P denote the convex hull of characteristic vectors of matchable sets of a bipartite graph $G = (V, E)$ with bipartition $\{A, B\}$. Prove that if $x \in P$ then x satisfies

$$x(C) - x(N(C)) \leq 0 \text{ for all } C \subseteq A$$
$$x(A) - x(B) = 0$$
$$0 \leq x_v \leq 1, \text{ for all } v \in V.$$

6.6. Prove that in the previous exercise, $x \in P$ if and only if $x \leq 1$ and there exists $y \in \mathbf{R}^E$ such that

$$y(\delta(v)) = x_v, \text{ for all } v \in V$$
$$y_e \geq 0, \text{ for all } e \in E.$$

6.7. (Balas and Pulleyblank) Prove that P of Exercise 6.5 is the set of solutions of the system written there. Hint: Use a flow feasibility theorem and the result of Exercise 6.6.

6.8. Prove that the convex hull of characteristic vectors of b-factors of G is the set of solutions of

$$x(\delta(v)) = b_v, \text{ for all } v \in V$$
$$x(\gamma(S)) + x(F) \leq (b(S) + |F| - 1)/2, \text{ for all } S \subseteq V,$$
$$F \subseteq \delta(S), \ b(S) + |F| \text{ odd}$$
$$0 \leq x_e \leq 1, \text{ for all } e \in E.$$

6.9. Use the previous exercise to show that, if $w \in V$ and $b_v = 1$ for all $v \neq w$, and $b(V)$ is even, then the convex hull of characteristic vectors of b-factors of G is the set of solutions of

$$x(\delta(v)) = b_v, \text{ for all } v \in V$$
$$x(\gamma(S)) \leq (|S| - 1)/2, \text{ for all } S \subseteq V \setminus \{w\}, \ |S| \text{ odd}$$
$$x_e \geq 0, \text{ for all } e \in E.$$

6.10. Use the previous exercise to prove that the convex hull of characteristic vectors of matchings of size k in a graph G is the set of solutions of

$$x(\delta(v)) \leq 1, \text{ for all } v \in V$$
$$x(\gamma(S)) \leq (|S| - 1)/2, \text{ for all } S \subseteq V, \ |S| \text{ odd}$$
$$x(E) = k$$
$$x_e \geq 0, \text{ for all } e \in E.$$

6.3 FACETS

Consider the three polytopes given in Figure 6.4. Our intuitive notion is that the first polytope has dimension 0, the second has dimension 1, and the third dimension 2. This agrees with the formal definition given below. A finite

Figure 6.4. Polytopes of dimension 0, 1, and 2

set of vectors X is *affinely independent* if the only solution to the system of equations

$$\sum(x\lambda_x : x \in X) = 0 \tag{6.6}$$
$$\sum(\lambda_x : x \in X) = 0$$

is to set $\lambda_x = 0$ for all $x \in X$. The *dimension* of a set $K \subseteq \mathbf{R}^n$ (denoted by $\dim(K)$) is one less than the maximum cardinality of an affinely-independent set $X \subseteq K$.

As you would expect, the dimension of a set is invariant under translation, that is, if $K \subseteq \mathbf{R}^n$ has dimension k then so does $\{x - w : x \in K\}$ for any $w \in \mathbf{R}^n$. Note that this would not be true if we replaced affine independence by (the more familiar) linear independence; for example the vectors $(0, 1)$ and

$(1,2)$ are linearly independent, but the vectors $(1,1)$ and $(2,2)$ are linearly dependent (even though the second vectors are obtained by adding $(1,0)$ to each of the first). For affine independence we have the following simple proposition.

Proposition 6.14 *If a set $X \subseteq \mathbf{R}^n$ is affinely independent, then for any $w \in \mathbf{R}^n$ the set $\{x - w : x \in X\}$ is affinely independent.*

Proof: Suppose X is affinely independent and $w \in \mathbf{R}^n$. Let $\{\lambda_x : x \in X\}$ be a solution to the system

$$\sum((x - w)\lambda_x : x \in X) = 0$$
$$\sum(\lambda_x : x \in X) = 0.$$

From the first equation we have $\sum(x\lambda_x : x \in X) = w \sum(\lambda_x : x \in X)$, which, using the second equation, gives $\sum(x\lambda_x : x \in X) = 0$. So $\{\lambda_x : x \in X\}$ is also a solution to (6.6), and since X is affinely independent we have $\lambda_x = 0$ for all $x \in X$. ∎

A polyhedron $P \subseteq \mathbf{R}^n$ of dimension n is said to be of *full dimension*.

We say that an inequality $a_i^T x \leq b_i$ in a system $Ax \leq b$ is an *implied equation* if $a_i^T x = b_i$ for all solutions of $Ax \leq b$.

Proposition 6.15 *Let $P = \{x \in \mathbf{R}^n : Ax \leq b\}$ and let $\bar{A}x = \bar{b}$ be the system of implied equations in $Ax \leq b$. Then $\dim(P) = n - \rho$, where ρ is the rank of the matrix $\bar{A}$.* ∎

We leave the proof of this proposition as an exercise.

A maximal proper face of a polyhedron is called a *facet*. Facets are of special interest in polyhedral combinatorics since, in a sense that we will make precise, each facet corresponds to a distinct inequality that must be present in any defining system for the polyhedron.

In the plane, the facets are the edges of a polygon. In general, using Proposition 6.15 and Proposition 6.6, we have the following characterization (Exercise 6.13).

Proposition 6.16 *Let F be a nonempty proper face of a polyhedron P. Then F is a facet if and only if $\dim(F) = \dim(P) - 1$.* ∎

In the plane, this corresponds to the fact that polygons have dimension 2, while lines have dimension 1.

We say an inequality $w^T x \leq t$ *induces* a facet F of P if $F = \{x \in P : w^T x = t\}$. We will show that in a minimal defining system for a polyhedron, each inequality induces a distinct facet and each facet of the polyhedron corresponds to a distinct inequality. (A system of inequalities and equations is *minimal* if no inequality can be made into an equation without reducing the size of the solution set and no inequality or equation can be omitted without enlarging the solution set.)

Theorem 6.17 *Let $P = \{x \in \mathbf{R}^n : A'x = b', A''x \leq b''\}$ be a nonempty polyhedron. Then the defining system is minimal if and only if the rows of A' are linearly independent and for each row i of A'' the inequality $(a_i'')^T x \leq b_i''$ induces a distinct facet of P.*

Proof: Suppose that the system is minimal. If some row of A' were a linear combination of other rows, then either the corresponding equation can be deleted or P is empty, in either case a contradiction. So the rows of A' must be linearly independent.

Now let i be a row of A'' and let $\bar{A}''x \leq \bar{b}''$ be the system obtained by removing $(a_i'')^T x \leq b_i''$ from $A''x \leq b''$. Since $A'x = b', A''x \leq b''$ is minimal, there exists an x^1 with $A'x^1 = b', \bar{A}''x^1 \leq \bar{b}'', (a_i'')^T x^1 > b_i''$. Choose an $x^2 \in P$ such that $A'x^2 = b', A''x^2 < b''$. (See Exercise 6.14.) Now, examining these linear systems, it follows that there exists a vector z on the line segment from x^1 to x^2 such that $A'z = b', (a_i'')^T z = b_i'', \bar{A}''z < \bar{b}''$. So the implied equations for the face induced by $(a_i'')^T x \leq b_i''$ are precisely $A'x = b', (a_i'')^T x = b_i$. Therefore, the dimension of this face is $\dim(P) - 1$ (by Proposition 6.15) and hence $(a_i'')^T x \leq b''$ induces a facet F_i of P. Furthermore, the existence of the vector z implies that F_i is distinct from the facets corresponding to other rows in A''.

Conversely, suppose the rows of A' are linearly independent and for each row i of A'', the inequality $a_i'' x \leq b_i''$ induces a distinct facet of P. It follows that no inequality in $A''x \leq b''$ can be made into an equation without reducing the solution set of the system. Therefore, $A'x = b'$ are the only implied equations in $A'x = b', A''x \leq b''$ and if any one of these equations were removed the solution set would be larger (since, by Proposition 6.15, the dimension of the set would increase).

Finally, let i be a row of A'' and let F be the facet induced by $(a_i'')^T x \leq b_i''$. Since each inequality in $A''x \leq b''$ induces a distinct facet, the implied equations for F are precisely $A'x = b', (a_i'')^T x = b_i''$. Therefore, there exists an x^1 such that $A'x^1 = b', (a_i'')^T x^1 = b_i''$, and $(a_j'')^T x^1 < b_j''$ for all rows j of A'' other than i. Let x^2 be a vector in P such that $(a_i'')^T x^2 < b_i''$. Then for a sufficiently small positive ϵ, the vector $(1+\epsilon)x^1 - \epsilon x^2$ satisfies all equations and inequalities in $A'x = b', A''x \leq b''$ other than $(a_i'')^T x \leq b_i''$. So $(a_i'')^T x \leq b_i''$ cannot be removed from the system. ∎

If P is of full dimension, then by Proposition 6.16 any two inequalities that induce the same facet are positive multiples of one another. So we have the following consequence of Theorem 6.17.

Corollary 6.18 *A full-dimensional polyhedron has a unique (up to positive scalar multiples) minimal defining system.* ∎

Another direct consequence of the theorem is the result we mentioned earlier.

Corollary 6.19 *Any defining system for a polyhedron must contain a distinct facet-inducing inequality for each of its facets.* ∎

Digression: Matching Polyhedra

For a polyhedron P of full dimension, Theorem 6.17 implies that $Ax \leq b$ is a defining system if and only if $P \subseteq \{x : Ax \leq b\}$ and each facet of P is represented in $Ax \leq b$. This suggests the following method for proving that $Ax \leq b$ defines P: We suppose $w^T x \leq t$ induces a facet F, then using information about P we prove enough structural results about w to conclude that some inequality in $Ax \leq b$ also induces F. We illustrate this technique below.

The *matching polytope* of a graph G (denoted by $M(G)$) is the convex hull of the (characteristic vectors of) matchings of G. It is easy to check that $M(G)$ is contained in the solution set of the system

$$x(\delta(v)) \leq 1, \text{ for all } v \in V \tag{6.7}$$
$$x(\gamma(S)) \leq (\mid S \mid -1)/2, \text{ for all } S \subseteq V, \mid S \mid \geq 3 \text{ and odd.}$$
$$x_e \geq 0, \text{ for all } e \in E.$$

As with perfect matchings, the weighted matching algorithm outlined in Chapter 5 can be used to prove the following theorem of Edmonds [1965a].

Theorem 6.20 *(Matching Polytope Theorem)* *For any graph* $G = (V, E)$, *the matching polytope is defined by (6.7).*

A nice property of $M(G)$ is that it is of full dimension (since $\emptyset$ is a matching and for each edge $e \in E$ the singleton $\{e\}$ is a matching). Using this, Lovász [1979] applied the proof technique outlined above.

Proof of the Matching Polytope Theorem: Suppose

$$w^T x \leq t \tag{6.8}$$

induces a facet of $M(G)$ and let $\mathcal{M}^*$ denote the set of matchings for which (6.8) holds as an equation. We will show that some inequality in (6.7) induces the same facet as (6.8). To do this, we repeatedly use the following argument: If $\bar{w}^T x \leq \bar{t}$ is a valid inequality that every matching in $\mathcal{M}^*$ satisfies with equality, then $\bar{w}^T x \leq \bar{t}$ induces the same facet as (6.8). (This follows from the fact that a facet is a maximal proper face.)

First, suppose $w_e < 0$ for some $e \in E$. Then every matching in $\mathcal{M}^*$ must satisfy $x_e = 0$ (if $M \in \mathcal{M}^*$ had $e \in M$ then the matching obtained by removing e would violate (6.8)). Therefore, the inequality $x_e \geq 0$ induces the same facet as (6.8).

Next, suppose that for some node $v \in V$ every matching in $\mathcal{M}^*$ contains an edge that covers v. Then each matching in $\mathcal{M}^*$ satisfies $x(\delta(v)) = 1$, so $x(\delta(v)) \leq 1$ induces the same facet as (6.8).

Finally, suppose $w_e \geq 0$ for all $e \in E$ and for each node $v \in V$ there is some matching in $\mathcal{M}^*$ that misses v (that is, no edge in the matching covers

v). Let G' be the graph formed by those edges $e \in E$ having $w_e > 0$ and let S denote its node set. Notice that G' is connected. We will show that each matching in $\mathcal{M}^*$ satisfies

$$x(\gamma(S)) = (\mid S \mid -1)/2, \tag{6.9}$$

which implies that the corresponding inequality for S induces the same facet as (6.8). This will complete the proof of the theorem.

If $\mathcal{M}^*$ contains matchings that do not satisfy (6.9), then (removing edges having $w_e = 0$ if necessary) one of them must miss some two nodes u and v in S. Of all such matchings, choose the one, M_1, and the nodes u and v it misses, such that the distance in G' between u and v is as small as possible.

Since $w_e > 0$ for all edges of G', the nodes u and v cannot be adjacent in G'. So there exists a node z on the shortest path from u to v.

Let $M_2 \in \mathcal{M}^*$ be a matching that misses z. Notice that, by the choice of M_1, both u and v are covered by M_2. Furthermore, by the choice of u and v, M_1 must cover z. So in the graph G^* formed by the edges $M_1 \cup M_2$, the nodes u, v, and z all have degree 1. Therefore, we can assume (renaming u and v if necessary) that the connected component of G^* that contains u consists of a path P that does not cover z.

Consider the two matchings

$$\bar{M}_1 = (M_1 \setminus (M_1 \cap E(P))) \cup (M_2 \cap E(P))$$

and

$$\bar{M}_2 = (M_2 \setminus (M_2 \cap E(P))) \cup (M_1 \cap E(P))$$

obtained from M_1 and M_2, respectively, by flipping the edges of P in and out of the matchings. Since $\bar{M}_1 \cup \bar{M}_2 = M_1 \cup M_2$, both $\bar{M}_1$ and $\bar{M}_2$ are in $\mathcal{M}^*$. But $\bar{M}_2$ misses z and u, contradicting our choice of the matching M_1. ∎

Facet-inducing Inequalities

We describe below some common techniques for proving that a given inequality induces a facet of a specified polyhedron. Motivation for this study comes from the fact that a complete description of the facets of a polyhedron provides a "best possible" min-max theorem. Indeed, such a min-max result for the matching problem is the basis for our tutorial.

Proving Necessity

Suppose we have a polyhedron $P = \{x : Ax \leq b\}$ of full dimension. Then an "instant" proof that an inequality $a_i^T x \leq b_i$ from $Ax \leq b$ induces a facet of P is to display a vector $\bar{x}$ such that $a_i^T \bar{x} > b_i$ but $\bar{x}$ satisfies all other inequalities in $Ax \leq b$. This proof establishes that $a_i^T x \leq b_i$ must be included in any subset of inequalities from $Ax \leq b$ that still defines P. Since $Ax \leq b$ contains a minimal defining system for P, this means that $a_i^T x \leq b_i$ is present in a minimal system. So, by Theorem 6.17, $a_i^T x \leq b_i$ induces a facet of P. The same proof technique will work when P is not of full dimension, but we must first display a vector $x' \in P$ with $a_i^T x' < b_i$, to prove that $a_i^T x \leq b_i$ is not an implicit equation. (See Theorem 6.17.)

Although the existence of an appropriate vector $\bar{x}$ (and x', in the non-full-dimensional case) is guaranteed, it is normally readily available only for simple types of constraints.

We illustrate the method on the nonnegativity constraints for the matching problem. Let e be an edge of a graph $G = (V, E)$ and define $\bar{x}$ by $\bar{x}_e = -1$ and $\bar{x}_f = 0$ for all edges $f \in E \setminus \{e\}$. Then $\bar{x}$ satisfies all inequalities in the matching system (6.7), other than $x_e \geq 0$. So the constraint $x_e \geq 0$ induces a facet of $M(G)$.

Direct Construction

A second method for proving that an inequality $w^T x \leq t$ induces a facet of a full-dimensional polyhedron P is to display a set of $\dim(P)$ affinely-independent vectors in P, each of which satisfies $w^T x \leq t$ with equality. This proves that the face induced by $w^T x \leq t$ is a facet, since it has the correct dimension (one less than that of P). Again, if P is not of full dimension, we first need to establish that $w^T x \leq t$ is not an implicit equation, say by displaying a vector $x' \in P$ with $w^T x' < t$.

To construct the set of affinely-independent vectors, we should take advantage of the combinatorial structure of the inequality. A simple illustration of this is the following proof that most of the degree constraints in the matching system induce facets. Let $G = (V, E)$ be a graph and let V' denote the set of all nodes $v \in V$ that either have at least three distinct neighbors in G or exactly two neighbors, u and y, and uy is not an edge of G or exactly one neighbor and v is a node of a two-node connected component of G. It is easy to see that if $v \in V \setminus V'$, then $x(\delta(v)) \leq 1$ does not induce a facet of $M(G)$. (Exercise 6.16.) So suppose $v \in V'$. To prove that $x(\delta(v)) \leq 1$ induces a facet of $M(G)$, we must construct $|E|$ affinely-independent vectors. Consider the $|E|$ distinct matchings defined as follows: For each edge $e \in \delta(v)$ let $M_e = \{e\}$ and for each edge $e \in E \setminus \delta(v)$ let $M_e = \{e, f\}$, where f is an edge in $\delta(v)$ that does not meet e. The set of characteristic vectors of the matchings $\{M_e : e \in E\}$ is the desired affinely-independent set.

Verifying Maximality

As above, let $w^T x \leq t$ be a valid inequality for a polyhedron P of full dimension. Another method for proving that $w^T x \leq t$ is facet-inducing is to show that the face F it induces is not contained in any larger proper face. This has a nice interpretation if, as is normally the case, P is the convex hull of a finite set S of combinatorial objects. In this case, let $\mathcal{F} = \{s \in S : w^T x_S = t\}$. Then we need only to show that if $c^T x \leq d$ is any valid inequality for P such that $\mathcal{F} \subseteq \{x : c^T x = d\}$, then, in fact, $c^T x \leq d$ is a positive scalar multiple of $w^T x \leq t$. If P is not of full dimension, then, as before, we must first verify that $w^T x \le t$ is not an implicit equation. Furthermore, we must allow for the possibility that $c^T x \le d$ and $w^T x \le t$ differ not only by a positive scalar multiple, but also by a linear combination of the implicit equations for P.

The procedure normally goes step by step, using the properties of $\mathcal{F}$ to gather information about the coefficients of c. We illustrate the method with the following proof of Lovász (given in Lovász and Plummer [1986]), verifying that certain blossom inequalities define facets of $M(G)$.

Let $G = (V, E)$ be a graph and let $S \subseteq V$ be a set of nodes such that $G[S]$ is *two-connected*, that is, $G[S]$ is connected and has no node v such that $G[V \setminus \{v\}]$ is not connected, and $G[S]$ is *hypomatchable*, that is, for each $v \in S$, $G[S]$ has a matching that misses only v. We will show that the blossom inequality

$$x(\gamma(S)) \le (|S| - 1)/2 \tag{6.10}$$

induces a facet of $M(G)$. So let $\mathcal{F}$ be the set of matchings of G that satisfy (6.10) as an equation and suppose that the valid inequality

$$\sum (c_e x_e : e \in E) \le d \tag{6.11}$$

is also satisfied by every matching in $\mathcal{F}$ as an equation.

We first show that $c_e = 0$ for every edge $e \in E \setminus \gamma(S)$. Indeed, for such an edge e we can find a matching $M \in \mathcal{F}$ such that $M \cup \{e\}$ is also in $\mathcal{F}$. But since both M and $M \cup \{e\}$ must satisfy (6.11) with equality, we can conclude that $c_e = 0$. So we may suppose that if $c_e \ne 0$ then $e \in \gamma(S)$.

To finish the proof, we only need to show that for all edges e and f in $\gamma(S)$, we have $c_e = c_f$, since this implies that (6.10) and (6.11) differ by only a positive scalar multiple. So suppose this is not the case and let $v \in S$ be a node such that c_e takes on different values for edges in $\delta(v) \cap \gamma(S)$. Let the graph G' be obtained from $G[S]$ by splitting v into two nodes v' and v'' such that all of the edges $e \in \delta(v) \cap \gamma(S)$ for which c_e takes on the minimum value (of the edges in $\delta(v) \cap \gamma(S)$) are incident with v' and all of the others are incident with v''. Since $G[S]$ is two-connected, we know G' is connected. Using this, we can check that G' has a perfect matching. (If this were not the case, by Tutte's Theorem (Theorem 5.3), there would exist a nonempty subset Y of the nodes of G' whose removal creates more than $|Y|$ odd components. But then for any $u \in Y$, if we let $\bar{u}$ be the corresponding node in G, then the graph $G[S \setminus \{\bar{u}\}]$ does not have a perfect matching, a contradiction.) So let M be a perfect matching of G' and let e_1 and e_2 be the edges in M such that e_1 meets v' and e_2 meets v''. Then both $M \setminus \{e_1\}$ and $M \setminus \{e_2\}$ are in $\mathcal{F}$. But since $c_{e_1} < c_{e_2}$, we cannot have that both $M \setminus \{e_1\}$ and $M \setminus \{e_2\}$ satisfy (6.11) with equality. This contradiction completes the proof.

Combining this facet characterization with the matching results given above, we can give a complete description of the facet-inducing inequalities for G. Indeed, let V' be defined as above and let $\mathcal{B}$ denote the set of all subsets $S \subseteq V$ such that $|S| \ge 3$ and $G[S]$ is hypomatchable and two-connected. Then it is easy to check (using the Matching Polytope Theorem) that the system

$$x_e \geq 0, \text{ for all } e \in E \tag{6.12}$$
$$x(\delta(v)) \leq 1, \text{ for all } v \in V'$$
$$x(\gamma(S)) \leq (|S| - 1)/2, \text{ for all } S \in \mathcal{B}$$

defines the matching polytope, $M(G)$, of G. (Exercise 6.18.) Therefore, the results of this section can be combined into the following theorem.

Theorem 6.21 *Let $G = (V, E)$ be a graph. Then (6.12) is the unique (up to positive scalar multiples of the inequalities) minimal defining system for $M(G)$.* ∎

This characterization is due to Pulleyblank and Edmonds [1974]. Their proof that the blossom inequalities in (6.12) induce facets actually follows the direct construction method, making use of the nested structure of hypomatchable graphs.

Exercises

6.11. Show that a finite set $X \subseteq \mathbf{R}^n$ is affinely independent if and only if for any $\bar{x} \in X$ the set $\{x - \bar{x} : x \in (X \setminus \bar{x})\}$ is linearly independent.

6.12. Prove Proposition 6.15.

6.13. Prove Proposition 6.16.

6.14. Suppose that $A'x = b', A''x \leq b''$ is a minimal defining system for a polyhedron P. Show that there exists a vector $\bar{x} \in P$ such that $A''\bar{x} < b''$.

6.15. Let $G = (V, E)$ be a graph and let $e \in E$. Prove that $x_e \geq 0$ induces a facet of the matching polytope, $M(G)$, by constructing a set of $|E|$ affinely-independent vectors in $M(G)$, each of which satisfies $x_e = 0$.

6.16. Show that for a given graph $G = (V, E)$, if $v \in V \setminus V'$, then $x(\delta(v)) \leq 1$ does not induce a facet of $M(G)$. (Where V' is defined as in the text above.)

6.17. Let $G = (V, E)$ be a graph and suppose $v \in V'$. Prove that $x(\delta(v)) \leq 1$ induces a facet of $M(G)$ by displaying a vector $\bar{x}$ such that $\bar{x}(\delta(v)) > 1$ but $\bar{x}$ satisfies all other inequalities in (6.7).

6.18. Using the Matching Polytope Theorem, prove that the system of inequalities (6.12) defines the matching polytope of G.

6.4 INTEGRAL POLYTOPES

Throughout the remainder of this chapter, we will be restricting our attention to *rational polyhedra*, that is, polyhedra that can be defined by rational linear systems. In the context of combinatorial optimization, this restriction is not very harsh since we deal almost exclusively with integral objects.

A rational polyhedron is called *integral* if every nonempty face contains an integral vector. In this definition, it is clear that we need only consider minimal faces. So a pointed rational polyhedron is integral if and only if all its vertices are integral.

The applicability of polyhedral methods in combinatorial optimization (at least from a theoretical point of view) often comes down to our ability to prove that polyhedra are integral. An important tool in this effort is the following characterization of Hoffman [1974], which allows us to restrict our attention to the optimal objective values rather than the actual solutions to linear-programming problems over a polytope.

Theorem 6.22 *A rational polytope P is integral if and only if for all integral vectors w the optimal value of $\max\{w^T x : x \in P\}$ is an integer.*

Proof: The necessity of the condition is clear, since if P is integral then $\max\{w^T x : x \in P\}$ has an integral optimal solution.

To prove sufficiency, suppose that for all integral vectors w the optimal value of $\max\{w^T x : x \in P\}$ is an integer. Let $v = (v_1, \ldots, v_n)^T$ be a vertex of P (since P is polytope, P has vertices) and let w be an integral vector such that v is the unique optimal solution to $\max\{w^T x : x \in P\}$. (See Exercise 6.19.) By multiplying w by a large positive integer if necessary, we may assume $w^T v > w^T u + u_1 - v_1$ for all vertices u of P other than v. This implies that if we let $\bar{w} = (w_1 + 1, w_2, \ldots, w_n)^T$, then v is an optimal solution to $\max\{\bar{w}^T x : x \in P\}$. So $\bar{w}^T v = w^T v + v_1$. But, by assumption, $w^T v$ and $\bar{w}^T v$ are integers. Thus v_1 is an integer. Repeating this for each component of v, we can conclude that v is integral. ∎

It is straightforward to extend this result to the case of (unbounded) pointed polyhedra. A more difficult generalization is to remove the restriction that the polyhedron has vertices. But this can also be done, as was shown by Edmonds and Giles [1977].

The importance of Theorem 6.22 is that it guides us toward indirect proof techniques for establishing integrality; for example, one can imagine a proof by induction on the objective value. We will come back to this theorem in Section 6.6.

Exercises

6.19. Let $P = \{x : Ax \leq b\}$ be a rational polyhedron and let F be a face of P. Show that there exists an integral vector w and an integer t such that $F = \{x \in P : w^T x = t\}$.

6.20. Show that a face of an integral polyhedron is again an integral polyhedron.

6.21. Derive the Perfect Matching Polytope Theorem from the fact that the linear system (6.7) defines the convex hull of the matchings of a given graph G.

6.5 TOTAL UNIMODULARITY

Proving polyhedra are integral is often a difficult task, involving proofs along the lines we have illustrated with matching problems. This has led to the development of various sufficient conditions for integrality, stating that if A and b satisfy certain properties, then we know $\{x : Ax \leq b\}$ is integral. In terms of the number of applications, perhaps the most successful of these integrality conditions is one developed by Hoffman and Kruskal [1956], involving the determinants of submatrices of A. To introduce the concept, consider the following result.

Proposition 6.23 *Let A be an integral, nonsingular, m by m matrix. Then $A^{-1}b$ is integral for every integral vector $b \in \mathbf{R}^m$ if and only if $\det(A) = 1$ or -1.*

Proof: Suppose $\det(A) = \pm 1$. By Cramer's Rule (from linear algebra) we know that A^{-1} is integral, which implies $A^{-1}b$ is integral for every integral b. Conversely, suppose $A^{-1}b$ is integral for all integral vectors b. Then, in particular, $A^{-1}e_i$ is integral (where e_i is the ith unit vector) for all $i = 1, \ldots, m$. This means that A^{-1} is integral. So $\det(A)$ and $\det(A^{-1})$ are both integers. But, since $\det(A) \cdot \det(A^{-1}) = 1$, this implies that $\det(A) = \pm 1$. ∎

This connection between integral solutions and ± 1 determinants can be pushed quite a way in the context of polyhedral theory.

We call a matrix A of full row rank *unimodular* if A is integral and each basis of A has determinant ± 1. So the above result characterizes square, unimodular matrices. In general, Veinott and Dantzig [1968] proved the following nice relationship between unimodular matrices and certain integral polyhedra.

Theorem 6.24 *Let A be an integral m by n matrix of full row rank. Then the polyhedron defined by $Ax = b, x \geq 0$ is integral for every integral vector $b \in \mathbf{R}^m$ if and only if A is unimodular.*

Proof: Suppose A is unimodular. Let $b \in \mathbf{R}^m$ be an integral vector and let $\bar{x}$ be a vertex of $\{x : Ax = b, x \geq 0\}$. (The nonnegativity constraints imply that the polyhedron does indeed have vertices—Exercise 6.3.) Then there are n linearly independent constraints satisfied by $\bar{x}$ with equality. It follows that the columns of A corresponding to the nonzero components of $\bar{x}$ are linearly independent. Extending these columns to a basis B of A, we have that the nonzero components of $\bar{x}$ are contained in the integral vector $B^{-1}b$. So $\bar{x}$ is integral.

Conversely, suppose $\{x : Ax = b, x \geq 0\}$ is integral for all integral vectors b. Let B be a basis of A and let v be an integral vector in $\mathbf{R}^m$. By Proposition 6.23 it suffices to show that $B^{-1}v$ is integral. So let y be an integral vector such that $y + B^{-1}v \geq 0$, and let $b = B(y + B^{-1}v)$. Note that b is integral. Furthermore, by adding zero components to the vector $y + B^{-1}v$, we can obtain a vector $z \in \mathbf{R}^n$ such that $Az = b$. Then z is a vertex of $\{x : Ax = b, x \geq 0\}$, since z is in the polyhedron and satisfies n linearly independent constraints with equality: the m equations $Ax = b$ and the $n - m$ equations $x_i = 0$ for the columns i outside of B. So z is integral, and thus $B^{-1}v$ is integral. ∎

Proceeding further, we call a matrix *totally unimodular* if all of its square submatrices have determinant $0, 1$, or -1. So, in particular, every entry in a totally unimodular matrix is $0, 1$, or -1. Moreover, it is easy to see that an m by n matrix A is totally unimodular if and only if $[A\ I]$ is unimodular, where I is the m by m identity matrix. (Exercise 6.23.)

Using Theorem 6.24, we can give a short proof of the following well-known characterization of total unimodularity by Hoffman and Kruskal [1956].

Theorem 6.25 *(Hoffman-Kruskal Theorem) Let A be an m by n integral matrix. Then the polyhedron defined by $Ax \leq b, x \geq 0$ is integral for every integral vector $b \in \mathbf{R}^m$ if and only if A is totally unimodular.*

Proof: Applying the linear programming trick of adding slack variables, we have that for any integral b, the polyhedron $\{x : Ax \leq b, x \geq 0\}$ is integral if and only if the polyhedron $\{z : [A\ I]z = b, z \geq 0\}$ is integral. So the result follows from Theorem 6.24. ∎

This characterization requires b to vary: For a given vector b it may be true that $\{x : Ax \leq b, x \geq 0\}$ is an integral polyhedron even if A is not totally unimodular. (See Exercise 6.25.) On the other hand, the following result (also by Hoffman and Kruskal [1956]) shows that total unimodularity is sufficient to deduce integrality of polyhedra without the nonnegativity restriction of Theorem 6.25.

Theorem 6.26 *Let A be an m by n totally unimodular matrix and let $b \in \mathbf{R}^m$ be an integral vector. Then the polyhedron defined by $Ax \leq b$ is integral.*

Proof: Let F be a minimal face of $\{x : Ax \leq b\}$. Then, by Proposition 6.7, $F = \{x : A^\circ x = b^\circ\}$ for some subsystem $A^\circ x \leq b^\circ$ of $Ax \leq b$, with A° having full row rank. By reordering the columns if necessary, we may write A° as $[B\ N]$ where B is a basis of A°. It follows that

$$\bar{x} = \begin{pmatrix} B^{-1}b^\circ \\ 0 \end{pmatrix}$$

is an integral vector in F. ∎

This theorem is applied very often in polyhedral combinatorics. It is something that should always be kept in mind when trying to establish the integrality of polyhedra.

But how can we detect that a given matrix is totally unimodular? In view of Theorem 6.26 this is certainly an important question. In response, a number of characterizations were proposed in the 60s and 70s, but the general problem remained open until 1980 when Seymour [1980] proved a deep theorem which gives a polynomial-time test. His theorem shows that every totally unimodular matrix can be built in a specific way from certain "network matrices" (described below) and the two 5 by 5 matrices given in Figure 6.5. A detailed account of Seymour's work can be found in Schrijver [1986]. For

$$\begin{bmatrix} 1 & -1 & 0 & 0 & -1 \\ -1 & 1 & -1 & 0 & 0 \\ 0 & -1 & 1 & -1 & 0 \\ 0 & 0 & -1 & 1 & -1 \\ -1 & 0 & 0 & -1 & 1 \end{bmatrix} \qquad \begin{bmatrix} 1 & 1 & 1 & 1 & 1 \\ 1 & 1 & 1 & 0 & 0 \\ 1 & 0 & 1 & 1 & 0 \\ 1 & 0 & 0 & 1 & 1 \\ 1 & 1 & 0 & 0 & 1 \end{bmatrix}$$

Figure 6.5. Non-network totally unimodular matrices

our discussion, we will concentrate on network matrices, since, via Seymour, we know they form the backbone of all totally unimodular matrices and since most familiar cases of total unimodularity can be deduced directly from this class.

We begin with the following example, described by Poincaré [1900].

Theorem 6.27 *Let A be a 0, ± 1 valued matrix, where each column has at most one $+1$ and at most one -1. Then A is totally unimodular.*

Proof: Let N be a k by k submatrix of A. If $k = 1$, then $\det(N)$ is either 0 or ± 1. So we may suppose that $k \geq 2$ and proceed by induction on k. If N has a column having at most one nonzero, then expanding the determinant along this column we have that $\det(N)$ is either 0 or ± 1, by our induction hypothesis. On the other hand, if every column of N has both a $+1$ and a -1, then the sum of the rows of N is 0 and hence $\det(N) = 0$. ∎

This result may be restated as follows: Let $D = (V, E)$ be a directed graph and let A be its incidence matrix, that is, the rows of A are indexed by V and the columns by E, and the entry $A_{v,e}$ is $+1$ if v is the head of e, -1 if v is the tail of e, and 0 otherwise. Then A is totally unimodular. This can be used to give another proof of the max-flow min-cut theorem. (See Exercise 6.27.)

To extend this theorem, let $T = (V, E')$ be a spanning tree on the same set of nodes as D. Now consider the matrix M having rows indexed by E' and columns indexed by E, where, for $e = (u, v) \in E$ and $e' \in E'$

$$M_{e',e} = \begin{cases} +1 \text{ if the } (u, v)\text{-path in } T \text{ uses } e' \text{ in the forward direction} \\ -1 \text{ if the } (u, v)\text{-path in } T \text{ uses } e' \text{ in the backward direction} \\ 0 \text{ if the } (u, v)\text{-path in } T \text{ does not use } e'. \end{cases}$$

Then M is a *network matrix*.

Theorem 6.28 *Network matrices are totally unimodular.*

This important result was obtained by Tutte [1965]. In the proof we make use of Proposition 2.15, characterizing the bases of an incidence matrix. Namely, B is a basis of A if and only if the arcs corresponding to B form a spanning tree of D. Notice that this implies that the removal of any row i of A gives a matrix $\bar{A}$ of full row rank.

Proof of Theorem 6.28: Let M be a network matrix, arising from the tree $T = (V, E')$ and the directed graph $D = (V, E)$. Let N be the node-arc incidence matrix of D and let $\bar{N}$ be obtained from N by deleting a single row so that $\bar{N}$ has full row rank. Similarly, let L be the node-arc incidence matrix of T and let $\bar{L}$ be obtained by deleting a single row (corresponding to the same node that was deleted from N). By Theorem 6.27, the matrix $[\bar{L} \ \bar{N}]$ is totally unimodular. Now, by Proposition 2.15 we know that $\bar{L}$ is a basis of $[\bar{L} \ \bar{N}]$. It follows that the matrix $\bar{L}^{-1}\bar{N}$ is totally unimodular. (See Exercise 6.24.)

Carefully checking the definitions of L, M, and N, it can be seen that $LM = N$. This gives us $\bar{L}M = \bar{N}$ and hence $M = \bar{L}^{-1}\bar{N}$. So M is totally unimodular. ∎

We give several examples of network matrices.

Example 1: Not surprisingly, the matrices described in Theorem 6.27 are network matrices. This can be seen by considering trees T of the form given in Figure 6.6. ∎

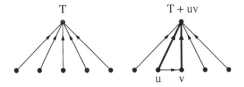

Figure 6.6. Tree for node-arc incidence matrices

Example 2: Let $G = (V, E)$ be a bipartite graph and let M be its node-edge incidence matrix: $M_{v,e} = 1$ if e meets v and 0 otherwise. Then M is a network matrix, as can be seen by considering trees of the form given in Figure 6.7.

Combining this with Theorem 6.26 gives another proof of Birkhoff's Theorem. ∎

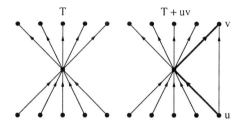

Figure 6.7. Tree for bipartite graphs

Example 3: If the $1's$ in each column of a $(0, 1)$-matrix M occur consecutively, then M is a network matrix. To see this, take a directed path as the tree T. ∎

Exercises

6.22. Let A be an m by n totally unimodular matrix. Show that A^T and $[A\ I]$ (where I is the m by m identity matrix) are also totally unimodular matrices.

6.23. Let A be an integral m by n matrix. Show that A is totally unimodular if and only if $[A\ I]$ is unimodular, where I is the m by m identity matrix.

6.24. Let A be a totally unimodular matrix of full row rank and let B be a basis of A. Prove that $B^{-1}A$ is totally unimodular. (Hint: Use Exercise 6.23.)

6.25. Give an example of a $\{0, 1\}$-valued matrix A and a vector b such that $\{x : Ax \leq b, x \geq 0\}$ is an integral polytope but A is not totally unimodular.

6.26. Let G be a graph and let A be its node-edge incidence matrix. Prove that A is totally unimodular if and only if G is bipartite.

6.27. Use total unimodularity to prove Theorem 3.5 (the Max-Flow Min-Cut Theorem).

6.28. Let E be a finite set and let $\mathcal{F}_1$, $\mathcal{F}_2$ be families of subsets of E. For $i = 1, 2$, suppose that if C and D are members of $\mathcal{F}_i$ then either $C \cap D = \emptyset$, $C \subseteq D$, or $D \subseteq C$. Show that the incidence matrix A of the family $\mathcal{F}_1 \cup \mathcal{F}_2$ is totally unimodular. (The rows of A are indexed by $\mathcal{F}_1 \cup \mathcal{F}_2$ and the columns of A are indexed by E. For $F \in \mathcal{F}_1 \cup \mathcal{F}_2$ and $e \in E$, the entry $A_{F,e} = 1$ if $e \in F$ and 0 otherwise.) (Hint: Row operations on subdeterminants will be useful.)

6.6   TOTAL DUAL INTEGRALITY

As we have emphasized, the basic theme of polyhedral combinatorics is the application of the linear-programming duality equation

$$\max \{w^T x : Ax \leq b\} = \min \{y^T b : y^T A = w^T, y \geq 0\} \qquad (6.13)$$

to combinatorial problems. This usually means proving that $Ax \leq b$ defines an integral polyhedron, whose vertices correspond to certain combinatorial objects. The end product, when we are successful, is a min-max theorem for the objects.

Such a min-max result can often be strengthened by imposing restrictions on the values of y in (6.13). The most natural of these restrictions is to insist that y be integer-valued. Indeed, several of the min-max theorems we presented in Chapters 3, 4, and 5 state precisely that for certain linear systems and objective vectors w, both sides of (6.13) have integral optimal solutions. Besides being more aesthetically pleasing, such integral min-max theorems also have practical implications. For example, the dual solutions may correspond to combinatorial objects of particular interest, such as "cuts" or "covers." Or the integrality may allow us to design a more efficient primal-dual algorithm by concentrating on discrete dual solutions, such as in the matching algorithms of Chapter 5.

Motivated by this, we define a rational linear system $Ax \leq b$ to be *totally dual integral* if the minimum in (6.13) can be achieved by an integral vector y for each integral w for which the optima exist. A nice feature of this definition is that we get primal integrality for free, provided b is integral, using the following result of Hoffman [1974].

Theorem 6.29   *Let $Ax \leq b$ be a totally dual integral system such that $P = \{x : Ax \leq b\}$ is a rational polytope and b is integral. Then P is an integral polytope.*

Proof: Since b is integral, the duality equation implies that $\max \{w^T x : x \in P\}$ is an integer for all integral vectors w. Thus, by Theorem 6.22, P is integral. ∎

Therefore, to prove an integral min-max theorem we need only show $Ax \leq b$ is totally dual integral. A generalization of this theorem to unbounded polyhedra was given by Edmonds and Giles [1977].

The qualification "provided b is integral" is very important, since the definition of total dual integrality can be abused in the following sense: For any rational system $Ax \leq b$, there exists a positive integer t such that $(1/t)Ax \leq (1/t)b$ is totally dual integral. (Exercise 6.30.) So total dual integrality in itself says nothing about the structure of a polyhedron. Moreover, our arguments for considering the dual problem center around the combinatorial nature of integral dual solutions. This is lost when A is nonintegral.

Therefore, we normally seek totally dual integral defining systems that involve only integral coefficients. Giles and Pulleyblank [1979] showed this is always possible.

Theorem 6.30 *Let P be a rational polyhedron. Then there exists a totally dual integral system $Ax \leq b$, with A integral, such that $P = \{x : Ax \leq b\}$. Furthermore, if P is an integral polyhedron, then b can be chosen to be integral.*

Proof: Let $P = \{x \in \mathbf{R}^n : Mx \leq d\}$, where M is integral. (This is where we are using the rationality of P.) Let $L = \{l \in \mathbf{Z}^n : l = y^T M, 0 \leq y \leq 1\}$ be the set of all integral vectors that can be written as a nonnegative combination of the rows of M, where no multiplier is greater than 1. Clearly L is finite since it is the set of integral points in a bounded set. For each $l \in L$, let $t(l) = \max \{l^T x : x \in P\}$.

We define the system $Ax \leq b$ as the inequalities $l^T x \leq t(l)$, where l ranges over L. Since each row of M is in L, we have $P = \{x : Ax \leq b\}$. Also, if P is integral, then $t(l)$ is an integer for each $l \in L$, so, in this case, b is integer-valued. We need only show that $Ax \leq b$ is totally dual integral.

Let w be an integral vector such that $\max \{w^T x : x \in P\}$ exists. We will construct an integral optimal solution to $\min \{y^T b : y^T A = w^T, y \geq 0\}$. To this end, let

$$z^* = \max \{w^T x : x \in P\} = \min \{y^T d : y^T M = w^T, y \geq 0\} \qquad (6.14)$$

and let y^* achieve the minimum in (6.14). Setting $\bar{w} = (y^* - \lfloor y^* \rfloor)^T M$ (where $\lfloor y^* \rfloor$ denotes the vector $(\lfloor y_1^* \rfloor, \lfloor y_2^* \rfloor \ldots)^T$), then $\bar{w} \in L$, and it is easy to check that $y^* - \lfloor y^* \rfloor$ is an optimal solution to $\min \{y^T d : y^T M = \bar{w}, y \geq 0\}$. So $t(\bar{w}) = (y^* - \lfloor y^* \rfloor)^T d$. Rearranging this, we have $z^* = t(\bar{w}) + \lfloor y^* \rfloor^T d$. But z^* is also the optimal value of $\min \{y^T b : y^T A = w^T, y \geq 0\}$. Hence, an integral optimal solution is obtained by setting the dual variables corresponding to the rows of M to $\lfloor y^* \rfloor$ and the dual variable corresponding to $\bar{w}x \leq t(\bar{w})$ to 1. ∎

This result is much more satisfying than the $(1/t)$ scaling. It sets up a nice connection between totally dual integral systems and integral polyhedra. This can be used in a procedure for proving integrality of polyhedra:

- Find an appropriate defining system $Ax \leq b$, with A and b integral.

- Prove $Ax \leq b$ is totally dual integral.

- Using Theorem 6.29, conclude that $\{x : Ax \leq b\}$ is an integral polyhedron.

With Theorem 6.30, we know this plan can be carried out. Its usefulness depends on methods for proving total dual integrality of systems. We will see an example of the process in Chapter 8. For further discussion, see Schrijver [1984].

Notice that the proof of Theorem 6.30 involved the addition of many inequalities that are redundant for the definition of P. This is sometimes unavoidable, even in "natural" combinatorial systems. Consider, for example, the system

$$x(\delta(v)) \leq 2 \quad \text{for all } v \in V \tag{6.15}$$
$$x_e \geq 0 \quad \text{for all } e \in E$$

for the graph, K_4, given in Figure 6.8. Let $w_e = 1$ for all $e \in E$ and consider

Figure 6.8. 2-matching problem

the dual problem

$$\text{Minimize } \sum(2y_v : v \in V) \tag{6.16}$$
$$\text{subject to}$$
$$y_u + y_v \geq w_e, \quad \text{for all } e = uv \in E$$
$$y_v \geq 0, \quad \text{for all } v \in V.$$

Since any solution must have y_v positive for at least three nodes, we know (6.16) has no integral optimal solution. But adding the single redundant constraint $\sum(x_e : e \in E) \leq 4$ makes (6.15) totally dual integral.

Certainly, we should always check any defining system we might find for the possibility that it is totally dual integral, since this gives an immediate improvement to our min-max result. However, if this is not the case, we must decide if the addition of redundant constraints to achieve integrality actually improves the min-max theorem or not.

Exercises

6.29. Let $P = \text{conv.hull}(\{(0,0),(1,2),(2,0)\})$. Find a totally dual integral system $Ax \leq b$, with A and b integral, such that $P = \{x : Ax \leq b\}$. Is your system minimal?

6.30. Let $Ax \leq b$ be a system of linear inequalities with A and b rational. Show that there exists a positive integer t such that $(1/t)Ax \leq (1/t)b$ is totally dual integral.

6.7   CUTTING PLANES

Cutting-Plane Proofs

Up to now, we have been discussing techniques that help us to make statements about the totality of objective functions for a given combinatorial problem. In many situations, however, we need to solve a problem only for a single, fixed, objective. So, although a description of the appropriate convex hull may be elusive, we may still be able to apply linear-programming techniques to the problem at hand. One method for carrying this out is the "cutting-plane" technique described below.

To set things up, we need to formulate the combinatorial problem as an integer programming problem

$$\max \{w^T x : Ax \leq b, x \text{ integer } \}. \tag{6.17}$$

This is normally an easy matter; for example, the maximum weight matching problem on a graph $G = (V, E)$ is just

$$\text{Maximize } \sum (w_e x_e : e \in E)$$
$$\text{subject to}$$
$$x(\delta(v)) \leq 1, \text{ for all } v \in V$$
$$x_e \geq 0, \text{ for all } e \in E$$
$$x_e \text{ integer, for all } e \in E.$$

Then, the problem of establishing the optimality of a solution (or at least providing an upper bound on the optimal value) is equivalent to proving $w^T x \leq t$ is valid for all integral solutions to $Ax \leq b$, where t is the maximum value in (6.17) (or a desired upper bound). If we did not have the integrality restriction, we could prove the validity of $w^T x \leq t$ with the help of Farkas' Lemma. Our goal is to establish a similar proof method for integral solutions.

Let's start with an example. Consider the linear system

$$\begin{array}{rcr}
2x_1 + 3x_2 & \leq & 27 \\
2x_1 - 2x_2 & \leq & 7 \\
-6x_1 - 2x_2 & \leq & -9 \\
-2x_1 - 6x_2 & \leq & -11 \\
-6x_1 + 8x_2 & \leq & 21.
\end{array} \tag{6.18}$$

The polytope defined by these inequalities is depicted in Figure 6.9. As can easily be seen, every integral solution to the system satisfies $x_2 \leq 5$. Note, however, that we cannot derive this directly with Farkas' Lemma, since there is a fractional vector, $(9/2, 6)$, with $x_2 = 6$. But we can derive the inequality

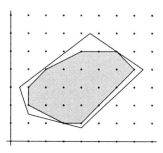

Figure 6.9. Polytope and its integer hull

in two Farkas-like steps. First, by multiplying the last inequality by $1/2$ we obtain the valid inequality

$$-3x_1 + 4x_2 \leq 21/2.$$

Since every integral (x_1, x_2) will make $-3x_1 + 4x_2$ an integer, we know that every integral solution to (6.18) will satisfy the stronger inequality

$$-3x_1 + 4x_2 \leq 10 \qquad\qquad (6.19)$$

obtained by rounding $21/2$ down to the nearest integer. Next, multiplying (6.19) by 2 and multiplying the first inequality, $2x_1 + 3x_2 \leq 27$, by 3, and adding the resulting inequalities we obtain

$$17x_2 \leq 101.$$

Multiplying this inequality by $1/17$ and rounding down the right-hand side, we can conclude that every integral solution to (6.18) satisfies $x_2 \leq 5$.

In general, suppose our system consists of the m inequalities

$$a_i^T x \leq b_i \quad (i = 1, \ldots, m). \qquad\qquad (6.20)$$

Let $y_1, \ldots, y_m$ be nonnegative real numbers and set

$$c = \sum (y_i a_i : i = 1, \ldots, m)$$

and

$$d = \sum (y_i b_i : i = 1, \ldots, m).$$

Trivially, every solution to (6.20) satisfies $c^T x \leq d$. Moreover, if c is integral, then, arguing as above, we know that all integral solutions to (6.20) also satisfy the stronger inequality

$$c^T x \leq \lfloor d \rfloor \qquad\qquad (6.21)$$

where $\lfloor d \rfloor$ denotes d rounded down to the nearest integer. We call (6.21) a *Gomory-Chvátal cutting plane*. The "cutting plane" part of the name comes from the fact that the rounding operation "cuts off" part of the polyhedron defined by (6.20), although it does not cut off any integral vectors. (*Cutting planes*, or *cuts*, generally refer to any inequality that is valid for all integral vectors in a polyhedron.) Gomory and Chvátal pioneered this approach to integer programming: Gomory [1960] with an algorithm based on generating inequalities like (6.21) and Chvátal [1973] with the "cutting-plane proof" method we describe below.

Gomory-Chvátal cutting planes can also be defined directly in terms of the polyhedron P defined by (6.20): We just take a valid inequality $c^T x \leq d$ for P with c integral and round down to obtain the cutting plane $c^T x \leq \lfloor d \rfloor$. The use of the nonnegative numbers y_i is to provide a *derivation* of $c^T x \leq \lfloor d \rfloor$ from (6.20). With the y_i's in hand, we are easily convinced that $c^T x \leq d$ and hence $c^T x \leq \lfloor d \rfloor$ are indeed valid.

Once we have derived a cut, we can add it to our system (6.20) and make use of it in deriving further inequalities. We call a sequence of such derivations a *cutting-plane proof*. That is, a cutting-plane proof of an inequality $w^T x \leq t$ from the system (6.20) is a sequence of inequalities

$$a_{m+k}^T x \leq b_{m+k} \quad (k = 1, \ldots, M)$$

together with nonnegative numbers

$$y_{kj} \quad (1 \leq k \leq M, 1 \leq j \leq m + k - 1)$$

such that for each $k = 1, \ldots, M$, the inequality $a_{m+k}^T x \leq b_{m+k}$ is derived from the system

$$a_i^T x \leq b_i \quad (i = 1, \ldots, m + k - 1)$$

using the numbers $y_{kj}, j = 1, \ldots, m + k - 1$, and such that the last inequality in the sequence is $w^T x \leq t$. Such a proof is a clean way to demonstrate that $w^T x \leq t$ is valid for all integral solutions to (6.20). A number of combinatorial examples of cutting-plane proofs can be found in Chvátal [1973, 1985], Chvátal, Cook, and Hartmann [1989], and Grötschel and Pulleyblank [1986].

Chvátal [1973] (and indirectly Gomory [1960]) showed that cutting-plane proofs are always available, provided P is a polytope.

Theorem 6.31 *Let $P = \{x : Ax \leq b\}$ be a rational polytope and let $w^T x \leq t$ be an inequality, with w integral, satisfied by all integral vectors in P. Then there exists a cutting-plane proof of $w^T x \leq t'$ from $Ax \leq b$, for some $t' \leq t$.*

An important special case of this theorem occurs when P contains no integral vectors.

Theorem 6.32 *Let $P = \{x : Ax \leq b\}$ be a rational polytope that contains no integral vectors. Then there exists a cutting-plane proof of $0^T x \leq -1$ from $Ax \leq b$.*

This result is analogous to the familiar version of Farkas' Lemma that a polyhedron $P = \{x : Ax \leq b\}$ is empty if and only if $0^T x \leq -1$ can be written as a nonnegative linear combination of the inequalities $Ax \leq b$.

Since proving $w^T x \leq t$ is valid is equivalent to proving $P \cap \{x : w^T x \geq t + 1\}$ contains no integral vectors (if t is not an integer, then replace $t + 1$ by t rounded up), Theorem 6.32 can be used to give "indirect" proofs of inequalities. In some instances, perhaps surprisingly, these proofs can be less complex than direct cutting-plane proofs.

In our proof of the two theorems, we proceed by induction on the dimension of the polytope (following Schrijver [1980]). The technique will be to push the inequality $w^T x \leq l$ into the polytope as far as possible, apply induction to prove that the face F induced by $w^T x \leq l$ contains no integral vectors, then push the inequality to $w^T x \leq l - 1$. Continuing this, we eventually reach $w^T x \leq t$, as desired. To carry this out, we use the following lemma, which allows us to translate the cutting-plane proof on the face F to a proof on the entire polytope. This "rotation" technique is illustrated in Figure 6.10.

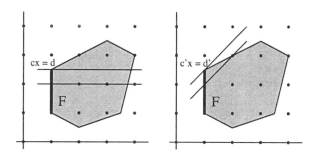

Figure 6.10. Rotating a cutting-plane for a face F

Lemma 6.33 *Let F be a face of a rational polytope P. If $c^T x \leq \lfloor d \rfloor$ is a Gomory-Chvátal cutting-plane for F, then there exists a Gomory-Chvátal cutting-plane $(c')^T x \leq \lfloor d' \rfloor$ for P such that*

$$F \cap \{x : (c')^T x \leq \lfloor d' \rfloor\} = F \cap \{x : c^T x \leq \lfloor d \rfloor\}.$$

Proof: Let $P = \{x : A'x \leq b', A''x \leq b''\}$, where A'' and b'' are integral and $F = \{x : A'x \leq b', A''x = b''\}$. We may assume $d = \max\{c^T x : A'x \leq b', A''x = b''\}$. By Farkas' Lemma, there exist vectors $y' \geq 0$ and y'' such that

$$(y')^T A' + (y'')^T A'' = c^T$$
$$(y')^T b' + (y'')^T b'' = d.$$

Due to the equality constraints, y'' need not be nonnegative. Thus, to obtain a Gomory-Chvátal cut for P we must replace y'' by a vector that is necessarily nonnegative. To this end, define c' and d' as

$$c' = c - (\lfloor y'' \rfloor)^T A'' = (y')^T A' + (y'' - \lfloor y'' \rfloor)^T A''$$

$$d' = d - (\lfloor y'' \rfloor)^T b'' = (y')^T b' + (y'' - \lfloor y'' \rfloor)^T b''$$

where $\lfloor y'' \rfloor$ denotes the vector $(\lfloor y_1'' \rfloor, \lfloor y_2'' \rfloor, \ldots)^T$. Then c' is integral (since $(\lfloor y'' \rfloor)^T A''$ is integral) and $(c')^T x \le d'$ is a valid inequality for P (since $y'' - \lfloor y'' \rfloor$ is nonnegative). Furthermore, since $\lfloor d \rfloor = \lfloor d' \rfloor + (\lfloor y'' \rfloor)^T b''$, we have

$$F \cap \{x : (c')^T x \le \lfloor d' \rfloor\} =$$
$$F \cap \{x : (c')^T x \le \lfloor d' \rfloor, (\lfloor y'' \rfloor)^T A'' x = (\lfloor y'' \rfloor)^T b''\} =$$
$$F \cap \{x : c^T x \le \lfloor d \rfloor\}$$

and the result follows. ∎

We first prove Theorem 6.32, then, making use of it, we prove the more general Theorem 6.31.

Proof of Theorem 6.32: We use induction on the dimension of P. Since the theorem is trivial if $\dim(P) = 0$, we may assume $\dim(P) \ge 1$ and that the result holds for all rational polytopes of dimension less than $\dim(P)$.

Let $w^T x \le l$ be an inequality, with w integral, that induces a proper face of P and let $\bar{P} = \{x \in P : w^T x \le \lfloor l \rfloor\}$ be the polytope obtained by applying the Gomory-Chvátal cut $w^T x \le \lfloor l \rfloor$. If $\bar{P} = \emptyset$, then with Farkas' Lemma we can deduce $0^T x \le -1$ from $Ax \le b, w^T x \le \lfloor l \rfloor$. So suppose $\bar{P} \ne \emptyset$ and let $F = \{x \in \bar{P} : w^T x = \lfloor l \rfloor\}$. By the choice of $w^T x \le l$, we know that F has dimension less than $\dim(P)$. (Either F is a proper face of P (when l is integral) or P contains vectors that do not satisfy $w^T x = \lfloor l \rfloor$.) So, by the induction hypothesis, we may assume there exists a cutting-plane proof of $0^T x \le -1$ from $Ax \le b, w^T x = \lfloor l \rfloor$. Using Lemma 6.33, this gives us a cutting-plane proof, from $Ax \le b, w^T x \le \lfloor l \rfloor$, of an inequality $c^T x \le \lfloor d \rfloor$ such that $\bar{P} \cap \{x : c^T x \le \lfloor d \rfloor, w^T x = \lfloor l \rfloor\} = \emptyset$. Thus, after applying this sequence of cuts to $\bar{P}$, we have $w^T x \le \lfloor l \rfloor - 1$ as a Gomory-Chvátal cutting-plane. Since P is bounded, $\min\{w^T x : x \in P\}$ is finite. So, continuing in the above manner, letting $\bar{P} = \{x \in P : w^T x \le \lfloor l \rfloor - 1\}$, and so on, we finally will obtain a cutting-plane proof of some $w^T x \le t$ such that $P \cap \{x : w^T x \le t\} = \emptyset$. With Farkas' Lemma, we may then derive $0^T x \le -1$ from $Ax \le b, w^T x \le t$. ∎

Proof of Theorem 6.31: Suppose P contains no integral vectors. Then, from Theorem 6.32, there exists a cutting-plane proof of $0x \le -1$. Also, since P is bounded, $l = \max\{w^T x : x \in P\}$ is finite. So, with Farkas' Lemma, we can derive $w^T x \le \lfloor l \rfloor$. Adding an appropriate multiple of $0x \le -1$ to $w^T x \le \lfloor l \rfloor$, we obtain a cutting-plane proof of $w^T x \le t'$ for some $t' \le t$. Thus, the result holds in this case.

Now suppose P contains integral vectors. Again, let $l = \max\{w^T x : x \in P\}$ and let $\bar{P} = \{x \in P : w^T x \le \lfloor l \rfloor\}$ be the polytope obtained by applying the cut $w^T x \leq \lfloor l \rfloor$ to P. If $\lfloor l \rfloor \leq t$ we are finished, so suppose not. Consider the face $F = \{x \in \bar{P} : w^T x = \lfloor l \rfloor\}$ of $\bar{P}$ induced by $w^T x \leq \lfloor l \rfloor$. Since $w^T x \leq t$ is valid for all integral vectors in P, we know that F contains no integral vectors. So, by Theorem 6.32, there exists a cutting-plane proof of $0x \leq -1$ from $Ax \leq b, w^T x = \lfloor l \rfloor$. Using Lemma 6.33, this implies that there exists a cutting-plane proof of an inequality $c^T x \leq \lfloor d \rfloor$, from $Ax \leq b, w^T x \leq \lfloor l \rfloor$, such that $\bar{P} \cap \{x : c^T x \leq \lfloor d \rfloor, w^T x = \lfloor l \rfloor\} = \emptyset$. Thus, after applying the sequence of cuts to $\bar{P}$, we have $w^T x \leq \lfloor l \rfloor - 1$ as a Gomory-Chvátal cutting plane. Continuing in this fashion, we finally derive an inequality $w^T x \leq t'$ with $t' \leq t$. ∎

Chvátal Rank

Gomory-Chvátal cutting planes have an interesting connection with the general problem of finding linear descriptions of combinatorial convex hulls. In this context, we do not think of the cuts as coming sequentially, as in cutting-plane proofs, but rather in waves that provide successively tighter approximations to P_I, the convex hull of the integral vectors in P. This approximation process gives a finite procedure for obtaining a linear description of P_I.

We start by taking all possible Gomory-Chvátal cuts for P in a first wave. Although there appear to be infinitely many such cutting-planes, actually a finite subset imply the rest. A nice way to describe this is to let P' denote the set of all vectors in P that satisfy every Gomory-Chvátal cut for P. Then we have the following result of Schrijver [1980].

Theorem 6.34 *If P is a rational polyhedron, then P' is also a rational polyhedron.*

Proof: Let $P = \{x : Ax \leq b\}$ with A and b integral. Then P' is defined by $Ax \leq b$ and the set of all inequalities that can be written as

$$(y^T A)x \leq \lfloor y^T b \rfloor \qquad (6.22)$$

for some vector y such that $0 \leq y < 1$ and $y^T A$ is integral. To verify this, we will prove that every Gomory-Chvátal cut can be written as the sum of an inequality in (6.22) and a linear combination of inequalities from $Ax \leq b$. This will give us the result, since it means that all cuts other than those in the finite set (6.22) are redundant in the definition of P'. So let $w^T x \leq \lfloor t \rfloor$ be a Gomory-Chvátal cut, derived from $Ax \leq b$ with the nonnegative vector $\bar{y}$. Let $\bar{y}' = \bar{y} - \lfloor \bar{y} \rfloor$ be the fractional parts of $\bar{y}$. Then $w' = (\bar{y}')^T A = w - (\lfloor \bar{y} \rfloor)^T A$ is an integral vector and $t' = (\bar{y}')^T b = t - (\lfloor \bar{y} \rfloor)^T b$ differs from t by an integral amount. So the cut $(w')^T x \leq \lfloor t' \rfloor$ derived with $\bar{y}'$, together with the valid inequality $((\lfloor \bar{y} \rfloor)^T A)x \leq (\lfloor \bar{y} \rfloor)^T b$ (a nonnegative combination of $Ax \leq b$) sum to the cut $w^T x \leq t$. ∎

We can take as a second wave the set of all Gomory-Chvátal cuts for P', the third wave as all cuts for P'', and so on. Thus, letting $P^{(0)} = P$ and $P^{(i)} = P^{(i-1)'}$ for all positive integers i, we have a sequence of polyhedra

$$P = P^{(0)} \supseteq P^{(1)} \subseteq P^{(2)} \supseteq \cdots \supseteq P_I$$

generated by the waves of cuts. (See Figure 6.11.) Translating Theorem 6.31

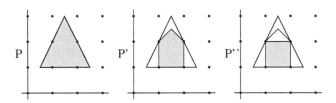

Figure 6.11. A polytope of Chvátal rank 2

to these terms, we obtain the following result.

Theorem 6.35 *If P is a rational polytope, then $P^{(k)} = P_I$ for some integer k.*   ∎

The least k for which $P^{(k)} = P_I$ is known as the *Chvátal rank* of P. (For an extension of Theorem 6.35 to unbounded polyhedra see Schrijver [1980].)

The notion of Chvátal rank provides a framework for attacking the problem of finding a linear description for P_I. The principal example of this is the fractional matching polytope, which has Chvátal rank 1. (This can be checked by examining the form of the blossom inequalities.) Other combinatorial examples can be found in Chvátal, Cook, and Hartmann [1989] and Gerards and Schrijver [1986].

Cutting-plane Algorithms

In practice, cutting-plane proofs are usually not applied to specific instances of a combinatorial problem, but rather are used to prove the validity of general classes of inequalities, such as the blossom constraints for matchings. These valid inequalities may then be applied in a cutting-plane based algorithm for the combinatorial problem. The general idea is that we have an integer-programming formulation of the problem, max $\{w^T x : x \in P, x$ integral $\}$ for some polyhedron P, and some classes of inequalities that we know are valid for all integral solutions. Using a linear-programming algorithm, we find an optimal solution x^* to the linear-programming relaxation max $\{w^T x : x \in P\}$. If x^* is integral, then it is also an optimal solution to our combinatorial problem. Otherwise, we search the classes of valid inequalities to find (we hope) some that are violated by x^*, that is, $c^T x^* > d$ (although $c^T x \leq d$ for all integral solutions). We then add the violated inequalities to our linear-programming relaxation and find a new optimal solution x^{**}. If x^{**} is integral we are done. Otherwise we search for inequalities that are violated by x^{**}, add them to our linear-programming relaxation, and so on. (See Figure 6.12.)

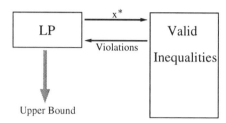

Figure 6.12. Cutting-plane algorithm

If we are lucky, we end up finding an integral solution to one of the linear-programming relaxations and thus solve the combinatorial problem. But, depending on the objective function and our classes of valid inequalities, the subroutine that searches for violations may come back empty-handed. So the whole process may terminate without reaching an integral optimal solution.

In any case, the optimal value of each linear-programming relaxation provides an upper bound on the optimal value of the combinatorial problem that is usually better than the previous such bound (and certainly no worse). The last such bound can be employed, for example, in a "branch and bound" search procedure for solving the problem, the branching coming about by considering a separate subproblem for each possible integral value of some selected variable x_i. (This is often called "branch and cut," a term coined by Padberg and Rinaldi [1991]. We illustrate this cutting-plane and branching procedure with the traveling salesman problem in Chapter 7.)

For the above method to be successful, the cutting-plane procedure must quickly arrive at a tight upper bound for the combinatorial problem. This, of course, depends on the choice of the classes of valid inequalities. An effective heuristic for finding good classes is to study those that induce facets of the underlying integral polyhedron. There is some theoretical justification for this since the facet-inducing inequalities imply all other valid inequalities for the problem (Theorem 6.17). Also, such inequalities tend to have some combinatorial meaning, which may be fundamental in deriving a good upper bound.

Keeping in mind that we will apply the classes of inequalities in a cutting-plane based algorithm (and so, for example, we must be able to check for violations efficiently), we may use the property of inducing a facet as a guide in our search for good classes.

Exercises

6.31. Let $P = \text{conv.hull}(\{(0,0),(1,0),(1/2,3)\})$. Find a system of linear inequalities that defines P'.

6.32. Let k be a positive integer and let

$$P = \text{conv.hull}(\{(0,0),(1,0),(1/2,k)\}).$$

Show that the Chvátal rank of P is at least k.

6.33. Let $Ax \le b$ be a totally dual integral system, with A integral, and let $P = \{x : Ax \le b\}$. Show that $P' = \{x : Ax \le \lfloor b \rfloor\}$.

6.34. Let $P = \{x : Ax \le b\}$, where A is an integral m by n matrix and b is an integral vector. Let σ denote the greatest (in absolute value) entry in A and b. Show that there exist an integral matrix A' and an integral vector b' such that $P' = \{x : A'x \le b'\}$ and the greatest (in absolute value) entry in A' and b' is at most $n\sigma$.

6.35. Let P be a (perhaps nonrational) polytope. Using Theorem 6.35, show that $P^{(k)} = P_I$ for some integer k.

Figure 6.13. 5-wheel

6.36. Let $G = (V,E)$ be a graph with no isolated nodes and consider the polytope P defined by the system of inequalities

$$x_u + x_v \le 1, \text{ for every edge } uv \in E \qquad (6.23)$$
$$x_v \ge 0, \text{ for every node } v \in V.$$

Let $\bar{x}$ be a $0-1$ vector in P. Then the set of nodes $v \in V$ such that $\bar{x}_v = 1$ is a *stable set* of G, that is, no two nodes in the set are adjacent in G. The convex hull of the incidence vectors of the stable sets of G is called the *stable set polytope* of G, and denoted by $S(G)$. Clearly, $S(G) = P_I$. For each of the following classes of valid inequalities for $S(G)$, give a cutting-plane proof starting from the system (6.23).

(i) Let $C \subseteq V$ be the node set of an odd circuit in G. Then the inequality $x(C) \le (|C| - 1)/2$ is valid for $S(G)$.

(ii) The graph illustrated in Figure 6.13 is called a *5-wheel*. If W is the node set of a 5-wheel in G and r is the center node, then $2x_r + x(W \setminus \{r\}) \leq 2$ is valid for $S(G)$.

(iii) A set of nodes $K \subseteq V$ is called a *clique* of G if each pair of nodes in K are adjacent in G. For any clique K, the inequality $x(K) \leq 1$ is valid for $S(G)$.

6.8   SEPARATION AND OPTIMIZATION

Recall the polyhedral combinatorics "plan" that we outlined in Section 6.2: Formulate a problem as optimizing over a finite set of vectors S, find a linear description of conv.hull(S), and apply the Duality Theorem of Linear Programming. The end result, a min-max relation for the combinatorial problem, can be used in designing an optimization procedure, perhaps in the spirit of the blossom algorithm for matchings, or perhaps along the lines of the cutting-plane algorithms described in the previous section. But in either case, the connection between the polyhedral description and an optimization algorithm seems somewhat tenuous, since we must overcome the fact that the size of the linear system may be far greater than the size of the original combinatorial problem. Although this appears to be a large hurdle, a powerful theorem of Grötschel, Lovász, and Schrijver [1988] states that we can always get over it. They show that good knowledge of the defining system for conv.hull(S) is equivalent to the existence of a polynomial-time algorithm for optimizing over S. So the "plan" has direct algorithmic consequences. We describe a precise version of this result below.

To begin, we need a general framework for describing linear systems that may be very large (relative to the size of our combinatorial problem). Simply listing the inequalities will not do the job. Instead, we adopt a cutting-plane approach and ask for an exact method for finding a violated inequality from the list. That is, we want to be able to solve the following *separation problem* in polynomial time.

Separation problem: Given a bounded rational polyhedron $P \subseteq \mathbf{R}^n$ and a rational vector $v \in \mathbf{R}^n$, either conclude that v belongs to P or, if not, find a rational vector $w \in \mathbf{R}^n$ such that $w^T x < w^T v$ for all $x \in P$.

To go along with this, it will be useful to state formally the corresponding *optimization problem*.

> *Optimization problem*: Given a bounded rational polyhedron $P \subseteq \mathbf{R}^n$ and a rational (objective) vector $w \in \mathbf{R}^n$, either find $x^* \in P$ that maximizes $w^T x$ over all $x \in P$, or conclude that P is empty.

The equivalence we mentioned above will translate to "separation $\equiv$ optimization," that is, we can solve one of the two problems in polynomial time if and only if we can also solve the other problem in polynomial time. (We have restricted ourselves to bounded polyhedra only to simplify the presentation; otherwise we must allow for the possibility that the optimization problem has an unbounded objective value.)

Following the development in Schrijver [1986], we consider classes of polyhedra $\mathcal{P} = \{P_t : t \in \mathcal{O}\}$ where $\mathcal{O}$ is some collection of objects and for each $t \in \mathcal{O}$, P_t is a bounded rational polyhedron. For example, $\mathcal{O}$ could be the collection of all graphs and P_t the perfect matching polytope for the graph t. We call the class $\mathcal{P}$ *proper* if for each object $t \in \mathcal{P}$ we can compute in polynomial time (with respect to the size of t) positive integers n_t and s_t such that $P_t \subseteq \mathbf{R}^{n_t}$ and such that P_t can be described by a linear system where each inequality has size at most s_t. (These technical conditions should be easy to verify in most applications that you will come across.) We say that the separation problem is *polynomially solvable* over the class $\mathcal{P}$ if there exists an algorithm that solves the separation problem for any $P_t \in \mathcal{P}$ and any rational vector $v \in \mathbf{R}^{n_t}$ in time polynomial in the sizes of t and v. Similarly, we say that the optimization problem is polynomially solvable over $\mathcal{P}$ if there exists a polynomial-time algorithm for solving any instance (P_t, w) of the optimization problem, where $t \in \mathcal{O}$ and w is a rational vector in $\mathbf{R}^{n_t}$. With these definitions, we can state a formal version of the "separation $\equiv$ optimization" theorem.

Theorem 6.36 *For any proper class of polyhedra, the optimization problem is polynomially solvable if and only if the separation problem is polynomially solvable.* ∎

This theorem is a consequence of the more general result of Grötschel, Lovász, and Schrijver [1988]. Its proof involves some advanced topics in linear programming, including the ellipsoid method for solving linear-programming problems.

At the present time, this equivalence of optimization and separation is only a theoretical tool, since the resulting algorithms (although polynomial-time) do not appear to be efficient in practice. We should view this equivalence as a guide for searching for more practical polynomial-time algorithms.

Digression: Matching Polyhedra

The collection of polyhedra $\mathcal{P} = \{PM(G) : G \text{ a graph}\}$ is easily seen to be a proper class (even if we did not know Edmonds' linear description of $PM(G)$). Thus, we can apply Theorem 6.36. Now what can we learn? First of all, using the blossom algorithm we know that the optimization problem over $\mathcal{P}$ is polynomially solvable. So we can conclude that the separation problem over $\mathcal{P}$ is also polynomially solvable. The nontrivial consequence is that given a graph $G = (V, E)$ and a rational vector $x^* \in \mathbf{R}^E$ such that $x^* \geq 0$ and $x^*(\delta(v)) = 1$ for all $v \in V$, we can find an odd set $S \subseteq V$ with $x^*(\delta(S)) < 1$ if such a set exists. A direct combinatorial algorithm for this *blossom separation* problem was first shown by Padberg and Rao [1982], some fifteen years after Edmonds' blossom algorithm. But we can also go the other way around. Padberg and Rao's algorithm (which is much simpler than the blossom algorithm), together with Theorem 6.36 and the Perfect Matching Polytope Theorem, implies the existence of a polynomial-time algorithm for solving the weighted perfect matching problem!

We will describe the blossom separation algorithm in a slightly more general form, which will be useful in Chapter 7. Let $T \subseteq V$ be an even cardinality subset of nodes, with $T \neq \emptyset$. Given a nonnegative vector $u \in \mathbf{R}^E$ of edge capacities, the *minimum T-cut problem* is to find a T-cut, $\delta(Q)$, with minimum capacity $u(\delta(Q))$. An algorithm for this problem can be used for blossom separation by taking $T = V$ and $u = x^*$. If the minimum T-cut has capacity less than 1, then it corresponds to a violated blossom inequality. Otherwise, we can conclude that all blossom inequalities for G are satisfied by x^*.

Our presentation of the minimum T-cut algorithm follows the description given in Grötschel, Lovász, and Schrijver [1988]. The algorithm begins by finding a cut $\delta(S)$ (not necessarily a T-cut) of minimum capacity that satisfies $S \cap T \neq \emptyset$ and $(V \setminus S) \cap T \neq \emptyset$. This step can be carried out by solving $|T| - 1$ maximum flow problems, using the technique described in Section 3.5. Since every T-cut, $\delta(Q)$, satisfies $Q \cap T \neq \emptyset$ and $(V \setminus Q) \cap T \neq \emptyset$, if $|S \cap T|$ happens to be odd, then $\delta(S)$ is a minimum T-cut and we are done. Otherwise, we make use of the following result.

Theorem 6.37 *Among the minimum T-cuts of the graph G, there exists one, $\delta(W)$, with either $W \subseteq S$ or $W \subseteq V \setminus S$.*

Proof: Let $\delta(Q)$ be a minimum T-cut, and suppose that $Q \cap S \neq \emptyset$ and $Q \cap (V \setminus S) \neq \emptyset$. We may assume that $|Q \cap S \cap T|$ is odd, since we can replace S by $V \setminus S$ without changing the conclusion of the theorem.

So $|Q \cap S \cap T|$ is odd and $|(V \setminus Q) \cap S \cap T|$ is odd. Now by replacing Q by $V \setminus Q$ if necessary, we may assume that both $(S \cup Q) \cap T \neq \emptyset$ and $(V \setminus (S \cup Q)) \cap T \neq \emptyset$. We claim that $Q \cap S$ is a minimum T-cut. Indeed, by the choice of S we have

$$u(\delta(S)) \leq u(\delta(S \cup Q)).$$

Combining this with the inequality (see Exercise 3.66)

$$u(\delta(S)) + u(\delta(Q)) \ge u(\delta(S \cup Q)) + u(\delta(S \cap Q))$$

we can conclude that

$$u(\delta(S \cap Q)) \le u(\delta(Q)).$$

∎

This allows us to reduce the problem to two problems on smaller graphs. To see this, let G^1 be the graph obtained by shrinking $V \setminus S$ to a single (new) node and let $T^1 = T \setminus S$ and $u_e^1 = u_e$ for all $e \in E(G^1)$. Similarly, let G^2 be the graph obtained by shrinking S to a single (new) node and let $T^2 = T \cap S$ and $u_e^2 = u_e$ for all $e \in E(G^2)$. Then the minimum T-cut in G can be found by solving the minimum T^1-cut problem in G^1 and the minimum T^2-cut problem in G^2.

We solve the two new problems with the same procedure, splitting them into further subproblems if necessary.

Since we can find S and build G^1 and G^2 in polynomial time, it follows by induction on $|T|$ that the whole procedure runs in polynomial time. (See Exercise 6.37.)

As one would expect, the minimum T-cut algorithm performs poorly in practice for larger test instances. A more efficient alternative (actually, the algorithm proposed by Padberg and Rao), works by computing a Gomory-Hu cut-tree, as described in Exercise 6.39.

Exercises

6.37. Show that the minimum T-cut algorithm runs in time $O(n^5)$. (Hint: Use induction and the fact that $|T| = |T^1| + |T^2|$.)

6.38. Suppose that we are given a vector $\bar{x} \in \mathbf{R}^V$ satisfying the initial valid inequalities (6.23) for the stable set polytope of $G = (V, E)$. Show how to reduce the separation problem for the odd circuit inequalities to the problem of finding a minimum weight odd circuit in a graph having nonnegative edge weights. Show that the latter problem can be solved using shortest path methods. (See Exercise 2.38).

6.39. Let $G = (V, E)$ be a graph, $T \subseteq V$ with $|T|$ even, and $u \in \mathbf{R}^E$ a nonnegative capacity function. Consider a Gomory-Hu cut-tree H with T as the set of terminals. Show that there exists an edge e of H such that the bipartition of V defined by the two components of $H \setminus e$ gives a minimum T-cut.

CHAPTER 7

The Traveling Salesman Problem

7.1 INTRODUCTION

In the general form of the traveling salesman problem, we are given a finite set of points V and a cost c_{uv} of travel between each pair $u, v \in V$. A *tour* is a circuit that passes exactly once through each point in V. The *traveling salesman problem* (TSP) is to find a tour of minimal cost.

The TSP can be modeled as a graph problem by considering a complete graph $G = (V, E)$, and assigning each edge $uv \in E$ the cost c_{uv}. A tour is then a circuit in G that meets every node. In this context, tours are sometimes called *Hamiltonian circuits*.

The TSP is one of the best known problems of combinatorial optimization. A nice collection of papers tracing the history and research on the problem can be found in Lawler, Lenstra, Rinnooy Kan, and Shmoys [1985].

Unlike the cases of matching or network flows, no polynomial-time algorithm is known for solving the TSP in general. Indeed, it belongs to the class of $\mathcal{NP}$-*hard* problems, which we describe in Chapter 9. Consequently, many people believe that no such efficient solution method exists, for such an algorithm would imply that we could solve virtually every problem in combinatorial optimization in polynomial time.

Nevertheless, TSPs do arise in practice, and relatively large ones can now be solved efficiently to optimality. In this chapter we discuss how. We will illustrate the methods on the 1173-node Euclidean problem depicted in Figure 7.1. Although this problem is smaller than the largest solved so far (as of August 1994, the record is 7397 nodes (Applegate, Bixby, Chvátal, Cook [1995])), it is still of a respectable size. The node coordinates for this instance are con-

241

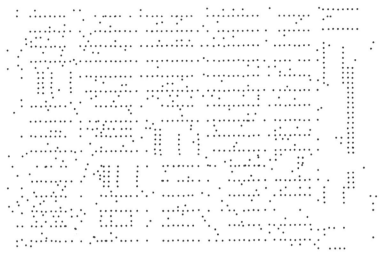

Figure 7.1. Sample TSP

tained in the "TSPLIB" library of test problems described in Reinelt [1991]. We encourage readers to try out some of their own methods.

Whereas in prior chapters, we were often able to describe polynomial-time algorithms that also performed well in practice, in this chapter we discuss algorithms which do work well empirically, but for which only very weak guarantees can be provided.

7.2 HEURISTICS FOR THE TSP

Heuristics are methods which cannot be guaranteed to produce optimal solutions, but which, we hope, produce fairly good solutions at least some of the time. For the TSP, there are two different types of heuristics. The first attempts to construct a "good" tour from scratch. The second tries to improve an existing tour, by means of "local" improvements. In practice it seems very difficult to get a really good tour construction method. It is the second type of method, in particular, an algorithm developed by Lin and Kernighan [1973], which usually results in the best solutions and forms the basis of the most effective computer codes.

Nearest Neighbor Algorithm

In Chapter 1 we described the Nearest Neighbor Algorithm for the TSP: Start at any node; visit the nearest node not yet visited, then return to the start node when all other nodes are visited. Applying it to our test problem, we obtain the tour of cost $67,822$ that is exhibited in Figure 1.1. Note that the tour includes some very expensive edges. In practice, this almost always seems to happen when we use the Nearest Neighbor Algorithm.

Johnson, Bentley, McGeoch, and Rothberg [1997] report that on problems in TSPLIB, the average costs of the tours found by the Nearest Neighbor Algorithm are about 1.26 times the costs of the corresponding optimal tours. Thus, for some applications, Nearest Neighbor may be an effective method: It is easy to implement, runs quickly, and usually produces tours of reasonable quality. It should be noted, however, that the "1.26 times optimal" estimate is an empirical observation, not a performance guarantee. Indeed, it is easy to construct problems on only four nodes for which the Nearest Neighbor Algorithm can produce a tour of cost arbitrarily many times that of the optimal tour. (See Exercise 7.2.)

To obtain a guaranteed bound, we need to assume that the edge costs are nonnegative and satisfy the triangle inequality:

$$c_{uv} + c_{vw} \geq c_{uw}, \text{ for all } u, v, w \in V.$$

In this case, Rosenkrantz, Stearns, and Lewis [1977] show that a Nearest Neighbor tour is never more than $\frac{1}{2}\lceil log_2 n \rceil + \frac{1}{2}$ times the optimum, where n is the cardinality of V.

This bound may seem very weak (particularly when compared to the 1.26 observed bound on the TSPLIB problems), but Rosenkrantz, Stearns, and Lewis [1977] proved that we cannot do much better. They showed this by describing a family of problems with nonnegative costs satisfying the triangle inequality and with arbitrarily many nodes, such that the Nearest Neighbor Algorithm can produce a tour of cost $\frac{1}{3}\lceil log_2(n+1) + \frac{4}{9} \rceil$ times the optimum. This result shows that if we wish to give a worst case bound on the performance of this heuristic, the bound we get is so bad that it is not of much practical interest.

The proofs of the two results are not hard, but they are technical and we refer the reader to the reference cited above.

Insertion Methods

Insertion methods provide a different set of tour construction heuristics. They start with a tour joining two of the nodes, then add the remaining nodes one by one, in such a way that the tour cost is increased by a minimum amount. There are several variations, depending on which two nodes are chosen to start, and more importantly, which node is chosen to be inserted at each stage.

In practice, usually the best insertion method is *Farthest Insertion*. In this case, we start with an initial tour passing through two nodes that are the ends of some high-cost edge. For each uninserted node v, we compute the minimum cost between v and any node in the tour constructed thus far. Then we choose as the next node to be inserted the one for which this cost is *maximum.*

At first, this may seem counterintuitive. However, in practice, it often works well. This is probably because a rough shape of the final tour to be produced is obtained quite early, and in later stages, only relatively slight modifications are made.

Nearest Insertion is a heuristic which, at each stage, chooses as the next node to insert the one for which the cost to any node in the tour is minimum.

Another variant is *Cheapest Insertion*. In this case, the next node for insertion is the one that increases the tour cost the least.

Usually the solutions produced by Nearest Insertion and Cheapest Insertion are inferior to those produced by Farthest Insertion. In Figures 7.2 and 7.3 we show the results of applying Nearest Insertion and Farthest Insertion to our test problem. On the TSPLIB problems, Johnson, Bentley, McGeoch, and Rothberg [1997] report that, on average, Farthest Insertion found tours of length about 1.16 times that of the optimal tours.

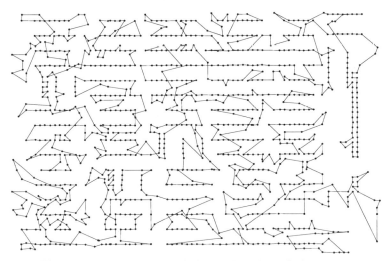

Figure 7.2. Sample TSP and Nearest Insertion solution: 72337

An extensive worst-case analysis of various insertion heuristics is provided in Rosenkrantz, Stearns, and Lewis [1977]. They prove that any insertion heuristic produces a solution whose value is at most $\lceil log_2 n \rceil + 1$ times the optimum, for a problem with n nodes (for which the edge costs are nonnegative and satisfy the triangle inequality). They show, further, that Cheapest Insertion and Nearest Insertion always produce solutions whose costs are at most twice the optimum. They also give an example which shows that this bound is essentially tight. See the above reference for details.

Interestingly, no examples are known which force any insertion method to construct a tour of cost more than four times that of the optimum. Also, in spite of the fact that Farthest Insertion usually produces the best solutions of any insertion method, no better worst case bound has been established for it than for insertion methods in general.

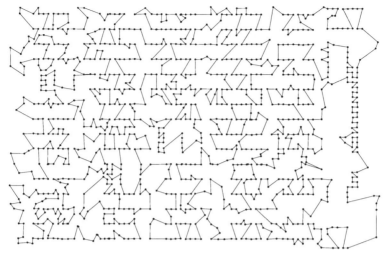

Figure 7.3. Sample TSP and Farthest Insertion solution: 65980

Christofides' Heuristic

We describe one more tour construction method, due to Christofides [1976]. It has the best worst case bound of any known method: It always produces a solution of cost at most $\frac{3}{2}$ times the optimum (assuming that the graph is complete and the costs are nonnegative and satisfy the triangle inequality).

The algorithm begins by finding a minimum-cost spanning tree, T, of G using, for example, Kruskal's Algorithm. The edges in T will be used in a search for a good tour.

Let W be the set of nodes which have odd degree in T, and find a perfect matching M of $G[W]$, the subgraph of G induced by W, which is of minimum cost with respect to c.

Now let J consist of $E(T) \cup M$, where, if some edge is in both T and M, we take two copies of the edge. Then J is the edge-set of a connected graph with node-set V for which each node has even degree. If all nodes have degree 2, then J is the edge-set of a tour and we terminate with it. If not, let v be any node of degree at least 4 in (V, J). Then there are edges uv and vw such that if we delete these edges from J and add the edge uw to J, then the subgraph remains connected. Moreover, the new subgraph has even degree at each node. (This is because the subgraph induced by J has an Euler tour; we choose uv and vw to be consecutive edges of the tour.) Make this "shortcut" and repeat this process until all nodes are incident with two edges of J.

Theorem 7.1 *Suppose we have a TSP with nonnegative costs satisfying the triangle inequality. Then any tour constructed by Christofides' Heuristic has cost at most $\frac{3}{2}$ times the cost of an optimal tour.*

Proof: Let H^* be an optimal tour. Removing any edge from H^* yields a spanning tree, so the cost $c(T)$ of a minimum-cost spanning tree T is at most $c(H^*)$. We can define a circuit C on the set W of odd nodes of T by joining these nodes in the order they appear in H^*. Note that $|W|$ is even and the edge-set of C partitions into two perfect matchings of $G[W]$. Since c satisfies the triangle inequality, each edge of these matchings has cost no greater than the corresponding subpath of H^*. Therefore one of these matchings has cost at most $c(H^*)/2$. This implies that the cost of the minimum-cost perfect matching M of $G[W]$ is at most $c(H^*)/2$. Thus $c(J) \leq \frac{3}{2} \cdot c(H^*)$. Since shortcutting can only improve $c(J)$, the final tour produced also has cost at most $\frac{3}{2} \cdot c(H^*)$, as required.   ∎

In the Johnson, Bentley, McGeoch, and Rothberg [1997] tests on the TSPLIB problems, Christofides' Heuristic produced tours that were about 1.14 times the optimum. They also made the interesting discovery that if at each short-cut step the best shortcut for the given node is chosen, then the performance of the algorithm improves to 1.09 times the optimum.

Tour Improvement Methods: 2-opt and 3-opt

There are several standard methods for attempting to improve an existing tour T. The simplest is called *2-opt*. It proceeds by considering each nonad-jacent pair of edges of T in turn. If these edges are deleted, then T breaks up into two paths T_1 and T_2. There is a unique way that these two paths can be recombined to form a new tour T'. If $c(T') < c(T)$, then we replace T with T' and repeat. This process is called a *2-interchange*. See Figure 7.4. If $c(T') \geq c(T)$ for every choice of pairs of nonadjacent edges, then T is *2-optimal* and we terminate.

Figure 7.4. 2-interchange

For a general cost function, in order to check whether a tour is 2-optimal we check $O(|V|^2)$ pairs of edges. For each pair, the work required to see if the switch decreases the tour cost can be performed in constant time. Thus the amount of time required to check a tour for 2-optimality is $O(|V|^2)$. But this does not mean that we can transform a tour into a 2-optimal tour in polynomial time. Indeed, Papadimitriou and Steiglitz [1977] show that if we make unfortunate choices, we may in some cases perform an exponential number of interchanges, before a 2-optimal tour is found.

The 2-opt algorithm can be generalized naturally to a k-opt algorithm, wherein we consider all subsets of the edge-set of a tour of size k, or size at most k, remove each subset in turn, then see if the resulting paths can be recombined to form a tour of lesser cost. The problem is that the number of subsets grows exponentially with k, and we soon reach a point of diminishing return. For this reason, k-opt for $k > 3$ is seldom used.

Johnson, Bentley, McGeoch, and Rothberg [1997] report that on the TSPLIB problems, 2-opt produces tours about 1.06 times the optimum and 3-opt about 1.04 times the optimum.

Tour Improvement Methods: Lin-Kernighan

Lin and Kernighan [1973] developed a heuristic which works extremely well in practice. It is basically a k-opt method with two novel features. First, the value of k is allowed to vary. Second, when an improvement is found, it is not necessarily used immediately. Rather the search continues in hopes of finding an even greater improvement. In order to describe it, we require several definitions.

A δ-*path* in a graph G on n nodes is a path containing n edges and $n + 1$ nodes all of which are distinct except for the last one, which will appear somewhere earlier in the path. See Figure 7.5. (The name comes from the shape of the path.)

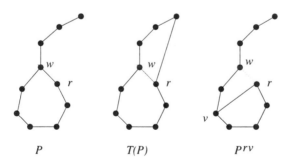

Figure 7.5. δ-path

Note that a tour is a δ-path for which the last node is the same as the first node. If P is a δ-path which is not a tour, then we can obtain a tour $T(P)$ as follows. Let w be the last node of the path (which also appears earlier in the path). Let wr be the first edge of the subpath of P between the two occurrences of w. By removing the edge wr and adding the edge joining r to the first node of the path, we obtain the edge set of a tour. See Figure 7.5.

Suppose that P is a δ-path which is not a tour. Again, let w be the last node and let wr be the first edge of the subpath of P between the two occurrences of w. If we remove wr we obtain the edge-set of a path ending at r. If we then add one more edge rv and node v, we obtain a new δ-path P^{rv} ending at v. We call this operation an *rv-switch*. Note that $c(P^{rv}) = c(P) + c_{rv} - c_{wr}$. Again, see Figure 7.5.

The Lin-Kernighan heuristic starts with a tour and then constructs a sequence of non-tour δ-paths, each obtained from the preceding one by an *rv*-switch. For each δ-path P so produced, it computes the cost of $T(P)$. If this is better than the best tour known, then it is "remembered." When the scan is complete, it replaces the starting tour with the best tour found in the scan. A full description of the *Core Lin-Kernighan Heuristic* is given below.

Step 1 [Outer loop: node/edge pairs]. For each node v of G, for each of the two edges uv of T incident with v in turn, perform Steps 2 through 5 in an attempt to obtain an improvement. This process is called an *edge scan*.

Step 2 [Initialize edge scan]. Initially, the best tour found is T. Let $u_0 = u$. Remove edge $u_0 v$ and add an edge $u_0 w_0$, for some $w_0 \neq v$, provided that such a w_0 can be found for which $c_{w_0 u_0} \leq c_{u_0 v}$. If no such w_0 can be found, then this scan is complete and we go on to the next node/edge pair.

We now have a δ-path P^0 (with last edge $u_0 w_0$) and $c(P^0) \leq c(T)$. Set $i = 0$ and proceed to Step 3.

Step 3 [Test tour]. Construct the tour $T(P^i)$. If $c(T(P^i))$ is less than the cost of the best tour found so far, then store this as the new best tour found so far. In either case, proceed to the next step.

Step 4 [Build next δ-path]. Let u_{i+1} be the neighbor of w_i in P^i which belongs to the subpath joining w_i to u_i. If the edge $w_i u_{i+1}$ was an edge added to a δ-path in this iteration, then go to Step 5 and stop this scan. Otherwise, try to find a node w_{i+1} such that $u_{i+1} w_{i+1}$ is not in T and when we perform the $u_{i+1} w_{i+1}$-switch, the new δ-path P^{i+1}, with last edge $u_{i+1} w_{i+1}$, we obtain has cost no greater than that of T. Again, if no such w_{i+1} can be found, we go to Step 5 and stop this scan. But if we are successful, then we set $i = i + 1$ and go back to Step 3. (See Figure 7.6.)

Step 5 [End of node/edge scan]. If we have found a tour whose cost is less than that of T, replace T with the minimum cost such tour found. If there remain untested node/edge combinations, then return to Step 2 and try the next.

Now let us make a few comments. First, note that in the process of a single node/edge scan, for any given edge, we can either add it to a δ-path or remove it from a δ-path but not both. Thus it makes sense to speak of *added* and *removed* edges. Each successive δ-path generated will have cost no greater than that of T, the initial tour. This is equivalent to saying that the sum of the costs of removed edges minus the sum of the cost of added edges is kept nonnegative. This difference is sometimes called the *gain sum*.

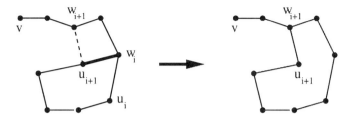

Figure 7.6. Construction of next δ-path

Notice that when a better tour $T(P^i)$ is found, we do not immediately abandon the process. Rather we continue looking for an even better completion.

In Steps 2 and 4 we chose a node w_{i+1}, subject to certain conditions. In general there will be many possible choices for these nodes. It may be too time consuming to try all possibilities at each stage, so Lin and Kernighan suggest the following compromise, intended to limit the amount of back-tracking. For each candidate w, compute $l(w) = c_{wu_{i+2}} - c_{u_{i+1}w}$, where u_{i+2} is the node which would be selected the next time through this step. Note that u_{i+2} is completely determined. When choosing each of w_0 and w_1, we consider in turn each of the five candidates w for which $l(w)$ is maximum. For all subsequent iterations, we consider only the best candidate. Thus in the process of scanning associated with a single node/edge pair, we will in fact consider as many as 25 choices for the first two edges to be switched in. For each we completely follow its chain of switches. If a better tour is found, then T is replaced and we start over. If not, we go on to the next.

They recommend one additional modification to the above core. The first time Step 4 is executed, when we choose u_1, we consider a second alternative. This is the neighbor of w_0 in the path back to v. Now removing the edge u_1w_0 yields a circuit and a path joining v and u_1. By joining u_1 to a node w_1 in the circuit, we obtain once again a δ-path starting from v. (We may orient the δ-path in either direction around the circuit.) We choose the best such w_1, according to the above criterion. See Figure 7.7(a). Going one step further in this situation, they also allow w_1 to be a node in the path joining v and u_1 (rather than in the circuit). In this case, we let u_2 be the first node on the subpath from w_1 to u_1, delete the edge w_1u_2, and form a δ-path by letting w_2 be a node in the circuit. See Figure 7.7(b).

This completes the description of the core algorithm. There are a wide range of possible modifications to the core that can be considered. Some interesting variants are described in Johnson and McGeoch [1997], Mak and Morton [1993], and Reinelt [1994].

To produce a very good quality tour, we need to embed the core algorithm into a larger search procedure. Lin and Kernighan propose running the core repeatedly, starting from many different tours. They also propose several

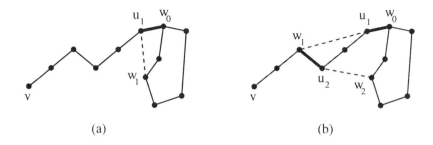

(a)   (b)

Figure 7.7. Extra first step

different methods for reducing the total amount of work required by these many runs of the core routine.

An alternative has been proposed by Martin, Otto, and Felten [1992] which seems to work very well in practice. Each time we complete a run of the core routine and hence have a "locally optimum" tour, T, we apply to it a "kick" that will perturb the tour so that it is likely to no longer be locally optimal. We then rerun the core routine from this new tour. If the core routine produces a new tour T' that is cheaper than T, then we replace T by T', and repeat the kicking process with this new tour. Otherwise, we go back and repeat the process with our best tour T.

One kick that they propose is a 4-interchange that the core routine is incapable of performing. It consists of randomly choosing four nonadjacent edges $u_0v_0, u_1v_1, u_2v_2, u_3v_3$ of the tour where we assume that the nodes appear in the above order on the tour. We remove these edges and add the edges $u_0v_2, u_1v_3, u_2v_0, u_3v_1$. See Figure 7.8. This becomes our new starting tour. Its cost will probably be much worse than that of the old local optimum, but it does provide a new starting point to rerun the core routine.

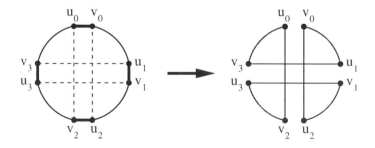

Figure 7.8. 4-interchange to restart core

This procedure is called *Chained Lin-Kernighan*. Martin and Otto [1996] describe it in a general context for search procedures for combinatorial optimization. One point that they make is that it may be helpful to replace T

by T' even when T' is slightly more expensive than T. This added flexibility may allow the procedure to break away from a locally optimal tour that does not seem to permit good kicks. The rule they suggest is to replace T by T' with a certain probability that depends on the difference in the costs of the two tours and on the number of iterations of the procedure that have already been carried out.

Notice that Chained Lin-Kernighan is not a finite algorithm, since we have not provided any stopping rules. We would normally let it run until we see a long period with no improvement. Sometimes, however, we have available a good lower bound on the optimum solution cost, which will permit us to stop sooner. Obtaining such bounds is the subject of the next two sections.

The result of applying this method to our test problem is shown in figure 7.9. On the TSPLIB problems, Johnson, Bentley, McGeoch, and Roth-

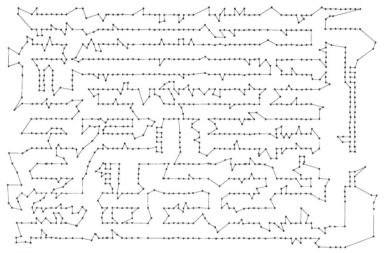

Figure 7.9. Sample TSP and Chained Lin Kernighan solution: 56892

berg [1997] report that Lin-Kernighan produces tours about 1.02 times the optimum, and Chained Lin-Kernighan under 1.01 times the optimum.

Running Times

The methods described above can all be implemented to run efficiently, even for quite large problems. We refer the reader to the extensive treatment of this subject by Johnson, Bentley, McGeoch, and Rothberg [1997]. They report, for example, that on a randomly generated 10,000-node Euclidean problem, running times on a fast workstation (that is, fast in 1994), are 0.3 seconds for Nearest Neighbor, 7.0 seconds for Farthest Insertion, 41.9 seconds for Christofides' Heuristic, 3.8 seconds for 2-opt, 4.8 seconds for 3-opt, and 9.7 seconds for Lin-Kernighan.

Exercises

7.1. Let $G = (V, E)$ be a graph and $c \in \mathbf{R}^E$. Show that the problem of finding a minimum-cost Hamiltonian circuit in G can be formulated as a TSP.

7.2. Show that if we have a TSP on n nodes for which the triangle inequality does not hold, then the ratio of the cost of the solution produced by the Nearest Neighbor Algorithm to the cost of an optimal tour can be arbitrarily large.

7.3. A variant on the TSP permits a node to be visited more than once if it results in a better solution. Show that if the cost function satisfies the triangle inequality, then there is always an optimum solution which visits each node exactly once. Show that if the triangle inequality does not hold, this problem can be solved by solving a TSP for which the triangle inequality does hold.

7.4. Show that if we have a Euclidean TSP, then a tour is 2-optimal only if it never crosses itself, but the converse is false.

7.5. Use the following example to show that the factor $3/2$ in Theorem 7.1 cannot be decreased. Let $V = \{v_1, v_2, \ldots, v_n\}$ and define $c_{v_i v_j}$ to be $\lfloor (|i - j| + 1)/2 \rfloor$ for $i \neq j$.

7.6. (Van Leeuwen and Schoone) Let V be any set of n nodes in the Euclidean plane and let T be any tour on these points. Suppose we attempt to obtain a noncrossing tour by choosing any pair of edges which cross and then performing a 2-interchange to uncross them. Show that the total number of crossings can be increased by such an operation. Show that after at most $|V|^3$ uncrossings, we necessarily obtain a noncrossing tour. (Hint: For each edge, viewed as a line segment, count the number of [infinite] lines that can be drawn through two cities so as to intersect that segment. Show that the removal of a crossing always reduces the total count, for all edges, by at least one.)

7.7. Show that if we consider all choices for w_{i+1} in Step 4 of the Core Lin-Kernighan Heuristic, then the resulting solution will always be 3-optimal but need not be 4-optimal. Show that the version described here (core only) need not be 3-optimal.

7.3 LOWER BOUNDS

We have emphasized the use of min-max relations in procedures for computing optimal solutions to combinatorial problems. The lower bound provided by the "max" side of the relation gives a proof of the optimality of the solution given in the "min" side. Unfortunately, for the TSP and many other problems known to be just as difficult as the TSP (see Chapter 9), no such min-max relation is known. Nonetheless it is important in some practical situations to give lower bounds as a measure of the quality of a proposed solution.

Held and Karp

A classic approach to lower bounds for the TSP involves the computation of minimum-cost spanning trees. The general technique is standard: To obtain a lower bound on a difficult problem we relax its constraints until we arrive at a problem that we know how to solve efficiently. In this case, the idea is that if we remove from a given tour the two edges incident with a particular node, then we are left with a path running through the remaining nodes. Although we do not in general know how to compute such a "spanning path" of minimum cost (it is just as hard as the TSP), we do know how to compute a minimum-cost spanning tree, and this will give us a lower bound on the cost of the path.

More precisely, suppose we have a graph $G = (V, E)$ with edge costs (c_e : $e \in E$) and a tour $T \subseteq E$. Let $v_1 \in V$, let e and f be the two edges in T that are incident with v_1, and let P be the edge-set of the path obtained by removing e and f from T. The cost of T can be written as $c_e + c_f + c(P)$. So if we have numbers A and B such that $A \leq c_e + c_f$ and $B \leq c(P)$, then $A + B$ will be a lower bound on the cost of the tour T. Our goal is to define A and B so that they are valid for all choices of T. In this way, we will obtain a lower bound for all tours.

So, what can we say about the edges e and f? Since all we know is that they both are incident with v_1, we can do no better than to set A equal to the sum of the costs of the two cheapest edges in E incident with that node.

The interesting part is the bound on $c(P)$. Notice that P is a spanning tree (albeit of a special form) for the graph $G \setminus v_1$ we get by deleting v_1 and the incident edges from G. Thus, if we let B be the minimum cost of a spanning tree in $G \setminus v_1$ (which we can compute with the methods described in Chapter 2), then we know that P must have cost at least B. So we have our bound, $A + B$. This is commonly called the *1-tree bound* and a set of edges consisting of two edges incident with node v_1 plus a spanning tree of $G \setminus v_1$ is called a *1-tree*. The name comes from the usual practice of denoting the node that we delete as node v_1 or node "1." We can summarize this discussion as follows.

1-tree Bound

Let $G = (V, E)$ with edge costs ($c_e : e \in E$) and let $v_1 \in V$. Now let $A = \min\{c_e + c_f : e, f \in \delta(v_1), e \neq f\}$ and let B be the cost of a minimum spanning tree in $G \setminus v_1$. Then $A + B$ is a lower bound for the TSP on G.

A minimum-cost 1-tree for the 1173-node problem is illustrated in Figure 7.10. (The node v_1 is in the lower right-hand corner of the figure, and is drawn larger than the other nodes.) Its value is approximately 90.5% of the cost of the best tour reported above. Although this is quite respectable, it leaves a rather large gap between the upper and lower bounds. To improve

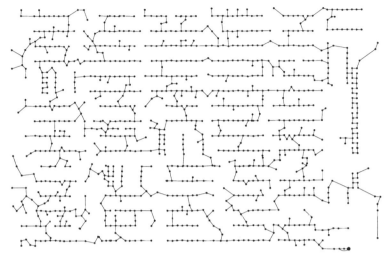

Figure 7.10. Sample TSP and 1-tree bound: 51488

this we might try several different nodes as "node v_1," but for larger problems, like our test problem, this will not result in a significant gain. The key to obtaining a real improvement can be found by examining the structure of the 1-tree in Figure 7.10. What catches your eye is that the optimal 1-tree does not at all resemble a tour: Many nodes do not have degree 2. We can use this to our advantage.

To see clearly what is happening, consider the small example given in Figure 7.11. With v_1 chosen as indicated, the 1-tree bound is 0. As you can

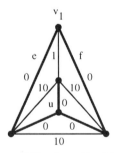

Figure 7.11. An optimal 1-tree

see, however, the cheapest tour has cost 10. What went wrong is that the spanning tree in G' can use all three of the 0 cost edges incident with node u, whereas a tour can only make use of two of them. There is a way around this problem, although at first it may seem like a sleight of hand.

What would happen if we added 10 to the cost of each of the edges incident with node u? Every tour uses exactly two of these edges, so the cost of every tour increases by precisely 20. So, as far as the TSP goes, we have not really changed anything: The old tour is still optimal, its cost now being 30. But what has happened to the 1-tree bound? A simple computation shows that it also has value 30. So we have a simple proof that no tour in the altered graph can have cost less than 30. This means that no tour in the original graph can have cost less than 10! By this simple transformation we have therefore increased the lower bound from 0 to 10. The point is that although the transformation does not alter the TSP, it does fundamentally alter the minimum spanning tree computation.

In the above example, we say that we "assigned u the node number -10." That is, we refer to the process of subtracting k from the cost of each edge incident with a given node v as assigning v the *node number* k. Notice that we could assign several node numbers at once. The change in the cost of any tour will just be twice the sum of the assigned numbers. With this fact, we can formally state a lower bounding technique, introduced by Held and Karp [1970].

Held-Karp Bound

Let $G = (V, E)$ be a graph with edge costs $(c_e : e \in E)$, let $v_1 \in V$, and for each node $v \in V$ let y_v be a real number. Now for each edge $e = uv \in E$ let $\bar{c}_e = c_e - y_u - y_v$ and let C be the 1-tree bound for G with respect to the edge costs $(\bar{c}_e : e \in E)$. Then $2\sum(y_v : v \in V) + C$ is a lower bound for the TSP on G (with respect to the original edge costs $(c_e : e \in E)$).

With a good set of node numbers, the difference between the Held-Karp lower bound and the 1-tree lower bound can be dramatic, as we indicate in Figure 7.12 for our 1173-node test problem. The 56349 bound we obtained in this way implies that the 56892-cost tour we found in the previous section is no more than 1% above the cost of an optimal tour.

For the important computational issue of how to find a good set of numbers, Held and Karp [1971] proposed a simple iterative scheme. The main step is the following. Suppose we have computed an optimum 1-tree T with respect to the altered edge costs $(c_{uv} - y_u - y_v : v \in V)$ for some set of node numbers $(y_v : v \in V)$. For each node $v \in V$, let $d_T(v)$ denote the number of edges of T that are incident with v. Based on our above discussion, if $d_T(v)$ is greater than 2 then we should decrease y_v and if $d_T(v)$ is equal to 1 (it cannot be less than 1) we should increase y_v. This is just what Held and Karp tell us to do: For each node v, replace y_v by

$$y_v + t(2 - d_T(v))$$

for some positive real number t (the *step size*).

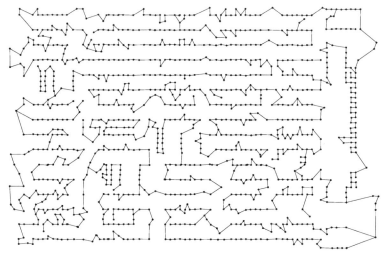

Figure 7.12. Sample TSP and Held-Karp bound: 56349

By iterating this step, we obtain a sequence of Held-Karp lower bounds. Although it is not true that the bound improves at each iteration, under certain natural conditions on the choice of the step sizes it can be shown that the bound will converge to the optimum Held-Karp bound (that is, the maximum Held-Karp bound over all choices of node numbers). (See Held, Wolfe, and Crowder [1974].) Unfortunately, no conditions are known that guarantee the convergence will occur in polynomial time. But all is not lost. As reported in Grötschel and Holland [1988], Holland [1987], Smith and Thompson [1977], and elsewhere, several ways of selecting the step sizes have shown good performance in practice. We describe one such method.

Motivated by geometrical considerations, at the kth iteration Held and Karp [1971] suggest the step size

$$t^{(k)} = \alpha^{(k)}(U - H)/\sum((2 - d_T(v))^2 : v \in V)$$

where U is some target value (an upper bound on the cost of a minimum tour), H is the current Held-Karp bound, and $\alpha^{(k)}$ is a real number satisfying $0 < \alpha^{(k)} \le 2$. Following Held, Wolfe, and Crowder [1974], we start with $\alpha^{(0)} = 2$ and decrease $\alpha^{(k)}$ by some fixed factor after every block of iterations, where the size of a block depends on the number of nodes in the TSP we are solving and the amount of computation time we are willing to spend on obtaining the lower bound.

The entire method is summarized in the box below.

Held-Karp Iterative Method

Input

Graph: $G = (V, E)$ with edge costs $(c_e : e \in E)$ and $v_1 \in V$

Real number: U (a target value)

Positive real number: ITERATIONFACTOR (for
example, 0.015)

Positive integer: MAXCHANGES (for example, 100)

Initialization

$y_v = 0 \ \forall v \in V$, $H^* = -\infty$, TSMALL = 0.001, $\alpha = 2$, $\beta = 0.5$

NUMITERATIONS = ITERATIONFACTOR $\times |V|$

Algorithm

For $i = 1$ to MAXCHANGES

    For $k = 1$ to NUMITERATIONS

        Let T be the optimum 1-tree with respect to the
edge costs $(c_{uv} - y_u - y_v : uv \in E)$ and let H
be the corresponding Held-Karp bound;

        If $H > H^*$, then set $H^* = H$ (improvement);

        If T is a tour, then STOP.

        Let $t^{(k)} = \alpha(U - H)/\sum((2 - d_T(v))^2 : v \in V)$;

        If $t^{(k)} < TSMALL$, then STOP.

        Replace y_v by $y_v + t^{(k)}(2 - d_T(v))$ for all $v \in V$;

    Replace α by $\beta\alpha$;

If you carry out experiments with this method, you will notice that the bound you obtain for a given graph and edge costs will depend on the choices of the input parameters (particularly the target value U). Only experimenting with the settings will allow you to optimize the method for a particular class of problems. Also, you should keep in mind that many other choices for the step sizes are possible, and experiments may suggest an alternative scheme that performs better on your problems.

Linear Programming

Dantzig, Fulkerson, and Johnson [1954] proposed to attack the TSP with linear-programming methods. The approach they outlined is still the most effective method known for computing good lower bounds for the TSP. Their work also plays a very important role in the history of combinatorial optimization since it was the first time cutting-plane methods were used to solve a combinatorial problem. We will discuss their method further in the next section, but here we present their linear-programming relaxation to the TSP.

Let x be the characteristic vector of a tour. Then x satisfies

$$x(\delta(v)) = 2, \text{ for all } v \in V$$

Box 7.1: *Lagrangean Relaxation*

The Held-Karp lower bounding technique is an instance of a general integer programming method known as *Lagrangean relaxation* (Held and Karp [1971]). The method is appropriate for integer programming problems max $\{w^T x : Ax \leq b, x$ integral $\}$ for which the constraints $Ax \leq b$ can be split into two parts, $A_1 x \leq b_1$ and $A_2 x \leq b_2$, in such a way that "relaxed" problems of the form max $\{c^T x : A_2 x \leq b_2, x$ integral $\}$ can be solved efficiently. The method is based on the observation that for any vector $y \geq 0$ (where the number of components of y matches the number of inequalities in $A_1 x \leq b_1$) the value

$$L(y) = y^T b_1 + \max \{(w - y^T A_1)^T x : A_2 x \leq b_2, x \text{ integral }\}$$

is an upper bound on the original integer programming problem (since $y b_1 \geq y A x$). By assumption, for any given vector y we can easily compute $L(y)$, so what we need is a way to find a good y (that is, one that gives a strong upper bound $L(y)$). This can be accomplished with a general iterative technique called *subgradient optimization* (Polyak [1967], Held and Karp [1971], Held, Wolfe, and Crowder [1974]). The kth step of the subgradient method is the following. Having the vector $y^{(k)}$, we compute an optimal solution $x^{(k)}$ to the problem

$$\max\{(w - y^{(k)^T}) A_1^T x : A_2 x \leq b_2, x \text{ integral }\}.$$

Now, for a specified step size $t^{(k)}$, we let

$$y^{(k+1)} = y^{(k)} - t^{(k)}(b_1 - A_1 x^{(k)})$$

and go on to the $(k+1)$st step. Polyak [1967] has shown that if the step sizes $t^{(0)}, t^{(1)}, \ldots$ are chosen so that they converge to 0 but not too fast (namely $\lim_{k \to \infty} t^{(k)} = 0$ and $\sum_{k=0}^{\infty} t^{(k)} = \infty$), then the sequence of upper bounds produced by the subgradient method converges to the optimal $L(y)$ bound.

$$0 \leq x_e \leq 1, \text{ for all } e \in E.$$

It is not true, however, that tours are the only integer solutions to this system, since every 2-factor (that is, disjoint union of circuits meeting all of the nodes) will appear in the solution set. To forbid these non-tours, we can add the inequalities

$$x(\delta(S)) \geq 2, \text{ for all } \emptyset \neq S \neq V$$

since any tour must both enter and leave such a set S, and thus will contain at least two edges from $\delta(S)$. These inequalities are known as *subtour constraints* since they forbid small circuits (or "subtours").

The Dantzig, Fulkerson, and Johnson relaxation of the TSP is

$$\text{Minimize } \sum(c_e x_e : e \in E) \qquad (7.1)$$
$$\text{subject to}$$
$$x(\delta(v)) = 2, \text{ for all } v \in V$$
$$x(\delta(S)) \geq 2, \text{ for all } \emptyset \neq S \neq V$$
$$0 \leq x_e \leq 1, \text{ for all } e \in E.$$

Note that any integral solution to (7.1) is a tour, so (7.1) can be used to create an integer-linear-programming formulation of the TSP. What is important for us, however, is that the optimal value of (7.1) is a lower bound on the cost of any tour. We call this the *subtour bound* for the TSP. We show below that the subtour bound is equal to the optimal Held-Karp bound!

To start off, choose some node $v_1 \in V$ and notice that we can restrict the set of subtour constraints to those sets S that do not contain v_1, since the constraints for S and $V \setminus S$ are identical. Furthermore, making use of the equations $x(\delta(v)) = 2$, we can write the subtour constraints in the "inside" form

$$x(\gamma(S)) \leq |S| - 1$$

(see Exercise 7.11). Also, the equation

$$x(\gamma(V \setminus \{v_1\})) = |V| - 2$$

is implied by the equations $x(\delta(v)) = 2$, and thus can be added as a redundant constraint to (7.1).

With these modifications, we can write (7.1) as

$$\text{Minimize } \sum(c_e x_e : e \in E) \qquad (7.2)$$
$$\text{subject to}$$
$$x(\delta(v)) = 2, \text{ for all } v \in V \qquad (7.3)$$
$$x(\gamma(S)) \leq |S| - 1, \text{ for all } S \subseteq V, \ v_1 \notin S \qquad (7.4)$$
$$x(\gamma(V \setminus \{v_1\})) = |V| - 2 \qquad (7.5)$$
$$0 \leq x_e \leq 1, \text{ for all } e \in E. \qquad (7.6)$$

These constraints look similar to the constraints of a linear-programming problem we saw in Chapter 2. Indeed, if we remove the equations (7.3) and the variables corresponding to the edges in $\delta(v_1)$, then we are left with the defining system for the convex hull of the spanning trees in the graph $G[V \setminus \{v_1\}]$. This is the connection with the Held-Karp bound. More directly, we can conclude that the cost of a minimum 1-tree in G is equal to

$$\min \{c^T x : x \text{ satisfies (7.4), (7.5), (7.6), and } x(\delta(v_1)) = 2\}. \tag{7.7}$$

To see how the node numbers come in, consider the form of the dual linear-programming problem of (7.2). We have a dual variable y_v for all $v \in V \setminus \{v_1\}$, together with a dual variable for each constraint in the 1-tree formulation (7.7). Now suppose we have an optimal solution to this dual linear-programming problem, and let $(y_v^* : v \in V \setminus \{v_1\})$ be the values of the variables $(y_v : v \in V \setminus \{v_1\})$. If we fix these variables at their values y_v^*, then the remaining variables constitute an optimal solution to the dual linear-programming problem of

$$\text{Minimize } \sum((c_{uv} - y_u^* - y_v^*)x_{uv} : uv \in E)$$
$$\text{subject to}$$
$$x \text{ satisfies } (7.4), (7.5), (7.6)$$
$$x(\delta(v_1)) = 2.$$

But this is a 1-tree problem. So the optimal value of (7.2) is equal to the Held-Karp bound obtained using the node numbers $(y_v^* : v \in V \setminus \{v_1\})$ (setting the node number on v_1 to 0).

Conversely, the arguments show that for any set of node numbers we can construct a feasible solution to the dual of (7.2) with objective value equal to the corresponding Held-Karp bound. Thus, we have shown the following result.

Theorem 7.2 *The subtour bound is equal to the optimal Held-Karp bound.*

∎

The Held and Karp procedure can therefore be viewed as a heuristic for approximating the subtour bound. Direct methods for computing this bound will be discussed in the next section.

Exercises

7.8. Modify the edge costs in the graph given in Figure 7.11 so that they satisfy the triangle inequality, keeping the fact that the 1-tree bound is not equal to the optimal value of the TSP, but the best Held-Karp bound is equal to the optimum TSP value.

7.9. Give a graph and edge costs such that the best Held-Karp bound is not equal to the optimum value of the TSP.

7.10. Let $G = (V, E)$ be a graph with edge costs $(c_e : e \in E)$, and let T be a minimum spanning tree of G. Show that if v is a leaf of T, then an optimum 1-tree with v as node v_1 can be obtained be adding to the edge-set of T the edge joining v to its second nearest neighbor. Give an example of G, c, and T, where the best choice of v_1 (that is, the one giving the greatest 1-tree bound) is not a leaf of T.

7.11. Let $G = (V, E)$ be a graph with edge costs $c \in \mathbf{R}^E$. Show that the linear-programming problem (7.1) is equivalent to the linear-programming problem (7.2).

7.4 CUTTING PLANES

In the last section, we described an intuitively motivated procedure for constructing what is usually a good lower bound on the cost of an optimal solution to a TSP. We showed that this procedure was in fact a heuristic for obtaining a good feasible solution to the "subtour" linear-programming problem (7.1), which we restate here:

$$\text{Minimize } \sum(c_e x_e : e \in E) \tag{7.8}$$

$$\text{subject to}$$

$$x(\delta(v)) = 2, \text{ for all } v \in V \tag{7.9}$$

$$x(\delta(S)) \geq 2, \text{ for all } S \subseteq V, \ S \neq V, S \neq \emptyset \tag{7.10}$$

$$0 \leq x_e \leq 1, \text{ for all } e \in E. \tag{7.11}$$

Suppose we tried to solve this linear-programming problem directly. What problems would we encounter? One big obstacle is that the number of inequalities (7.10) is about the same as the number of distinct subsets of cities, or about $2^{|V|}$. Even if we notice that we do not need inequalities for both S and $V \setminus S$, and hence can limit ourselves to sets S satisfying $|S| \leq |V|/2$, we still need about $2^{|V|-1}$ of these inequalities.

Dantzig, Fulkerson, and Johnson overcame this obstacle by solving the linear-programming problem using the cutting-plane approach described in Section 6.7. We describe their approach in this section.

We begin by solving the linear-programming problem (7.8), (7.9), and (7.11). If the optimal solution happens to be the characteristic vector of a tour, then we can stop since this must be the solution to the TSP. If not, we will try to find some subtour constraints (7.10) violated by the optimal solution. We add these inequalities to our starting set and solve the resulting linear program.

We perform this process over and over. If we ever obtain a solution which does not violate any subtour constraints, then we have solved the original linear-programming problem. If not, we add some violated subtour constraints to obtain the next problem.

If this iterative process is to work, there are two problems to solve. First, we must have an efficient method of checking an optimum solution to a relaxed problem to see whether it violates any subtour constraints from (7.10). Second, we must have an efficient way of solving the linear-programming problems that arise.

The first problem can be solved using results from Chapter 3, as we describe below.

For small TSPs the second problem is easily handled by any commercial-quality simplex-based linear-programming code. For larger TSPs, however, we will run into the difficulty of having to deal with linear-programming problems with a large number of variables. For example, the problems for our 1173-node sample TSP will have 687368 variables. In such a case, it is probably not a good idea to solve the problem directly. Instead, we handle the variables in a manner similar to the way we handle cutting planes: Start out with a linear-programming problem that contains only a subset of the variables and add in the remaining ones as they are needed. We need to explain what we mean by "as they are needed."

Suppose we select a set $E' \subseteq E$, such that the linear-programming problem

$$\text{Minimize } \sum(c_e x_e : e \in E') \tag{7.12}$$

$$\text{subject to}$$

$$x(\delta(v)) = 2, \text{ for all } v \in V$$

$$x(\delta(S)) \geq 2, \text{ for all } \emptyset \neq S \neq V$$

$$0 \leq x_e \leq 1, \text{ for all } e \in E'.$$

has a feasible solution. (A common choice is to take the union of a small number [say 10] of tours produced by the Core Lin-Kernighan Heuristic.) An optimal solution, x', to (7.12) can be extended to a feasible solution, x^*, to (7.8) by setting $x_e^* = 0$ for all $e \in E \setminus E'$. The trouble is that x^* may not be an optimal solution to (7.8).

To check optimality, let y', Y' be an optimal solution to the dual linear-programming problem of (7.12):

$$\text{Maximize } \sum(2y_v : v \in V) + \sum(2Y_S : S \subseteq V, S \neq V, S \neq \emptyset) \tag{7.13}$$

$$\text{subject to}$$

$$y_u + y_v + \sum(Y_S : uv \in \delta(S), S \subseteq V, S \neq V, S \neq \emptyset) \leq c_{uv}, \tag{7.14}$$

$$\text{for all } uv \in E'$$

$$Y_S \geq 0, \text{ for all } S \subseteq V, S \neq V, S \neq \emptyset. \tag{7.15}$$

If y', Y' is also feasible for the dual linear-programming problem of (7.8) (that is, the problem we obtain by replacing E' by E in (7.14)) then we know by linear-programming duality that x^* is indeed optimal for the original linear-programming problem (7.8). Otherwise, we can add those edges $e \in E \setminus E'$ for which the corresponding constraint (7.14) is violated to our set E', resolve the linear-programming problem (7.12), and repeat the process.

This is another example of column generation. It is similar to the methods we described in Chapter 3 for multicommodity flows and in Chapter 5 for solving minimum-weight perfect matching problems on dense graphs. Combining column generation with the cutting-plane approach allows us to solve linear-programming problems that are both "long" and "wide."

Using this combined method, suppose that we eventually solve the linear-programming problem (7.8). Generally this solution will not be the characteristic vector of a tour. What then? We could stop with a lower bound which is usually pretty good. We could go on to branch-and-bound, as discussed in the next section. Or we could try to find some other class of cutting planes to add which would permit this process to continue.

We now discuss the cutting-plane generation in some detail.

Handling Subtour Constraints

Suppose x^* is a feasible solution to the (initial) linear-programming problem

$$\text{Minimize } \sum(c_e x_e : e \in E) \tag{7.16}$$

$$\text{subject to}$$

$$x(\delta(v)) = 2, \text{ for all } v \in V \tag{7.17}$$

$$0 \leq x_e \leq 1, \text{ for all } e \in E. \tag{7.18}$$

We wish to determine whether all subtour constraints (7.10) are satisfied, and if not, find one or more that are violated.

If the solution falls apart into several components (that is, the graph with node-set V and edge-set $\{e \in E : x_e^* > 0\}$ is disconnected), then the node-set S of each component violates (7.10). This situation is easy to detect.

After several waves of cutting-plane addition, we will in general not have a disconnected solution. In this case, we need a more sophisticated separation algorithm.

For each edge e of G, define its *capacity* u_e to be x_e^*. Then the value $x^*(\delta(S))$ for any set S of nodes is precisely the same as the capacity of the cut in $\delta(S)$ in G. So we can apply the minimum cut methods we discussed in Section 3.5: There exists a set S of nodes that violates (7.10) if and only if some cut in G has capacity less than two.

Now we are in a good position. We solve the initial linear-programming relaxation. Then we can add violated subtour constraints as long as the solution is not sufficiently connected. Then we can use a minimum-cut algorithm to ensure that there are no violated subtour constraints at all. Each time we add more subtour constraints, we can use the simplex algorithm to obtain a new optimal solution.

For our 1173-node sample TSP, the optimal value of (7.8) is 56361. As we would expect, this is slightly better than the bound we found using the Held-Karp method.

Suppose we terminate with a solution such as in Figure 7.13. It is an optimal solution to the linear-programming problem (7.8) if all costs are Euclidean, but it is not a tour. Below, we describe a class of cutting planes that is very useful in improving the lower bound in situations like this.

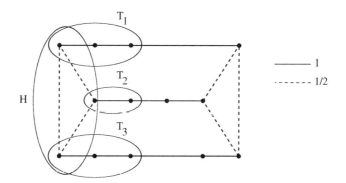

Figure 7.13. A fractional solution and violated comb

Comb Inequalities

A *comb* is defined by giving several subsets of nodes of the graph: We need one nonempty *handle* $H \subseteq V, H \neq V$ and $2k + 1$ pairwise disjoint, nonempty *teeth* $T_1, T_2, ..., T_{2k+1} \subseteq V$, for k at least 1. (So the number of teeth is odd and at least 3.) We also require each tooth to have at least one node in common with the handle and at least one node that is not in the handle. See Figure 7.13.

Chvátal [1973a] and Grötschel and Padberg [1979] proved the following result.

Theorem 7.3 *Let C be a comb with handle H and teeth $T_1, T_2, ..., T_{2k+1}$ for $k \geq 1$. Then the characteristic vector x of any tour satisfies*

$$x(\gamma(H)) + \sum_{i=1}^{2k+1} x(\gamma(T_i)) \leq |H| + \sum_{i=1}^{2k+1} (|T_i| - 1) - (k + 1).$$

Proof: Let x be a tour. Then x satisfies all the constraints (7.9)–(7.11). Add up the following constraints.

Equations (7.9) for the nodes in H.

Constraints (7.10) for the teeth T_i in the "inside form" $x(\gamma(T_i)) \leq |T_i| - 1$, (See Exercise 7.11).

Constraints $x_j \geq 0$ for the edges in $\delta(H)$ but not belonging to any tooth, but in the form $-x_j \leq 0$.

Constraints (7.10) for the sets $T_i \setminus H$ in the "inside form" $x(\gamma(T_i \setminus H)) \leq |T_i \setminus H| - 1$.

Constraints (7.10) for the sets $T_i \cap H$ (for those i such that T_i intersects H in more than one node) again in the "inside form" $x(\gamma(T_i \cap H)) \leq |T_i \cap H| - 1$.

We obtain

$$2x(\gamma(H)) + 2\sum_{i=1}^{2k+1} x(\gamma(T_i)) \leq 2|H| + 2\sum_{i=1}^{2k+1}(|T_i| - 1) - (2k+1).$$

Now, dividing through by 2, we obtain

$$x(\gamma(H)) + \sum_{i=1}^{2k+1} x(\gamma(T_i)) \leq |H| + \sum_{i=1}^{2k+1}(|T_i| - 1) - \frac{2k+1}{2}.$$

Since the left-hand side is integer-valued, we can round down the right-hand side, and get the desired result.   ∎

Another way of stating this theorem is that

$$x(\gamma(H)) + \sum_{i=1}^{2k+1} x(\gamma(T_i)) \leq |H| + \sum_{i=1}^{2k+1}(|T_i| - 1) - (k+1) \qquad (7.19)$$

is a valid cutting plane. These are called *comb inequalities*.

Often when we see a fractional solution x in which there exists an odd circuit all of whose edges have the value $1/2$, the node-set of the circuit forms the handle of a comb giving rise to a cutting plane that is violated by x. Again, see Figure 7.13. At the current time, there is no polynomial-time algorithm known for deciding in general whether a given (nonnegative) vector x violates some comb inequality.

When each tooth of a comb has exactly two nodes, we call the corresponding inequality a *blossom inequality*. This name comes from a connection to the 2-factor problem. (See Exercise 7.15.) For this special class of comb inequalities, Padberg and Rao [1982] showed that there is a polynomial-time separation routine. We describe their method below.

Let $G = (V, E)$ be a graph. A blossom inequality can be specified by a handle $H \subseteq V$ and a set of edges $A \subseteq \delta(H)$, with $|A|$ odd and at least 3. So the ends of the edges in A are the teeth of the corresponding comb.

When we are solving a TSP, we assume that the nonnegative vector x satisfies the equations (7.9). So we may write the blossom inequality

$$x(\gamma(H)) + x(A) \leq |H| + \frac{|A| - 1}{2} \tag{7.20}$$

in the form

$$x(\delta(H) \setminus A) - x(A) \geq 1 - |A|. \tag{7.21}$$

(See Exercise 7.16.) If we rewrite this inequality as

$$x(\delta(H) \setminus A) + (|A| - x(A)) \geq 1,$$

then the left-hand side looks similar to the capacity of the cut $\delta(H)$, except that the edges in $e \in A$ contribute $1 - x_e$ instead of the usual x_e. It is perhaps not surprising that our separation routine will make use of the minimum T-cut algorithm (see Section 6.8), as we now describe.

To speed up our computations, we begin by deleting from E all those edges e such that $x_e = 0$.

Now define a new graph G' by subdividing each edge $e \in E$ with two new nodes v'_e and v''_e, that is, if e has ends v and w then we replace e by the three edges vv'_e, $v'_e v''_e$, and $v''_e w$. Let T be the set of all new nodes v'_e, v''_e for all $e \in E$, and define edge weights $u \in \mathbf{R}^{E(G')}$ by setting $u_{vv'_e} = x_e$, $u_{v'_e v''_e} = 1 - x_e$, and $u_{v''_e w} = x_e$ for all edges $e = vw \in E$. See Figure 7.14.

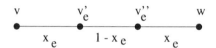

Figure 7.14. Subdivided edge

Suppose that the blossom inequality corresponding to $H \subseteq V$ and $A \subseteq \delta(A)$ is violated by x. Let $S \subset V(G')$ consist of H, together with the two new nodes v'_e, v''_e for all edges $e \in \gamma_G(H)$, and for each edge $e \in A$ the new node v'_e or v''_e that is joined by an edge in G' to a node in H. Then $S \cap T$ is odd and the capacity of $\delta_{G'}(S)$ is precisely

$$x(\delta(H) \setminus A) + (|A| - x(A)). \tag{7.22}$$

So $\delta_{G'}(S)$ is a T-cut in G' with capacity less than 1.

Conversely, let $\delta_{G'}(S)$ be a minimum T-cut in G' and suppose that its capacity is less than 1. Let $H = S \cap V$, and let A consist of those edges $e \in \delta_G(H)$ such that exactly one of the two new nodes v'_e or v''_e is in S and that node is adjacent to a node in H.

Using the fact that for any new node $t \in V(G') \setminus V$, the sum of the capacities of the two edges in $E(G')$ that are incident with t is exactly 1, it is easy to check that $|A|$ is odd and that the capacity of $\delta_{G'}(S)$ is at least (7.22). So the blossom inequality corresponding to H and A is violated by x.

The blossom separation problem can thus be solved by finding a minimum T-cut in G' using the algorithm described in Section 6.8. If the T-cut has capacity less than 1 then we can extract a violated blossom inequality. Otherwise, we conclude that no such inequality exists.

Using this method to optimize over the linear-programming problem we obtain by adding all blossom inequalities to (7.8), we obtain a lower bound of 56785 for our 1173-node sample TSP. This is a very good bound. It implies that the tour we found with Chained Lin-Kernighan in Section 7.3 has cost no more than 0.2% above that of an optimal tour.

A number of heuristics have been proposed for finding violated comb inequalities. One of the common ideas is to shrink certain subsets of nodes and look for a violated blossom inequality in the shrunk graph that corresponds to a violated comb in the original graph. (See Exercise 7.19.)

There are many more classes of valid cutting planes known for the TSP. We refer the reader to Grötschel and Padberg [1985], Jünger, Reinelt, and Rinaldi [1995], and Naddef [1990] for further discussions.

Exercises

7.12. Show how to solve linear program (7.16)–(7.18) as a minimum-cost flow problem. Use your construction to prove that there exists an optimal solution to (7.16)–(7.18) for which all variables have value $0, 1/2$, or 1.

7.13. Let x be a feasible solution to the linear program (7.16)–(7.18). Let F be the set of all $e \in E$ for which $x_e \neq 0$ or 1.

(a) Show that if F contains the edge-set of an even circuit then x can be expressed as a convex combination of two other feasible solutions, and so is not a vertex of the polyhedron defined by (7.17),(7.18).

(b) Show that if any connected component of F contains two or more odd circuits, then x is not a vertex of (7.17),(7.18).

(c) Show that if every component of F consists of a single odd circuit, then there exists a set of inequalities (7.18) that can be set to equations such that x is the unique vector satisfying these equations plus (7.17).

(d) State a necessary and sufficient condition, based on (a)–(c), for a vector x to be a vertex of (7.17),(7.18).

7.14. Prove that if x is a vertex of (7.17),(7.18), then at most $|V|$ components of x can have nonintegral values. Construct an example that shows that this bound can be attained.

7.15. Show that the blossom inequalities are satisfied by the characteristic vectors of the 2-factors of a graph G.

7.16. Show that in the presence of the equations (7.9), the blossom inequality (7.20) can be written as (7.21). What is the connection with the system (5.40)?

7.17. Show that in the presence of the equations (7.9), the comb inequality (7.19) can be written as

$$(x(\delta(H)) - 2) + \sum_{i=1}^{2k+1} (x(\delta(T_i)) - 2) \geq 2k. \tag{7.23}$$

7.18. (Karger) Let k be a fixed positive integer and let x be a vector that satisfies (7.9), (7.10), and (7.11). Use Exercise 7.17 and the result of Exercise 3.65 (on page 85) to give a polynomial-time algorithm that with probability at least $1 - \frac{1}{n}$ will find some violated comb inequality having at most $2k + 1$ teeth if such an inequality exists.

7.19. Let $G = (V, E)$ be a graph and $x \in \mathbf{R}^E$ a nonnegative vector. Suppose that $S \subset V$ has the property that $x(\gamma(S)) = |S| - 1$ and consider the graph, G', we obtain by shrinking S to a single node, v, and replacing the parallel edges $e_1, \ldots, e_k$ between v and any other node by a single edge e having $x_e = x_{e_1} + \cdots + x_{e_k}$. Show that any violated comb inequality in G' corresponds to a violated comb inequality in G.

7.5   BRANCH AND BOUND

Cutting-plane methods can provide a very good lower bound on a TSP. Combining this with a tour produced by Chained Lin-Kernighan will typically leave only a small gap between the cost of the tour and the value of the bound. But suppose the gap is too large for a given application. How can we proceed further? The *branch and bound* method we present below is a common approach for doing just this. We will describe it in terms of the TSP, but the same principles apply to virtually any combinatorial optimization problem. Our description follows the TSP algorithm of Padberg and Rinaldi [1991].

Suppose we have a graph $G = (V, E)$ with edge costs $(c_e : e \in E)$ and let $\mathcal{T}$ denote the set of all tours of G. A lower bound on the TSP is a number B such that $c(T) \geq B$ for all $T \in \mathcal{T}$. A lower bounding technique is a method for producing such a number B. Now suppose we split $\mathcal{T}$ into two sets $\mathcal{T}_0$ and $\mathcal{T}_1$ such that $\mathcal{T}_0 \cup \mathcal{T}_1 = \mathcal{T}$. If we can produce numbers B_0 and B_1 such that $c(T) \geq B_0$ for all $T \in \mathcal{T}_0$ and $c(T) \geq B_1$ for all $T \in \mathcal{T}_1$, then the minimum of B_0 and B_1 is a lower bound on the TSP. The point of splitting $\mathcal{T}$ is that the extra structure in $\mathcal{T}_0$ and $\mathcal{T}_1$ may allow our lower bounding technique to perform better than it did on the entire set $\mathcal{T}$. This is the basis of branch and bound methods: We successively split the solution set and apply our lower bounding algorithm to each part. To see how this works, we describe how to use the cutting-plane lower bound in a branch and bound framework.

In this context, a natural way to partition the set of tours is to select an edge e and let $\mathcal{T}_0$ be those tours that do not contain e and let $\mathcal{T}_1$ be those that do contain e. So if we let P denote the original TSP, then we can work with this partition by considering a new problem P_0 obtained by setting $x_e = 0$ and a new problem P_1 obtained by setting $x_e = 1$.

Suppose that we have applied our cutting-plane methods to obtain a linear-programming relaxation, LP, of the original TSP. Then we can immediately write linear programs LP_0 and LP_1 (corresponding to the new problems) by adding the equations $x_e = 0$ and $x_e = 1$ to LP. If e is chosen carefully, we may obtain an immediate improvement in the lower bound by simply solving LP_0 and LP_1. Moreover, we can apply our cutting-plane generation routines to strengthen each of these linear programs, obtaining the relaxations LP_0' and LP_1'. Our lower bound will then be the minimum of the optimal values of LP_0' and LP_1'.

If we have not already established the optimality of our best tour, we can repeat the above process by taking one of the problems, say P_1, and some edge f, and creating the problems P_{10} and P_{11} by setting $x_f = 0$ and $x_f = 1$. Again, we can apply the cut generation routines to each of the new problems, obtaining the linear programs LP_{10}' and LP_{11}'. A bound on our TSP is then the minimum of the optimal values of LP_0', LP_{10}', and LP_{11}'. And we can go further, creating two new problems from either P_0, P_{10}, or P_{11}, and so on.

A general stage of the process can be described by a tree, where the nodes represent problems. (See Figure 7.15.) Each node Q that is not a leaf of the tree has two children, corresponding to the problems Q_0 and Q_1 we created from Q. (See Figure 7.15.) At any point, a lower bound for the original

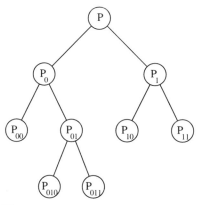

Figure 7.15. A branch and bound tree

TSP can be obtained by taking the minimum of the lower bounds we have computed for the problems corresponding to the leaves of the tree. We stop the procedure whenever this bound is greater than or equal to the cost of our best tour (in which case we have proven that our tour is optimal) or if we have established a bound that is strong enough for our given application.

Notice that while working on some problem Q, we might well discover a tour that has cost less than the cost of our current best tour. In such a case, we should let this new tour be our best tour and continue the process.

The procedure we have described is the branch and bound method. The "branching" is the process of choosing a problem Q (from the leaves of the tree) to split into Q_0 and Q_1, and the edge e that determines the split. We still need to specify how these choices are made.

Since our goal is to improve the lower bound, at any point we could choose to process a problem Q whose linear-programming value is equal to the minimum over all leaves of the branch and bound tree. This will lead to a direct improvement in the lower bound. Other strategies may be adopted (for example, a depth first search of the branch and bound tree, where we always process one of the most recently created problems), but this choice is simple and has proved to work well in practice.

Box 7.2: *Branch and Bound for General Integer Programming*
The most successful methods that have been developed for solving general integer programming problems max $\{wx : Ax \le b, x \text{ integer }\}$ are based on branch and bound techniques. Branch and bound is a general scheme that requires two main decisions: how to branch and how to bound. The standard bounding method for integer programming is to solve the linear-programming (LP) relaxation of the current subproblem. This is used almost uniformly in commercially available integer programming codes. In some cases the LP relaxations are strengthened by the addition of cutting-planes derived from the structure of the given matrix A. On the branching side, many different schemes have been proposed. A common one is to choose some variable x_i that takes on a fractional value x_i^* in the optimal solution to the current LP relaxation, and create one new subproblem with the additional constraint $x_i \le \lfloor x_i^* \rfloor$ and a second new subproblem with the additional constraint $x_i \ge \lceil x_i^* \rceil$. The rule for selecting the variable x_i often depends on user-specified priorities, the simplest being to choose the first variable x_i that takes on a fractional value. (So the user would input the problem with the variables in the order of their "importance" to the model.) For a detailed discussion of general integer programming methods see Nemhauser and Wolsey [1988].

Once we have selected a problem Q, what is a good choice for a "branching edge" e? If x^* is an optimal solution to the linear-programming relaxation for Q, then an obvious choice for e is some edge such that x_e^* is close to .5, since then both $x_e = 0$ and $x_e = 1$ will hopefully force the linear program to move far away from the current optimal solution (and cause the optimal value to increase). Along the same lines, since we want to increase the objective function, we prefer more expensive edges e over cheaper edges. So, one proposal for a branching choice is to examine all edges e such that x_e^* is in some fixed interval surrounding .5, and select that edge having the greatest cost c_e.

We now have a rudimentary branch and bound scheme for the TSP. Of the many enhancements that can be made, we would like to mention one that seems particularly useful in practice. This enhancement concerns the generation of cutting-planes. Since we are using the inequalities we described in Section 4, all of the cuts we find while processing problem Q are actually valid inequalities for all problems in the branch and bound tree. So we can save the cuts in a pool, and search the pool for violated inequalities during any of our cut generation steps. The use of a pool is especially important when our generation routines are not exact separation methods, but rather heuristics for finding cuts in a particular class. In this case, the pool not only speeds up the search, it actually gives us a chance to find cutting planes that our heuristic would miss.

Using this type of branch and bound scheme, Applegate, Bixby, Chvátal, and Cook [1995] showed that the tour for the 1173-node problem we reported in Section 2 is in fact optimal. Their branch and bound tree contained 25 nodes. Moreover, they have solved a 7397-node problem to optimality with this approach. We have, of course, skimmed over all of the implementation details, and we refer the reader to the papers of Padberg and Rinaldi [1991] and Jünger, Reinelt, and Thienel [1994] for discussions of their realizations of these techniques.

Exercises

7.20. A collection of TSPs (many coming from industrial applications) can be found in Reinelt [1991]. Develop a computer implementation of some of the techniques described in this chapter and apply your code to one of these "TSPLIB" problems.

CHAPTER 8

Matroids

8.1 MATROIDS AND THE GREEDY ALGORITHM

Let $G = (V, E)$ be a connected graph and let $c \in \mathbf{R}^E$. In Chapter 2 we saw that Kruskal's algorithm finds a maximum weight spanning tree. (There we treated minimum spanning trees, but here we discuss the equivalent maximization problem.)

Kruskal's Algorithm for Maximum Weight Spanning Trees

Set $J = \emptyset$;
While there exists $e \notin J$ with $J \cup \{e\}$ a forest
    Choose such e with c_e maximum;
    Replace J by $J \cup \{e\}$.

A slight variant (Exercise 2.6) finds a maximum weight forest. It is convenient to denote by $\mathcal{I}$ the set $\{J \subseteq E : J \text{ is a forest}\}$.

Greedy Algorithm

Set $J = \emptyset$;
While there exists $e \notin J$ with $c_e > 0$ and $J \cup \{e\} \in \mathcal{I}$
    Choose such e with c_e maximum;
    Replace J by $J \cup \{e\}$.

The above algorithm is the one we will refer to throughout this chapter as the *Greedy Algorithm*. In particular, we will be interested in applying it in other situations. Let us call the members of $\mathcal{I}$ "independent" sets. The family $\mathcal{I}$ of forests of a graph has the property that the Greedy Algorithm finds a maximum-weight independent set. We might ask whether other families $\mathcal{I}$ have this property. For example, suppose that, instead of declaring the forests of G to be the independent sets, we declare the matchings of G to be the independent sets. Then the Greedy Algorithm can be applied to the problem of finding a maximum weight matching. However, it is not hard to see that it does not always work. An example is in Figure 8.1. Here the Greedy Algorithm will find the matching $\{pq, rs\}$, but $\{ps, qr\}$ is better. So for some families $\mathcal{I}$ the Greedy Algorithm works and for some it does not.

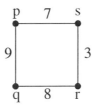

Figure 8.1. Greedy Algorithm for optimal matching

Families like forests for which the Greedy Algorithm always returns an optimal solution are called "matroids." In fact, matroids are plentiful enough that the Greedy Algorithm has a number of applications. In addition, other optimization problems on matroids (not just the optimal independent set problem) can be solved efficiently. This leads to the solution of some old problems (like bipartite matching) and some new ones (like finding an optimal directed spanning tree in a digraph).

Matroid Axioms and Examples

Let S be a finite set and $\mathcal{I}$ be a family of subsets of S, called *independent* sets. We say that $M = (S, \mathcal{I})$ is a *matroid* if the following axioms are satisfied:

(M0) $\emptyset \in \mathcal{I}$.

(M1) If $J' \subseteq J \in \mathcal{I}$, then $J' \in \mathcal{I}$.

(M2) For every $A \subseteq S$, every maximal independent subset of A has the same cardinality.

The first two axioms ("empty set is independent," "subsets of independent sets are independent") are usually easy to check, but (M2) is less easy. Let us first explain the connection between these axioms and the Greedy Algorithm for a special case. Namely, suppose that each c_e is 0 or 1. Let A be the set of elements of weight 1. Now in this case the weight $c(J)$ of an independent set J is just $|J \cap A|$. Moreover, if J is independent, then $J \cap A$ is also independent, and $c(J \cap A) = |J \cap A|$. In other words, a maximum cardinality independent subset of A will solve the problem of finding a maximum-weight independent set. The Greedy Algorithm will begin with $J = \emptyset$ and add elements of A until J is maximal independent, but then (M2) implies that J is a maximum weight independent set. So the fact that the Greedy Algorithm works for matroids when the weights are all 0 or 1 is an immediate consequence of the definition.

We need some terminology. A maximal independent subset of a set $A \subseteq S$ is called a *basis* of A. Its size (which depends only on A, by (M2)) is the *rank* of A, denoted $r(A)$. The bases of S are usually referred to simply as the bases of M. We begin by giving some examples of matroids. First, let us verify that the forests of a graph do define a matroid. So let $G = (V, E)$ be a graph, let $S = E$, and let $\mathcal{I} = \{J \subseteq S : J \text{ is a forest of } G\}$. It is obvious that $\emptyset$ is a forest, and that any subset of a forest is a forest. So we need to prove that (M2) holds. Take $A \subseteq S$ and let J be a basis of A. Then J is a maximal forest of the subgraph $G' = (V, A)$, so it consists of a spanning tree of every component of G'. This means that, if G' has connected components with vertex-sets $V_1, V_2, \ldots V_k$ then $|J| = \sum_{i=1}^{k}(|V_i| - 1)$. This does not depend on J, so (M2) is proved, and $(S, \mathcal{I})$ is a matroid. The matroid determined by G as above is called the *forest matroid* of G. Notice that the proof also established the rank function of a forest matroid. It is given by $r(A) = |V| -$ (# connected components of the graph (V, A)), for any $A \subseteq S$.

As a second example, we describe the important class of *linear* matroids. Let F be a field and N be a matrix over F with columns indexed by elements of S. Let $\mathcal{I} = \{J \subseteq S : \text{the columns indexed by elements of } J \text{ are linearly independent}\}$. We will show that $(S, \mathcal{I})$ is a matroid. By convention the empty set of vectors is linearly independent. Also every subset of a linearly independent set is linearly independent. So (M0), (M1) are true. Now consider $A \subseteq S$. For any basis J of A, the corresponding columns form a basis of the subspace generated by all the columns indexed by elements of A. All such sets have the same size, by a basic theorem of linear algebra.

Our next example is the class of *uniform* matroids, which are very simple. We have a set S and an integer k, and we define $\mathcal{I}$ to be $\{J \subseteq S : |J| \leq k\}$. Again, it is obvious that $(S, \mathcal{I})$ satisfies (M0) and (M1). Now suppose $A \subseteq S$

and let J be a basis of A. If $|A| \leq k$, then $J = A$ and if $|A| > k$, then $|J| = k$. So $|J| = \min(|A|, k)$, which does not depend on J. So (M2) is satisfied, and $(S, \mathcal{I})$ is a matroid. Its rank function is given by $r(A) = \min(|A|, k)$ for every $A \subseteq S$.

We have seen that the Greedy Algorithm does not work for optimal matching. Indeed, the example of Figure 8.1 shows that the matchings of a graph do not define the independent sets of a matroid. ($\{pq, rs\}$ and $\{qr\}$ are both bases of $\{pq, qr, rs\}$.) However, there is a different way to define a matroid based on matching. Let $G = (V, E)$ be a graph, let $S = V$, and let $\mathcal{I} = \{J \subseteq S :$ there is a matching M' of G covering all the elements of $J\}$. (Notice that M' may cover other nodes in addition.)

Proposition 8.1 $M = (S, \mathcal{I})$ *defined above is a matroid.*

Proof: Obviously the empty set is covered by the empty matching, and if a matching covers a set J, it covers any subset of J. So (M0) and (M1) are satisfied. Now consider $A \subseteq V$ and let J_1, J_2 be two bases of A, with $|J_1| < |J_2|$. Let M_1, M_2 be matchings covering the elements of J_1, and J_2, respectively. Consider the subgraph $G' = (V, M_1 \triangle M_2)$. As we have seen, the connected components of G' correspond to alternating paths and circuits, each of whose interior nodes are covered by both M_1 and M_2. Since $|J_1 \backslash J_2| < |J_2 \backslash J_1|$, there must be a component of G' that is a path joining a node $v \in J_2 \backslash J_1$ to a node $w \notin J_1$. (It could be that $w \in J_2 \backslash J_1$ or that $w \notin A$.) If we change M_1 by swapping edges on this path, the new matching M_1' covers all nodes in $J_1 \cup \{v\}$. This contradicts the choice of J_1.   ∎

Correctness of the Greedy Algorithm

There are some more examples of matroids in the exercises, and in the next section. Now we prove our earlier claim about the Greedy Algorithm. The result is due to Rado [1957]. It was rediscovered by Edmonds [1970], and he seems to be responsible for the "greedy" descriptive. First, we give a short, rather tricky proof. Later we give a polyhedral approach which also proves the theorem.

Theorem 8.2 *For any matroid* $M = (S, \mathcal{I})$ *and any* $c \in \mathbf{R}^S$, *the greedy algorithm finds a maximum-weight independent set.*

Proof: Suppose the theorem is false. Let J be the independent set delivered by the Greedy Algorithm and let J' be a maximum-weight independent set. Let $J = \{e_1, e_2, \dots, e_m\}$, where the e_i are chosen by the algorithm in that order. Then it is easy to see that $c_{e_1} \geq c_{e_2} \geq \dots \geq c_{e_m}$. Let $J' = \{q_1, q_2, \dots, q_l\}$, where $c_{q_1} \geq c_{q_2} \geq \dots \geq c_{q_l}$. Of course, J and J' may have elements in common. Choose the least index k such that $c_{q_k} > c_{e_k}$. If none exists, then we must have $l > m$; in this case let $k = m + 1$. In either case we know that the

Greedy Algorithm did not add any of $q_1, \ldots, q_{k-1}, q_k$ in step k. Since what it did choose (if anything) has smaller weight, it must be that q_i, for $1 \le i \le k$, has the property either that $q_i \in \{e_1, \ldots, e_{k-1}\}$ or that $\{e_1, \ldots, e_{k-1}, q_i\} \notin \mathcal{I}$. In other words, $\{e_1, \ldots, e_{k-1}\}$ is a basis of $A = \{e_1, \ldots, e_{k-1}, q_1, \ldots, q_k\}$. But this contradicts (M2), since $\{q_1, \ldots, q_k\}$, being a subset of J', is also an independent subset of A and is larger. ∎

We can apply the Greedy Algorithm to any $(S, \mathcal{I})$ satisfying (M0) and (M1). Such a pair is sometimes called an *independence system*. So $(E, \mathcal{I})$ is an independence system, where $\mathcal{I}$ is the set of matchings of $G = (V, E)$, but it is not in general a matroid. Suppose that we have an independence system $(S, \mathcal{I})$ that is not a matroid, and suppose that $A \subseteq S$ violates (M2). Choose c to be the characteristic vector of A and apply the Greedy Algorithm. If J' is a basis of A that is not of maximum cardinality, then it is quite possible for the Greedy Algorithm to terminate with $J = J'$. So for every independence system that is not a matroid it is possible for the Greedy Algorithm to deliver a nonoptimal independent set. If we combine this observation with Theorem 8.2, we get the following.

Theorem 8.3 *Let $(S, \mathcal{I})$ be an independence system. Then the Greedy Algorithm finds an optimal independent set for every $c \in \mathbf{R}^S$ if and only if $(S, \mathcal{I})$ is a matroid.* ∎

Matroid Algorithms

We want to claim that the Greedy Algorithm is efficient, but how do we estimate the work involved in deciding whether $J \cup \{e\} \in \mathcal{I}$? It is easy to see that the Greedy Algorithm is a polynomial-time algorithm if and only if such independence-testing questions can be answered in polynomial time. When analyzing algorithms that work on matroids we assume that each matroid is represented by a subroutine that answers questions about independence of sets in the matroid. We do not need to know the details of how the subroutine actually works, only to know that it answers such questions correctly. For this reason it may be better to imagine an *oracle*, rather than a subroutine.

We say that a matroid algorithm runs in polynomial time if the number of oracle questions asked is bounded by a polynomial in $|S|$ and the size of any other inputs, and the amount of additional computation is also bounded by a polynomial in $|S|$ and the size of any other inputs. (Other inputs might include, for example, weights of elements.) By this definition the Greedy Algorithm is a polynomial-time matroid algorithm. Notice that the definition has a nice consequence when we apply a matroid algorithm to a concrete class of matroids, for example, when we apply the Greedy Algorithm to the forest matroid of a graph. Then the input will be given to us by an input file, in this case a representation of a graph and a vector of weights. We can replace the independence oracle by an actual algorithm that tests whether a set of edges forms a forest. Since there exists a polynomial-time algorithm to do this, we immediately get a polynomial-time algorithm to find a maximum-weight forest of a graph. More generally, if we have a polynomial-time matroid algorithm and a class of matroids arising from input strings for which we have a polynomial-time algorithm to test independence, then we have a polynomial-time algorithm (in the usual sense) for the special class.

It is natural to ask whether all of this is necessary. That is, why not measure matroid algorithms in the same way we measure other algorithms, on the worst-case number of steps relative to input size? The key here is that we would need a general way to represent matroids (as we do, say, for graphs). One possibility would be just to list all the independent sets. But this would distort our measure of complexity, because $|\mathcal{I}|$ will generally be exponential in $|S|$. We need a representation for matroids that requires space polynomial in $n = |S|$. However, this is impossible! The reason is that there are so many different matroids on S, more than $2^{g(n)}$ where g is exponential in n. (See Welsh [1976].) On the other hand, there are only $2^{k+1} - 1$ binary strings of length at most k, so that if we were to have a different string for each matroid on S, some of the strings would have exponential length.

In Section 2 we make some further remarks on matroid algorithms.

Matroid Polytopes

Let $(S, \mathcal{I})$ be a matroid with rank function r, and let $J \in \mathcal{I}$. If x^0 is the characteristic vector of J, then for any $A \subseteq S$,

$$x^0(A) = |J \cap A| \leq r(A).$$

(The last inequality follows from the fact that $J \cap A$ is an independent subset of A.) It follows that every such vector x^0 is a feasible solution of the linear programming problem

$$\text{Maximize } \sum(c_e x_e : e \in S) \qquad (8.1)$$
$$\text{subject to}$$
$$x(A) \leq r(A), \text{ for all } A \subseteq S$$
$$x_e \geq 0, \text{ for all } e \in S.$$

We will prove the following strengthening of Theorem 8.2, due to Edmonds [1970].

Theorem 8.4 *Let $(S, \mathcal{I})$ be a matroid with rank function r, let $c \in \mathbf{R}^S$, and let x^0 be the characteristic vector of the set J found by the Greedy Algorithm. Then x^0 is an optimal solution of (8.1).*

Of course, by the results of Chapter 6, we get a polyhedral characterization as a consequence, which we call the Matroid Polytope Theorem.

Theorem 8.5 *Let $(S, \mathcal{I})$ be a matroid with rank function r. The convex hull of characteristic vectors of independent sets is $\{x \in \mathbf{R}^S : x \geq 0, \; x(A) \leq r(A)$ for all $A \subseteq S\}$.*∎

Proof of Theorem 8.4: We exhibit a feasible solution to the dual linear-programming problem of (8.1) that satisfies the complementary slackness conditions with x^0. The dual of (8.1) is

$$\text{Minimize} \sum (r(A) y_A : A \subseteq S) \qquad (8.2)$$
$$\text{subject to}$$
$$\sum (y_A : e \in A \subseteq S) \geq c_e, \text{ for all } e \in E$$
$$y_A \geq 0, \text{ for all } A \subseteq S.$$

The complementary slackness conditions are

$$x_e > 0 \text{ implies } \sum (y_A : e \in A \subseteq S) = c_e, \text{ for all } e \in S \qquad (8.3)$$
$$y_A > 0 \text{ implies } x(A) = r(A), \text{ for all } A \subseteq S. \qquad (8.4)$$

We construct our solution y^0 of (8.2) by a procedure called the *Dual Greedy Algorithm*. Order S as $\{e_1, \ldots, e_n\}$ so that $c_{e_1} \geq \ldots \geq c_{e_m} > 0 \geq c_{e_{m+1}} \geq \ldots \geq c_{e_n}$ and, if $c_{e_i} = c_{e_j}$, $e_i \in J$, $e_j \notin J$, then $i < j$. Define T_i, $1 \leq i \leq m$, by $T_i = \{e_1, \ldots, e_i\}$. (So $T_0 = \emptyset$.) Now put

$$y_A^0 = \begin{cases} c_{e_i} - c_{e_{i+1}}, & \text{if } A = T_i, \; 1 \leq i \leq m - 1; \\ c_{e_m}, & \text{if } A = T_m; \\ 0, & \text{for all other } A \subseteq S. \end{cases}$$

It is obvious that $y_A^0 \geq 0$ for all $A \subseteq S$. Now consider any $e_j \in S$. If $j > m$, then $\sum (y_A^0 : e_j \in A \subseteq S) = 0$. If $j \leq m$, then

$$\sum (y_A^0 : e_j \in A \subseteq S) = \sum_{i=j}^{m} y_{T_i}^0 = \sum_{i=j}^{m-1} (c_{e_i} - c_{e_{i+1}}) + c_{e_m} = c_{e_j}.$$

Hence y^0 is feasible to (8.2). Moreover, if $j \leq m$ then the corresponding constraint of (8.2) holds with equality. Since $x_e^0 > 0$ implies $e \in \{e_1, \ldots, e_m\}$, this proves (8.3). Now suppose that $y_A^0 > 0$. Then A is one of $T_1, \ldots, T_m$. So we must prove that $x^0(T_i) = r(T_i)$ for $1 \leq i \leq m$, that is, that $J \cap T_i$ is a basis of T_i for $1 \leq i \leq m$. Suppose this is not true. Then there is an element $e_k \in T_i \backslash J$ such that $(J \cap T_i) \cup \{e_k\} \in \mathcal{I}$. But in iteration k of the Greedy Algorithm, e_k was not added to J, so this is impossible, and (8.4) is proved.∎

Notice that we have proved that the Dual Greedy Algorithm finds an optimal solution of (8.1). We can also derive other useful characterizations of maximum-weight independent sets. (A different proof of the next result is the subject of Exercise 8.9.)

Corollary 8.6 *Let $M = (S, \mathcal{I})$ be a matroid, let $c \in \mathbf{R}^S$, and let $J \in \mathcal{I}$. Then J is a maximum-weight independent set with respect to c if and only if*

(a) $e \in J$ implies $c_e \geq 0$;

(b) $e \notin J$, $J \cup \{e\} \in \mathcal{I}$ imply $c_e \leq 0$;

(c) $e \notin J$, $f \in J$, $(J \cup \{e\}) \setminus \{f\} \in \mathcal{I}$ imply $c_e \leq c_f$.

Proof: It is obvious that the conditions are necessary for J to have maximum weight. Now suppose that the conditions are satisfied. We show that the characteristic vector x of J satisfies the complementary slackness conditions with the vector y^0 delivered by the Dual Greedy Algorithm. By (a), we know that (8.3) is satisfied. If (8.4) is violated, there exists i such that $c_{e_i} > c_{e_{i+1}} > 0$, and $J \cap T_i$ is not a basis of T_i. So there exists $e \in T_i \setminus J$ such that $(J \cap T_i) \cup \{e\} \in \mathcal{I}$. If $J \cup \{e\} \in \mathcal{I}$, this contradicts (b). If $J \cup \{e\} \notin \mathcal{I}$, extend $(J \cap T_i) \cup \{e\}$ to a basis J' of $J \cup \{e\}$. Then $|J'| = |J|$, so $J' = (J \cup \{e\}) \setminus \{f\}$ for some $f \notin T_i$. This contradicts (c). It follows from Theorem 8.4 that J' has maximum weight, and therefore, so does J. ∎

The next result will be useful in Section 5. It follows from Corollary 8.6 by adding a large constant to all the weights.

Corollary 8.7 *Let $M = (S, \mathcal{I})$ be a matroid, let $c \in \mathbf{R}^S$, and let J be a basis of M. Then J is a maximum weight basis with respect to c if and only if*

$$e \notin J, \ f \in J, \ (J \cup \{e\}) \setminus \{f\} \in \mathcal{I} \text{ imply } c_e \leq c_f.$$

∎

When we apply Theorem 8.5 to the forest matroid of a graph, we get a description of the convex hull of characteristic vectors of forests. With a little extra work, we can simplify the description. The proof is Exercise 8.12.

Theorem 8.8 *Let $G = (V, E)$ be a graph. The convex hull of characteristic vectors of forests of G is the set of all x satisfying*

$$x(\gamma(T)) \leq |T| - 1, \text{ for all } T, \ \emptyset \subset T \subseteq V \tag{8.5}$$
$$x_e \geq 0, \text{ for all } e \in E.$$

∎

A consequence of this result is the description of the convex hull of the spanning trees of G that is given in Theorem 2.8. (The proof is Exercise 8.13.)

Exercises

8.1. Consider the uniform matroid $M = (S, \mathcal{I})$, where $|S| = 4$ and $r(S) = 2$. Determine whether M is: (a) a forest matroid; (b) a linear matroid over the rational field; (c) a linear matroid over the binary field; (d) a matching matroid.

8.2. Let $G = (V, E)$ be a digraph, let N be its incidence matrix, and let M be the linear matroid defined by N (over the rational field, say). Prove that M is the forest matroid of the undirected graph corresponding to G. (Notice that this is the same as Exercise 2.28.)

8.3. Let N be a matrix whose columns are the seven distinct nonzero vectors of length three, each of whose components is 0 or 1. Let M be the corresponding linear matroid over the binary field. Show that M is not a linear matroid of any matrix over the rational field.

8.4. Prove that the rank function r of the matching matroid of a graph $G = (V, E)$ is given by

$$r(A) = \min(|A| + |S| - oc_A(G \setminus S) : S \subseteq V),$$

where $oc_A(G \setminus S)$ is the number of odd components of $G \subseteq S$, each of whose nodes is in A. (Hint: Modify the maximum matching algorithm to find a matching of G covering the maximum number of nodes of A, and hence get a min-max formula for this maximum.)

8.5. Let $G = (V, E)$ be a graph and let $\mathcal{I} = \{J \subseteq E$: each component of the subgraph (V, J) contains at most one circuit$\}$. Prove that $(E, \mathcal{I})$ is a matroid.

8.6. Let $G = (V, E)$ be a graph and let $\mathcal{I} = \{J \subseteq E$: each component of the subgraph (V, J) contains at most one circuit, and no even circuit$\}$. Prove that $(E, \mathcal{I})$ is a matroid.

8.7. In the previous exercise if the definition is changed to "at most one circuit, and no odd circuit," do we still get a matroid? Explain.

8.8. Let $G = (V, E)$ be a graph, and let $\mathcal{I} = \{J \subseteq E : G - J$ has the same number of connected components as $G\}$. Prove that $(E, \mathcal{I})$ is a matroid.

8.9. Prove Corollary 8.6 directly from Theorem 8.3.

8.10. Show that Corollary 8.6 is not generally true when $\mathcal{I}$ is the set of matchings of a graph.

8.11. Give an algorithm to find a maximum weight basis of S.

8.12. Prove Theorem 8.8.

8.13. Prove that the intersection of the polytope of independent sets of a matroid with the hyperplane $\{x \in \mathbf{R}^S : x(S) = r(S)\}$ is the convex hull of characteristic vectors of bases of the matroid. Hence prove Theorem 2.8.

8.14. Show that the Greedy Algorithm can be implemented so that it uses at most $n = |S|$ independence tests.

8.15. Show that, if G has no loops, then the forest polytope of G has full dimension. Also show that $x(\gamma(T)) \leq |T| - 1$ is facet-inducing if and only if $G(T, \gamma(T))$ is connected and has no cut node.

8.16. Use the proof technique introduced in Section 6.3 to give a proof of Theorem 8.5.

8.2   MATROIDS: PROPERTIES, AXIOMS, CONSTRUCTIONS

In this section we discuss a few of the basic results and examples of matroid theory. The scope is limited mainly to things we will use later. The interested reader is referred to Oxley [1992] for a more extensive treatment.

A *circuit* of an independence system $(S, \mathcal{I})$ is a minimal dependent set (that is, a minimal set that is not independent). It is easy to see that the circuits of the forest matroid of G correspond exactly to the circuits of G, which explains the name. One simple property of forests, which we used in Chapter 2, is that adding an edge to a forest produces at most one circuit. This observation can be generalized to matroids.

Theorem 8.9 *Let $(S, \mathcal{I})$ be a matroid, let $J \in \mathcal{I}$, and let $e \in S$. Then $J \cup \{e\}$ contains at most one circuit.*

Proof: Suppose that $J \cup \{e\}$ contains two circuits C_1, C_2. Choose J as small as possible. Then $J \cup \{e\} = C_1 \cup C_2$. There exist $a \in C_1 \backslash C_2$, $b \in C_2 \backslash C_1$. Then $J' = (C_1 \cup C_2) \backslash \{a, b\} \in \mathcal{I}$. (Otherwise, J' contains a circuit C, and then $(J \backslash \{a\}) \cup \{e\}$ contains circuits C and C_2, contradicting the choice of J.) Then J' is a basis of $C_1 \cup C_2$, but so is J, and they have different cardinalities, a contradiction. ∎

The circuits of a matroid completely determine the matroid, because a set is independent if and only if it contains no circuit. Similarly, the rank function r determines the matroid, since $A \in \mathcal{I}$ if and only if $r(A) = |A|$. As yet another example, the bases of M determine M, since the independent sets are simply the subsets of the bases. So, while we have used properties (M0), (M1), (M2) of independent sets to define the concept of matroid, we could instead do so using properties of circuits, or of the rank function, or of bases. We illustrate this idea for circuits; other examples are in the exercises.

Theorem 8.10 *A set $\mathcal{C}$ of subsets of S is the set of circuits of a matroid if and only if*

(i) $\emptyset \notin \mathcal{C}$.

(ii) If $C_1, C_2 \in \mathcal{C}$ and $C_1 \subseteq C_2$, then $C_1 = C_2$.

(iii) If $C_1, C_2 \in \mathcal{C}$, $C_1 \neq C_2$, and $e \in C_1 \cup C_2$, then there exists $C \in \mathcal{C}$, $C \subseteq (C_1 \cup C_2) \backslash \{e\}$.

Proof: Suppose that $\mathcal{C}$ is the circuit family of the matroid $M = (S, \mathcal{I})$. Then the truth of (i) and (ii) are obvious. Now suppose that (iii) is violated. Then $J = (C_1 \cup C_2) \backslash \{e\} \in \mathcal{I}$. But then $J \cup \{e\}$ contains two circuits, which is impossible, by Theorem 8.9.

Now suppose that $\mathcal{C}$ satisfies (i), (ii), (iii) and let $\mathcal{I} = \{J \subseteq S : J \not\supseteq C$ for all $C \in \mathcal{C}\}$. We will show that $(S, \mathcal{I})$ is a matroid. It is obvious that (M0) and (M1) are satisfied. Suppose (M2) is not. Let J_1 be a basis of $A \subseteq S$ and let $J_2 \subseteq A$ with $J_2 \in \mathcal{I}$ and $|J_1| < |J_2|$. We may choose J_2 to be a basis of A. We may choose J_1, J_2 so that $J_1 \cap J_2$ is as large as possible. There exists $e \in J_1 \backslash J_2$, for otherwise J_1 is not maximal. Now, $J_2 \cup \{e\}$ contains some $C \in \mathcal{C}$, and there is just one such C. (If there were a second, say C', then $(C \cup C') \backslash \{e\} \in \mathcal{I}$, contradicting (iii).) Now $C \not\subseteq J_1$, since $J_1 \in \mathcal{I}$, so there is $f \in C \backslash J_1$. Moreover, $J_3 = (J_2 \cup \{e\}) \backslash \{f\} \in \mathcal{I}$, since C is the only member of $\mathcal{C}$ contained in $J_2 \cup \{e\}$. But $|J_3| > |J_1|$ and $|J_3 \cap J_1| > |J_2 \cap J_1|$, contradicting the choice of J_1, J_2. ∎

Matroid Oracles

We have seen that there are several equivalent ways to define matroids. It is natural, then, to think of the independence oracle as just one of many equivalent methods to represent matroids as inputs to matroid algorithms. In fact, these different oracles have quite different properties. As an example, suppose that we use a representation motivated by the circuit definition, so an algorithm asks questions of the form "Is A a circuit?" for a subset A of S. The problems that could be solved by polynomial-time matroid algorithms under this definition are different. To see this, consider the problem of deciding whether S is independent, which is trivial for the independence oracle. Notice that, for every nonempty subset A of S, there is a matroid M_A having A as its only circuit. To decide whether S is independent, an algorithm using a circuit oracle would have to ask a question for every nonempty subset; otherwise it would not be able to distinguish whether the matroid was some M_A, for which the answer would be "no," or was the matroid in which S is independent. Therefore, this problem cannot be solved in polynomial time using the circuit oracle.

Arguments like the one above can be used also to show that certain natural problems cannot be solved in polynomial time with the independence oracle. An example is the problem of deciding whether a given matroid is uniform. See Exercise 8.25. Since the different oracles one may think of are not equivalent, one may now ask why we chose the independence oracle. The answer is that we want to choose one that is powerful enough to do something with, but not so strong that it is not available for classes of matroids arising in practice. We have seen that the circuit oracle is too weak. An example of one that is too strong is to ask for a largest circuit contained in a given subset. (Why?) The independence oracle, like the baby bear's porridge, is just right.

Matroid Constructions

Another important source of examples of matroids is the modification or combination of other matroids. The simplest way to get a new matroid from a given one is to delete some elements. Namely, given $M = (S, \mathcal{I})$ and $B \subseteq S$, define $M' = (S', \mathcal{I}')$ by $S' = S \backslash B$ and $\mathcal{I}' = \{J \subseteq S' : J \in \mathcal{I}\}$. That is, we delete the elements in B, and subsets of remaining elements are independent if and only if they were before. It is easy to see that $(S', \mathcal{I}')$ satisfies (M0), (M1), (M2). We write $M' = M \backslash B$. This operation is very natural. In forest matroids it corresponds to deleting some edges of a graph, and in linear matroids to deleting some columns of a matrix. So in fact we are saying that deleting a subset from a forest matroid gives a forest matroid, and deleting a subset from a linear matroid gives a linear matroid.

If we apply deletion to a matching matroid, however, we get some new matroids. The most obvious instance is to begin with the matching matroid of a graph consisting of two nodes and a single edge, and then delete one element. The resulting matroid cannot be a matching matroid, because any basis of a matching matroid is the set of nodes covered by a maximum matching in some graph, and so has even cardinality. An important class of matroids, the *transversal* matroids, arises from matching matroids via deletion, as follows. Begin with a bipartite graph G with bipartition $\{S, T\}$ and delete T from the matching matroid of G. So $J \subseteq S$ is independent if and only if there is a matching of G covering all nodes in J. We have seen that not every transversal matroid is a matching matroid. Actually, the opposite is true—Edmonds and Fulkerson [1965] proved that every matching matroid is the transversal matroid of some bipartite graph.

Another simple construction is *truncation*. Let $M = (S, \mathcal{I})$ be a matroid, and let k be a nonnegative integer. Define $\mathcal{I}'$ to be $\{J \in \mathcal{I} : |J| \leq k\}$. We claim that $M' = (S, \mathcal{I}')$ is a matroid. It is obvious that M' satisfies (M0) and (M1). Now let A be a subset of S and J a maximal subset of A that is in $\mathcal{I}'$. Then if $k \geq r(A)$, then J is a basis of A in M, so $|J| = r(A)$. If $k < r(A)$, then clearly $|J| = k$. In either case, the size of J depends only on A, not on the choice of J, so M' satisfies (M2), and so is a matroid. Moreover, we have proved that its rank function r' satisfies $r'(A) = \min\{r(A), k\}$ for all subsets A of S.

Yet another simple construction is the *disjoint union* of two (or more) matroids. Let $M_1 = (S_1, \mathcal{I}_1)$, $M_2 = (S_2, \mathcal{I}_2)$ be matroids, where $S_1 \cap S_2 = \emptyset$. Their disjoint union is $M = M_1 \oplus M_2 = (S, \mathcal{I})$, where $S = S_1 \cup S_2$ and $\mathcal{I} = \{J_1 \cup J_2 : J_1 \in \mathcal{I}_1, J_2 \in \mathcal{I}_2\}$. Why is M a matroid? Again (M0) and (M1) are obviously satisfied. Consider $A \subseteq S$ and let J be an M-basis of A. Then $J \cap S_1 \in \mathcal{I}_1$. If $J \cap S_1$ is not an M_1-basis of $A \cap S_1$, then there exists $e \in (A \cap S_1) \backslash J$ such that $(J \cap S_1) \cup \{e\} \in \mathcal{I}_1$. But then $J \cup \{e\} \in \mathcal{I}$, contradicting the choice of J. So $|J \cap S_1| = r_1(A \cap S_1)$, where r_1 is the rank function of M_1. Similarly, $|J \cap S_2| = r_2(A \cap S_2)$, so $|J| = r_1(A \cap S_1) + r_2(A \cap S_2)$, which does not depend on J, proving (M2). So M is a matroid. In fact, we have also proved that its rank function r is given by $r(A) = r_1(A \cap S_1) + r_2(A \cap S_2)$ for all $A \subseteq S$.

A *partition matroid* is defined as follows. Suppose that we have a partition $\{S_1, S_2, \ldots, S_m\}$ of S. Define $\mathcal{I}$ by $\{J \subseteq S : |J \cap S_i| \leq 1 \text{ for } 1 \leq i \leq m\}$. It is now easy to see that $M = (S, \mathcal{I})$ is a matroid. Namely, take the rank-one uniform matroid M_i on S_i, $1 \leq i \leq m$; then M is the disjoint union of the M_i. Moreover, the rank function of the partition matroid is given by $r(A) = |\{i : A \cap S_i \neq \emptyset\}|$ for all $A \subseteq S$.

Next we describe an important construction that generalizes the notion of contracting edges in a graph. Let $M = (S, \mathcal{I})$ be a matroid and $B \subseteq S$. Choose a basis J of B and define $M' = (S', \mathcal{I}')$ by $S' = S \backslash B$ and $\mathcal{I}' = \{J' \subseteq S' : J' \cup J \in \mathcal{I}\}$. (Suppose that M is the forest matroid of G and we contract an edge e from G to form G'. Then the forests of G' are precisely the sets that form a forest with e in G. So in this special case, M' is the forest matroid of G'.)

Theorem 8.11 *M' defined above is a matroid that does not depend on J, and its rank function r' is given by $r'(A) = r(A \cup B) - r(B)$, for all $A \subseteq S \backslash B$.*

Proof: Since $\emptyset \in \mathcal{I}$, and since $J'' \subseteq J'$ and $J' \cup J \in \mathcal{I}$ implies $J'' \cup J \in \mathcal{I}$, therefore $(S', \mathcal{I}')$ satisfies (M0) and (M1). Now suppose $A \subseteq S \backslash B$ and J' is a maximal member of $\mathcal{I}'$ contained in A. We claim $J \cup J'$ is an M-basis of $A \cup B$. If not, there is $e \in A \cup B \backslash (J \cup J')$ such that $(J \cup J') \cup \{e\} \in \mathcal{I}$. If $e \in A$, then $J' \cup \{e\} \in \mathcal{I}'$, a contradiction. If $e \in B$, then $J \cup \{e\} \in \mathcal{I}$, again a contradiction. Therefore $J \cup J'$ is an M-basis of $A \cup B$, so $|J'| = r(A \cup B) - |J| = r(A \cup B) - r(B)$. This does not depend on J', so (M2) is satisfied, and therefore M' is a matroid. Moreover, we have proved that its rank function r' satisfies $r'(A) = r(A \cup B) - r(B)$ for all $A \subseteq S \backslash B$. Finally, it is clear from this property of r' that M' does not depend on the choice of J. ∎

We say that the matroid M' is obtained from M by *contracting B* and we denote M' by M/B. Here is an application of contraction and direct sum, which will be useful later. Let $B \subseteq S$ and let $M' = (M/B) \oplus (M \backslash \overline{B})$. Then M' is a matroid on S and its bases are exactly those bases of M that intersect B in a basis of B. We extend this idea further in the following result. It will be used in Section 8.5, and its proof is part of Exercise 8.23. $T_1 \subseteq T_2 \subseteq \cdots \subseteq T_\ell$ of subsets of S. For convenience, we put $T_0 = \emptyset$ and $T_{\ell+1} = S$. This result will be used in Section 8.5; its proof is part of Exercise 8.28.

Proposition 8.12 *Let $\emptyset = T_0 \subseteq T_1 \subseteq T_2 \subseteq \cdots \subseteq T_{\ell+1} = S$. The bases of T_ℓ in M that intersect T_i in a basis of T_i for $1 \leq i \leq \ell$ are the bases of T_ℓ in the matroid $N = N_0 \oplus N_1 \oplus \cdots \oplus N_\ell$, where for each i, $N_i = (M/T_i) \backslash \overline{T}_{i+1}$.* ∎

Another important construction, the "dual," is inspired by Exercise 8.8, which described another matroid determined by a graph G. Its independent sets are the sets of edges whose deletion does not increase the number of connected components. Exercise 8.34 shows a connection with planar duality. If $M = (S, \mathcal{I})$ is the forest matroid of G and r is its rank function, then this second matroid $M^* = (S, \mathcal{I}^*)$ can be defined by $\mathcal{I}^* = \{J \subseteq S : r(S \backslash J) = r(S)\}$. For any matroid M we define its *dual* M^* in this way.

Theorem 8.13 *For any matroid M, M^* is also a matroid.*

Proof: It is clear that $\emptyset \in \mathcal{I}^*$, and if $J' \subseteq J$, then $r(S) \geq r(S \backslash J') \geq r(S \backslash J)$, so M^* satisfies (M0) and (M1). Now let $A \subseteq S$, and let J be a maximal member of $\mathcal{I}^*$ contained in A. Choose an M-basis B of $S \backslash A$ and extend it to an M-basis B' of $S \backslash J$. (Then $|B'| = r(S)$.) We claim that $B' \supseteq A \backslash J$. If this is not true, there is some $e \in (A \backslash J) \backslash B'$. But then $J \cup \{e\} \in \mathcal{I}^*$, since its complement contains a basis of S. This is a contradiction. So $r(S) = |B'| = |A \backslash J| + |B|$. Therefore $|J| = |A| + r(S \backslash A) - r(S)$, which does not depend on J, so M^* is a matroid. ∎

Notice that we have also established that the rank function r^* of M^* is given by $r^*(A) = |A| + r(S \backslash A) - r(S)$, for all $A \subseteq S$. Exercise 8.33 shows a connection among deletion, contraction, and duality.

Exercises

8.17. Let G_1, G_2 be graphs and let M be the disjoint union of their forest matroids. Show that M is the forest matroid of a graph G. Can G be connected?

8.18. Show that every uniform matroid is a transversal matroid.

8.19. Show that the forest matroid of a complete graph on four nodes is not a transversal matroid.

8.20. Show that a set J of edges is independent in the matroid defined in Exercise 8.5 if and only if there is a way to direct the edges of the subgraph (V, J) so that each node is the head of at most one arc. Use this fact to show that the matroid is a transversal matroid.

8.21. Prove that the rank function of a matroid is *submodular:* $r(A) + r(B) \geq r(A \cup B) + r(A \cap B)$ for all $A, B \subseteq S$.

8.22. Prove that r is the rank function of a matroid if and only if it is submodular and also satisfies $r(\emptyset) = 0$ and $r(A \cup \{e\}) = r(A)$ or $r(A) + 1$ for all $A \subseteq S$, $e \in S$.

8.23. Prove that $\mathcal{B}$ is the basis family of a matroid if and only if for all $B_1, B_2 \in \mathcal{B}$ and $a \in B_1 \backslash B_2$ there exists $b \in B_2 \backslash B_1$ such that $(B_2 \cup \{a\}) \backslash \{b\} \in \mathcal{B}$.

8.24. Compare the strength of the independence oracle with the basis oracle, that is, one that tells whether a given set is a basis of S.

8.25. Show that there is no polynomial-time algorithm to decide whether a given matroid is uniform. (Hint: Given a uniform matroid $M = (S, \mathcal{I})$ and any basis B of M, $\mathcal{I} \backslash \{B\}$ is the set of independent sets of a matroid.)

8.26. Let M_1, M_2 be linear matroids over the same field F. Show that $M_1 \oplus M_2$ is also a linear matroid. What if M_1, M_2 arise from different fields?

8.27. Let $M = (S, \mathcal{I})$ be a matroid. For each $e \in S$, let d_e be a nonnegative integer, and let S_e be a set of cardinality d_e. Let $S' = \cup(S_e : e \in S)$ and let $\mathcal{I}' = \{A \subseteq S' : |A \cap S_e| \leq 1 \text{ for each } e \in S, \text{ and } \{e : A \cap S_e \neq \emptyset\} \in \mathcal{I}\}$. Prove that $(S', \mathcal{I}')$ is a matroid. Find its rank function.

8.28. Prove Proposition 8.12. Use it to prove that the family of c-optimal bases of a matroid forms the bases of another matroid.

8.29. Show for any matroid M, that $(M^*)^* = M$.

8.30. Show that the following algorithm finds a maximum-weight basis of S. Begin with $J = S$. While $J \notin \mathcal{I}$ find $e \in J$ with $r(J \backslash \{e\}) = r(S)$ and such that c_e is minimum, and replace J by $J \backslash \{e\}$. (Hint: Use duality.)

8.31. Show that for any matroid M, the polytope $\{x \in \mathbf{R}_+^S : x(S) = r(S), x(A) \geq r(S) - r(S \backslash A), \text{ for all } A \subseteq S\}$ is the convex hull of the bases of M. (Hint: Use Exercise 8.13 and duality.)

8.32. Let N be a matrix and $M = (S, \mathcal{I})$ the linear matroid determined by N. Assume that N is of the form $[I|P]$ for some P. Let $N' = [-P^T|I]$. Prove that N' determines the matroid M^*.

8.33. Let $M = (S, \mathcal{I})$ be a matroid and let $A \subseteq S$. Show that $(M \backslash A)^* = M^*/A$. (Hint: Use the rank formulas.)

8.34. Let G be a connected planar graph, and let M be its forest matroid. Prove that M^* is the forest matroid of the planar dual of G. You may use Proposition 5.27.

8.3 MATROID INTERSECTION

We are given matroids M_1, M_2 on the same set S. We want to find a maximum weight (or maximum cardinality) common independent set. Obviously this problem generalizes the optimal independent set problem, taking $M_1 = M_2$. There are many other applications. We point out just two of them here.

First, consider the optimal matching problem for bipartite graphs. Where $\{V_1, V_2\}$ is a bipartition of $G = (V, E)$, take $S = E$ and $\mathcal{I}_i = \{J \subseteq S:$ each $v \in V_i$ is incident with at most one element of $J\}$. Notice that both $M_1 = (S, I_1)$ and $M_2 = (S, I_2)$ are partition matroids. It is easy to see that the common independent sets are exactly the matchings of G. (For example, the sets $S_v = \delta(v)$ for each $v \in V_1$ form a partition of S that determines the partition matroid M_1.) Therefore, the maximum weight matching problem in a bipartite graph is nothing but the maximum weight matroid intersection problem where both the matroids are partition matroids.

Let $G = (V, E)$ be a digraph. A *branching* of G is a forest in which each node has indegree at most one. We would like to find a branching of maximum weight in G. Suppose that we take $S = E$ and let M_1 be the forest matroid of G (more precisely, of the underlying undirected graph). Now let $\mathcal{I}_2$ be

$$\{J \subseteq E : \text{each node is the head of at most one element of } J\}.$$

Then $M_2 = (S, \mathcal{I}_2)$ is a partition matroid, and the common independent sets of M_1 and M_2 are exactly the arc sets of branchings of G. Therefore the problem of finding a branching of maximum weight in G is a matroid intersection problem.

In this section we will treat the problem of finding a maximum-cardinality common independent subset of two matroids. This problem is already very interesting and has numerous applications, of course including maximum cardinality bipartite matching.

The Matroid Intersection Theorem

There is a beautiful min-max theorem that characterizes the maximum cardinality of a common independent set. It is suggested by a natural upper bound, as follows. Let $J \in \mathcal{I}_1 \cap \mathcal{I}_2$ and let $A \subseteq S$. Then $J \cap A \in \mathcal{I}_1$ and $J \cap \overline{A} \in \mathcal{I}_2$, so

$$|J| = |J \cap A| + |J \cap \overline{A}| \leq r_1(A) + r_2(\overline{A}).$$

Therefore, if we can find J and A satisfying this relationship with equality, then we know that this J is maximum. In fact, such a pair (J, A) always exists; this is the content of the Matroid Intersection Theorem of Edmonds [1970].

Theorem 8.14 *For matroids M_1, M_2 on S,*

$$\max\{|J| : J \in \mathcal{I}_1 \cap \mathcal{I}_2\} = \min\{r_1(A) + r_2(\overline{A}) : A \subseteq S\}.$$

Before proving the theorem we show that König's Theorem 3.14 is an easy consequence. So we want to show that in a bipartite graph G, there exist a matching and a cover of equal size. It follows from the Matroid Intersection Theorem and the formula for the rank function of a partition matroid that there exist a matching M and a set $A \subseteq S$ such that

$$|M| = |\{v \in V_1 : A \cap \delta(v) \neq \emptyset\}| + |\{v \in V_2 : \overline{A} \cap \delta(v) \neq \emptyset\}|.$$

It is easy to see that the set of nodes counted in the right-hand side of the above formula form a cover of G, as required.

The Matroid Intersection Theorem follows from the algorithm described later, but we first give a short nonconstructive proof.

Proof of the Matroid Intersection Theorem: We have already observed that max $\leq$ min. The proof of the harder part is by induction on $|S|$. Let k be the minimum of $r_1(A) + r_2(\overline{A})$, and choose $e \in S$ with $\{e\} \in \mathcal{I}_1 \cap \mathcal{I}_2$. (If none exists, $k = 0$. This means that for every $e \in E$, $\{e\} \notin \mathcal{I}_1$ or $\{e\} \notin \mathcal{I}_2$, so we can take A to be $\{e \in E : r_1(\{e\}) = 0\}$, and we are done.) If the minimum of $r_1(A) + r_2(S' \backslash A)$ over subsets A of $S' = S \backslash \{e\}$ is k, then we are finished, by induction. If M_i' denotes $M_i / \{e\}$ and the minimum of $r_1'(B) + r_2'(S' \backslash B)$ over subsets B of S' is at least $k - 1$, then induction gives a common independent set of M_1', M_2' of size $k - 1$; adding e to it gives the desired J. We conclude that, if there is no common independent set of size k, then there exist subsets A, B of S' such that

$$r_1(A) + r_2(S' \backslash A) \leq k - 1$$

and

$$r_1(B \cup \{e\}) - 1 + r_2\left((S' \backslash B) \cup \{e\}\right) - 1 \leq k - 2.$$

Adding and applying submodularity, we have

$$r_1(A \cup B \cup \{e\}) + r_1(A \cap B) + r_2\left(S \backslash (A \cap B)\right) + r_2\left(S \backslash (A \cup B \cup \{e\})\right) \leq 2k - 1;$$

it follows that the sum of the middle two terms, or the sum of the other two terms, is at most $k - 1$, a contradiction. $\blacksquare$

We point out that the above proof gives no hint of how to find the desired subsets J and A efficiently.

The Matroid Intersection Algorithm

The Matroid Intersection Algorithm uses an augmenting path approach that generalizes a version of the matching algorithm for bipartite graphs. (In fact, the version we have in mind is the one that grows a forest of trees rooted only at the exposed nodes of V_2. This works because, in a bipartite graph, any augmenting path will join a node in V_1 to a node in V_2.) It maintains $J \in \mathcal{I}_1 \cap \mathcal{I}_2$, at each step either finding a larger such J, or finding A giving equality with J in the min-max formula, proving that J is maximum. It is not too hard to see that in the bipartite matching application the edges of an augmenting path form a sequence $e_1, f_1, e_2, f_2, \ldots, e_m, f_m, e_{m+1}$ such that

- $e_i \notin J$, $1 \leq i \leq m + 1$;
- $f_i \in J$, $1 \leq i \leq m$;
- $J \cup \{e_1\} \in \mathcal{I}_2$;
- $J \cup \{e_{m+1}\} \in \mathcal{I}_1$;
- $(J \cup \{e_i\}) \backslash \{f_i\} \in \mathcal{I}_1$, $1 \leq i \leq m$;
- $(J \cup \{e_{i+1}\}) \backslash \{f_i\} \in \mathcal{I}_2$, $1 \leq i \leq m$.

This suggests requiring the same conditions of "augmenting sequences" in the general case. In order to make the search for them more convenient, we define an auxiliary digraph $G = G(M_1, M_2, J)$. It has node-set $S \cup \{r, s\}$, and arcs:

- es for every $e \in S \backslash J$ such that $J \cup \{e\} \in \mathcal{I}_1$;
- re for every $e \in S \backslash J$ such that $J \cup \{e\} \in \mathcal{I}_2$;
- ef for every $e \in S \backslash J, f \in J$ such that $J \cup \{e\} \notin \mathcal{I}_1, (J \cup \{e\}) \backslash \{f\} \in \mathcal{I}_1$;
- fe for every $e \in S \backslash J, f \in J$ such that $J \cup \{e\} \notin \mathcal{I}_2, (J \cup \{e\}) \backslash \{f\} \in \mathcal{I}_2$.

Notice that if es is an arc, then e is the tail of no other arc and if re is an arc, then e is the head of no other arc. Similarly, if $ef \in E(G)$ with $f \in J$, $e \notin J$, then $es \notin E(G)$ and if $fe \in E(G)$ with $e \notin J$, then $re \notin E(G)$. Notice also that the arcs of this digraph can be divided into two types, those that are determined by M_1 and those that are determined by M_2. We will call them M_1-*arcs* and M_2-*arcs*, respectively. It is also worth noting that the construction is not symmetric in M_1 and M_2, but if we reverse the direction of all arcs of G and exchange r with s, then this has the effect of exchanging M_1 with M_2. In Figures 8.2 and 8.3 we show examples of the construction of auxiliary digraphs. In each case we show graphs G_1, G_2 whose forest matroids are M_1 and M_2, respectively, and we indicate on the graphs the common independent set J. Then we show the digraph $G(M_1, M_2, J)$. Notice that in the first case there is no (r, s)-dipath in the auxiliary digraph, and in the second case there is. Moreover, it is fairly easy to see that in the first case J is a maximum-cardinality common independent set, whereas in the second case it is not. These observations generalize to the following "augmenting path" theorem, which is the basis for the algorithm.

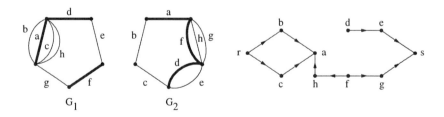

Figure 8.2. An auxiliary digraph where J is maximum

Theorem 8.15 *(Augmenting Path Theorem)*

    (a) If there exists no (r, s)-dipath in G, then J is maximum; in fact, if $A \subseteq S$ and $\delta(A \cup \{r\}) = \emptyset$, then $|J| = r_1(A) + r_2(\overline{A})$.

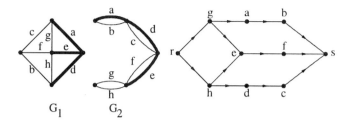

Figure 8.3. An auxiliary digraph where J is not maximum

(b) *If there exists an (r, s)-dipath in G, then J is not maximum; in fact, if $r, e_1, f_1, \ldots, e_m, f_m, e_{m+1}, s$ is the node-sequence of a chordless (r, s)-dipath, then $J \triangle \{e_1, f_1, \ldots, e_m, f_m, e_{m+1}\} \in \mathcal{I}_1 \cap \mathcal{I}_2$.*

The conclusion of part (a) of the Augmenting Path Theorem can be illustrated by the example of Figure 8.2. Here the set $A = \{a, b, c\}$ has the property that no arc leaves $A \cup \{r\}$, and one easily checks that $r_1(A) + r_2(S \backslash A) = 3$, proving that J is maximum. (Is A the only set with this property?) Part (b) of the theorem has a surprising extra condition attached to it—that the dipath be chordless. (A *chord* of a dipath P having node-sequence $v_0, v_1, \ldots, v_k$ is an arc of the form $v_i v_j$ where $i < j - 1$.) Although this requirement is not needed in the special case of bipartite matching, it is indeed needed in the general case—see Exercise 8.37. Note that a simple way to find a chordless dipath is to find a shortest one. Here is a statement of the algorithm implicit in Theorem 8.15.

Matroid Intersection Algorithm

Set $J = \emptyset$;
Loop
    Construct $G = G(M_1, M_2, J)$;
    If there is an (r, s)-dipath P in G
        Let $r, e_1, f_1, \ldots, e_m, f_m, e_{m+1}, s$ be the
        node-sequence of a chordless (r, s)-dipath P;
        Replace J by $J \triangle \{e_1, f_1, \ldots, e_m, f_m, e_{m+1}\}$;
    Else
        Let $A = \{e \in S : \text{there is an} (r, e)\text{-dipath in } G\}$;
        Stop.

Here is a proof of (a) in the Augmenting Path Theorem 8.15. Consider $e \in A \backslash J$. Since es is not an arc, $J \cup \{e\}$ contains an M_1-circuit C. Since there is no arc ef with $f \in S\backslash A$, we have $C \subseteq (A \cap J) \cup \{e\}$. Therefore, $J \cap A$ is an M_1-basis of A. Similarly, $J \cap \overline{A}$ is an M_2-basis of $\overline{A}$. Hence

$$|J| = |J \cap A| + |J \cap \overline{A}| = r_1(A) + r_2(\overline{A}).$$

To prove part (b) of Theorem 8.15, we need the following technical lemma. The *span* in a matroid $M = (S, \mathcal{I})$ of a set $A \subseteq S$ is the set $A \cup \{e \in S\backslash A :$ $e \in C \subseteq A \cup \{e\}$ for some circuit $C\}$, which we denote by span(A). (In a linear matroid span(A) corresponds to the columns in the subspace generated by the columns indexed by elements of A.)

Lemma 8.16 *Let $M = (S, \mathcal{I})$ be a matroid, let $J \in \mathcal{I}$, and let*

$$p_1, q_1, \ldots, p_k, q_k$$

be a sequence of distinct elements of S such that

(a) $p_i \notin J$, $q_i \in J$ for $1 \le i \le k$;

(b) $J \cup \{p_i\} \notin \mathcal{I}$, $(J \cup \{p_i\})\backslash\{q_i\} \in \mathcal{I}$ for $1 \le i \le k$;

(c) $(J \cup \{p_i\})\backslash\{q_j\} \notin \mathcal{I}$ for $1 \le i < j \le k$.

Let $J' = J \triangle \{p_1, q_1, \ldots, p_k, q_k\}$. Then $J' \in \mathcal{I}$ and span(J') = span(J).

Before proving the lemma, let us show how it allows us to prove part (b) of the Augmenting Path Theorem. With respect to the matroid M_1, the set J, and the sequence $e_1, f_1, \ldots, e_m, f_m$, the conditions of the lemma are satisfied. (Namely, it follows from the fact that there is no chord e_is that (b) is satisfied, and from the fact that there is no chord e_if_j for $i < j$ that (c) is satisfied.) Thus we can conclude that $J' = (J \cup \{e_1, \ldots, e_m\})\backslash\{f_1, \ldots, f_m\} \in \mathcal{I}_1$ and that $\text{span}_1(J') = \text{span}_1(J)$. Now we know that $e_{m+1} \notin \text{span}_1(J)$, and therefore, $e_{m+1} \notin \text{span}_1(J')$, so $J' \cup \{e_{m+1}\} \in \mathcal{I}_1$. Now we can apply the previously observed symmetry between M_1 and M_2 to conclude that the same set is in $\mathcal{I}_2$.

Proof of Lemma 8.16: The proof is by induction on k. Let us first check the case where $k = 1$. First, notice that $J' \in \mathcal{I}$, by (b). Now suppose that $g \in \text{span}(J)$, but $g \notin \text{span}(J')$. Then J is a basis of $J \cup \{p_1, g\}$, but $J' \cup \{g\}$ is another basis of larger size, a contradiction. Similarly, if $g \in \text{span}(J')$, but $g \notin \text{span}(J)$, then J' is a basis of $J' \cup \{q_1, g\}$, but $J \cup \{g\}$ is another basis of larger size, a contradiction.

Now suppose that $k \ge 2$ and that the result is true for smaller values. We use this assumption in two ways. First, we apply it to the set J and the sequence $p_1, q_1, \ldots, p_{k-1}, q_{k-1}$. Second, we apply it to the set $J\backslash\{q_k\}$ and the same sequence. We conclude that

(i) $((J' \cup \{q_k\})\backslash\{p_k\}) \in \mathcal{I}$ and span($(J' \cup \{q_k\})\backslash\{p_k\}$) = span($J$);

(ii) $J'\backslash\{p_k\} \in \mathcal{I}$ and $\mathrm{span}(J'\backslash\{p_k\}) = \mathrm{span}(J\backslash\{q_k\})$.

Since $p_k \notin \mathrm{span}(J\backslash\{q_k\})$, therefore $p_k \notin \mathrm{span}(J'\backslash\{p_k\})$, so $(J'\backslash\{p_k\})\cup\{p_k\} = J' \in \mathcal{I}$. To prove that $\mathrm{span}(J') = \mathrm{span}(J)$, it is enough to show that $\mathrm{span}(J') = \mathrm{span}((J'\cup\{q_k\})\backslash\{p_k\})$. We apply the case $k = 1$ to J' and q_k, p_k; to do this, we need to know that $J'\cup\{q_k\} \notin \mathcal{I}$ and that $(J'\cup\{q_k\})\backslash\{p_k\} \in \mathcal{I}$. Both facts follow from (i). ∎

Since the Matroid Intersection Algorithm will terminate after at most $n = |S|$ augmentations, the dipath or set A as in Theorem 8.15 can be found with standard methods, and each auxiliary digraph can be constructed with $O(n^2)$ independence tests, it follows that the Matroid Intersection Algorithm is a polynomial-time matroid algorithm. For concrete classes of matroids, one can usually obtain better bounds than arise from simply multiplying the number of oracle calls by the oracle complexity. For example, for matroids arising from two matrices each having at most n rows, using Gaussian elimination for each independence test, one obtains a total complexity of $O(n^6)$. But there is a straightforward implementation of the algorithm in time $O(n^4)$. (See Exercise 8.41.)

Exercises

8.35. Show that the problem of finding a minimum-cost simple directed (r, s)-dipath through all the nodes of a digraph G can be reduced to the problem of finding a maximum-weight common independent set of three matroids.

8.36. For the matching problem of Figure 3.5 and the matching illustrated there, construct the auxiliary digraph for the matroid intersection algorithm and use it to find a minimum cover.

8.37. For the matroid intersection problem on the matroids defined by the graphs of Figure 8.4, and the set J indicated there, construct the auxiliary digraph. Find an (r, s)-dipath for which the corresponding augmentation does *not* yield a common independent set.

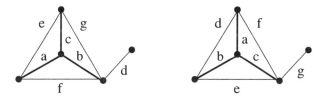

Figure 8.4. Necessity of chordlessness

8.38. Let r_1, r_2 be rank functions of matroids on S, and define f by $f(A) = r_1(A) + r_2(S\backslash A)$ for all $A \subseteq S$. Prove that if A_1 and A_2 both minimize f, then so also do $A_1 \cup A_2$ and $A_1 \cap A_2$.

8.39. Suppose that J is a common independent set of M_1 and M_2 and that $A \subseteq S$ with $|J| = r_1(A) + r_2(\overline{A})$. Prove that, in the corresponding auxiliary digraph G, we have $\delta(A \cup \{r\}) = \emptyset$.

8.40. Use the Matroid Intersection Theorem together with a formula for the rank function of a transversal matroid to solve Exercise 3.26.

8.41. Let M_1, M_2 be matroids determined by matrices N_1, N_2, where each N_i has n columns and at most n rows. Show how the auxiliary digraph for any common independent set can be constructed using $O(n^3)$ arithmetic operations. Hence show that the matroid intersection algorithm requires at most $O(n^4)$ arithmetic operations in total when applied to these two matroids.

8.42. Find a maximum-cardinality common independent set and a minimizing A in the Matroid Intersection Theorem for the matroids defined on the set $\{1, 2, \ldots, 7\}$ by the matrices

$$
\begin{pmatrix}
1 & 1 & 1 & -1 & 0 & 1 & 2 \\
1 & 0 & 0 & 1 & 1 & 1 & 1 \\
1 & 1 & 0 & 0 & 1 & 1 & 1 \\
0 & 1 & 1 & 1 & -1 & 0 & 1 \\
1 & 1 & 0 & 1 & 1 & -1 & 1
\end{pmatrix},
\quad
\begin{pmatrix}
1 & 1 & 1 & 1 & 1 & 0 & 1 \\
0 & 1 & 1 & 2 & 1 & -1 & 1 \\
1 & 1 & 0 & 1 & 1 & 0 & 1 \\
1 & 1 & 1 & -1 & 0 & 2 & 1 \\
1 & 1 & 1 & 1 & 1 & 0 & 0
\end{pmatrix}.
$$

8.43. Show that there does not exist a set J that is the set of edges of a spanning tree in both of the graphs of Figure 8.5.

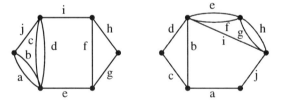

Figure 8.5. Is there a common spanning tree?

8.44. Prove that the graph $G = (V, E)$ has two edge-disjoint spanning trees if and only if there does not exist a partition of V into sets $V_0, V_1, \ldots, V_p$ such that the number of edges having ends in different V_i is less than $2p$. Hint: Apply the Matroid Intersection Theorem to M and M^*, where M is the forest matroid of G.

8.4 APPLICATIONS OF MATROID INTERSECTION

Matroid Partitioning

Many of the applications of matroid intersection are most easily derived through the theory of matroid partitioning. In fact, this theory is equivalent to that for matroid intersection, and actually was discovered earlier by Edmonds. Suppose that we are given matroids $M_i = (S, \mathcal{I}_i)$, $1 \leq i \leq k$. We say that a subset $J \subseteq S$ is *partitionable* with respect to $M_1, \ldots, M_k$ if $J = \cup(J_i : 1 \leq i \leq k)$ where $J_i \in \mathcal{I}_i$, $1 \leq i \leq k$; we call $(J_1, \ldots J_k)$ a *partitioning* of J with respect to the M_i. Obviously, we may assume that the J_i's are disjoint. The *matroid partitioning problem* is to find a maximum cardinality partitionable set. The main result of the theory is the following Matroid Partition Theorem. It is implicit in Edmonds and Fulkerson [1965]. (Actually, an earlier theorem of Rado [1942] can be shown to be equivalent.)

Theorem 8.17 *Let J be a maximum cardinality partitionable subset with respect to $M_i = (S, \mathcal{I}_i)$, $1 \leq i \leq k$. Then*

$$|J| = \min\{|\overline{A}| + \sum_{i=1}^{k} r_i(A) \ : \ A \subseteq S\}.$$

∎

As usual we can observe that for any such J and A, we have

$$|J| = |J \backslash A| + |J \cap A|$$
$$\leq |S \backslash A| + \sum |J_i \cap A|$$
$$\leq |\overline{A}| + \sum r_i(A).$$

The matroid partitioning problem is reduced to a matroid intersection problem as follows. (This construction, and the reverse one described in Exercise 8.49, are due to Edmonds [1970].) Make k disjoint copies

$$S_1, S_2, \ldots, S_k$$

of S, and imagine M_i as being defined on S_i rather than S. Let N_a be the direct sum of the M_i and let N_b be the (partition) matroid on $S' = \cup S_i$ in which a set is independent if and only if it contains at most one copy of e for each $e \in S$. It is easy to see that there is a correspondence between partitionable sets with respect to $M_1, \ldots, M_k$ and common independent sets of N_a, N_b. It is also easy to see that a set $B \subseteq S'$ that minimizes $r_a(B) + r_b(S' \backslash B)$ can be chosen to consist of all the copies of elements of A, for some $A \subseteq S$. It follows from the Matroid Intersection Theorem that the maximum size of a partitionable set is min $\{|\overline{A}| + \sum_{i=1}^{k} r_i(A) : A \subseteq S\}$, proving the Matroid Partition Theorem.

There is an important strengthening of the Matroid Partition Theorem, namely that in its statement, "maximum cardinality" can be replaced by "maximal." This follows from the observation that, whenever a copy of e is deleted from the common independent set of N_a, N_b by the intersection algorithm, it is replaced by another copy of e, so that no element is ever deleted from the partitionable set. Since the same argument could be applied to maximal partitionable subsets of an arbitrary subset B, we conclude that every maximal partitionable subset of B has the same cardinality. A consequence is the following result, which appeared explicitly for the first time in Nash-Williams [1966].

Theorem 8.18 *The subsets of S that are partitionable with respect to matroids $M_1, \ldots, M_k$ form the independent sets of a matroid. Its rank function r is given by $r(B) = \min\{(|B \backslash A| + \sum_{i=1}^{k} r_i(A)) : A \subseteq B\}$.* ∎

There is a neater description of the partitioning algorithm, obtained by identifying all copies of each element e of S in the auxiliary digraph for the intersection algorithm. The resulting digraph has node-set $S \cup \{r, s\}$ and has

- an arc re for each $e \in S \backslash J$;
- an arc es for each $e \in S$ such that $J_i \cup \{e\} \in \mathcal{I}_i$ for some i;
- an arc ef for each $e, f \in S$ such that $f \in J_i$, $J_i \cup \{e\} \notin \mathcal{I}_i$, and $J_i \triangle \{e, f\} \in \mathcal{I}_i$ for some i.

At termination of the algorithm, any set $A \subseteq S$ such that $\delta(A \cup \{s\}) = \emptyset$ has the property that $J_i \cap A$ M_i-spans A, and so $|J| = |\overline{A}| + \sum_{i=1}^{k} r_i(A)$.

Exercises

8.45. Let $G = (V, E)$ be an undirected graph and let k be an integer. Show that E can be partitioned into the edge-sets of k forests if and only if there does not exist a subset B of V such that $|\gamma(B)| > k(|B| - 1)$.

8.46. Let $G = (V, E)$ be a graph and k be an integer. Prove that G has k edge-disjoint spanning trees if and only if there does not exist a partition of V into sets $V_0, V_1, \ldots, V_p$ such that the number of edges of G joining nodes in different V_i is less than kp. (This is a generalization of Exercise 8.44.)

8.47. Use the Matroid Partition Theorem to prove the Matroid Polytope Theorem (Theorem 8.5) as follows. Let x be a rational point in $\mathbf{R}^S$ such that $x \geq 0$ and $x(A) \leq r(A)$ for all $A \subseteq S$. Let D be an integer such that $d = Dx \in \mathbf{Z}^S$. Show that, with respect to the matroid M' constructed in Exercise 8.27, S' is the union of D independent sets. Hence show that x is a convex combination of characteristic vectors of independent sets of M.

8.48. Let B_1, B_2 be bases of matroid M on S, and let $X_1 \subseteq B_1$. Show that there exists $X_2 \subseteq B_2$ such that $(B_1 \backslash X_1) \cup X_2$ and $(B_2 \backslash X_2) \cup X_1$ are both bases of M. Hint: If X_2 has the required properties, then $B_2 \backslash X_2$, X_2 provides a partitioning of B_2 with respect to the matroids $M_1 = M/X_1$ and $M_2 = M/(B_1 \backslash X_1)$.

8.49. (Reduction of matroid intersection to matroid partition) Let M_1, M_2 be matroids on S, let B be a maximal partitionable subset with respect to M_1 and M_2^*, and let J_1, J_2 be an associated partitioning. Extend J_2 to a basis B_2 of M_2^*. Show that $B \backslash B_2$ is a maximum cardinality common independent set of M_1 and M_2.

8.50. Give necessary and sufficient conditions for the existence of a partitioning $J_1, \ldots, J_k$ of S such that $|J_i| \leq b_i$ for all i, where $b_1, \ldots, b_k$ are given.

8.51. How could it be determined whether there exists a partitioning $J_i, 1 \leq i \leq k$ of S satisfying $a_i \leq |J_i| \leq b_i$, for all i, where $a_1, b_1, \ldots, a_k, b_k$ are given?

8.5 WEIGHTED MATROID INTERSECTION

We are given matroids $M_1 = (S, \mathcal{I}_1)$ and $M_2 = (S, \mathcal{I}_2)$ and a weight vector $c \in \mathbf{R}^S$. The *maximum weight common independent set problem* is

$$\max c(J) \qquad (8.6)$$
$$\text{subject to}$$
$$J \in \mathcal{I}_1 \cap \mathcal{I}_2.$$

We have solved two important special cases in earlier sections. Namely, the maximum weight independent set problem is the case where $\mathcal{I}_1 = \mathcal{I}_2$, and the cardinality matroid intersection problem is the case where $c_e = 1$ for all $e \in S$. (Another special case is maximum weight matching in a bipartite graph.)

The weighted matroid intersection problem was posed and solved by Edmonds [1970], who provided both optimality conditions and an efficient algorithm. We begin our treatment of the problem by stating the optimality conditions. Then we prove them nonconstructively. In particular, the proof uses polyhedral techniques as well as results from Section 8.1. Later we describe efficient algorithms for (8.6) and give an independent proof of the optimality conditions. The methods of this section rely heavily on the results developed in Sections 8.1 and 8.3.

Matroid Intersection Polyhedra

Suppose that we attempt to generalize the linear-programming conditions of Section 8.1. Notice that the characteristic vector $\bar{x}$ of any common independent set J is a feasible solution to:

$$x(A) \leq r_1(A), \text{ for all } A \subseteq S \qquad (8.7)$$

$$x(A) \leq r_2(A), \text{ for all } A \subseteq S$$
$$x_e \geq 0, \text{ for all } e \in S.$$

Therefore, the maximum of $c^T x$ subject to (8.7) is at least the maximum weight of a common independent set. Amazingly enough, this inequality is actually an equation! This is the content of the following theorem of Edmonds [1970].

Theorem 8.19 *For every* $c \in \mathbf{R}^S$ *there is a common independent set* J *whose characteristic vector maximizes* $c^T x$ *subject to* (8.7).

Of course, there is an equivalent statement in terms of polyhedra. (The equivalence follows from results of Chapter 6, and the fact that a $\{0, 1\}$-valued vector cannot be in the convex hull of other $\{0, 1\}$-valued vectors.)

Theorem 8.20 *(Matroid Intersection Polytope Theorem) The convex hull of the characteristic vectors of common independent sets of* M_1 *and* M_2 *is precisely the set of feasible solutions of* (8.7).

Another way to state Theorem 8.20 is that the vertices of the intersection of the polyhedra P_1 and P_2 determined by the matroids M_1 and M_2 are *precisely the common vertices* of the two polyhedra. This is quite a surprising fact. Normally, the intersection of two polyhedra will have many vertices in addition to those that are common to both.

It is interesting to apply these results to bipartite matching. We know how to express the matchings of a bipartite graph as the common independent sets of two matroids. Moreover, the corresponding rank functions are given by $r_i(A) = |\{v \in V_i : v \text{ is incident to at least one edge of } A\}|$, for $i = 1$ and 2. Now consider the inequality $x(A) \leq r_1(A)$ from (8.7). We will show that many of these inequalities are unnecessary. Namely, each such inequality is the sum of the inequalities $x(A \cap \delta(v)) \leq 1$ for each $v \in V_1$ such that $A \cap \delta(v) \neq \emptyset$. Moreover, each inequality $x(A \cap \delta(v)) \leq 1$ is the sum of the inequality $x(\delta(v)) \leq 1$ with the inequalities $-x_e \leq 0$ for $e \in \delta(v) \backslash A$. The same remarks apply to the inequalities $x(A) \leq r_2(A)$. Therefore, we have as a consequence of Theorem 8.20 that the convex hull of characteristic vectors of matchings of a bipartite graph is given by

$$x(\delta(v)) \leq 1, \text{ for all } v \in V$$
$$x_e \geq 0, \text{ for all } e \in E.$$

This result was proved in Chapter 6.

Now let us consider the dual of the linear-programming problem, $\max c^T x$ subject to (8.7). It is

$$\text{Minimize } \sum(r_1(A)y_A^1 + r_2(A)y_A^2 : A \subseteq S) \tag{8.8}$$

subject to

$$\sum(y_A^1 + y_A^2 : A \subseteq S, \ e \in A) \geq c_e, \text{ for all } e \in S$$
$$y_A^1, y_A^2 \geq 0, \text{ for all } A \subseteq S.$$

Choose an optimal solution $(\bar{y}^1, \bar{y}^2)$ of (8.8). Define $c^1, c^2 \in \mathbf{R}^S$ as follows. Let $c_e^1 = \sum(\bar{y}^1(A) : A \subseteq S, \ e \in A)$ for all $e \in S$, and let $c^2 = c - c^1$. Let Problem i denote (8.2) where c is replaced by c^i and M is replaced by M_i, for $i = 1$ and 2. Then we have the following useful observation.

Proposition 8.21 *For $i = 1$ and 2, $\bar{y}^i$ is optimal to Problem i. Conversely, if $\hat{y}^i$ is optimal to Problem i for $i = 1$ and 2, then $(\hat{y}^1, \hat{y}^2)$ is optimal to (8.8).* ∎

The second part of Proposition 8.21 will be used in the proof of Theorems 8.19 and 8.20. Its first part can be combined with Theorem 8.19 to give yet another version of the optimality condition. Call a pair c^1, c^2 of vectors a *weight-splitting* for c if $c = c^1 + c^2$. Suppose a weight-splitting and a common independent set J have the property that J is c^i-optimal in $\mathcal{I}_i$ for $i = 1$ and 2, and let J' be any other common independent set. Then $c(J) = c^1(J) + c^2(J) \geq c^1(J') + c^2(J') = c(J')$, so J is an optimal common independent set. In fact, such a weight-splitting always exists.

Corollary 8.22 *There exist a common independent set J and a weight-splitting c^1, c^2 for c such that J is c^i-optimal in $\mathcal{I}_i$ for $i = 1$ and 2.*

Proof: We define $(\bar{y}^1, \bar{y}^2)$ and c^1, c^2 as above. Now by Theorem 8.19, there is a common independent set J whose characteristic vector satisfies the optimality conditions of (8.8) with $(\bar{y}^1, \bar{y}^2)$. It is easy to see that the characteristic vector of J satisfies the optimality conditions of Problem i with $\bar{y}^i$ for $i = 1$ and 2. ∎

It remains to prove Theorem 8.20. It is obvious that it would be enough to prove that the set of solutions of (8.7) is an integral polytope. This in turn follows from the next result, via Theorem 6.29.

Theorem 8.23 *The system (8.7) is totally dual integral.*

Proof: Suppose that $c \in \mathbf{Z}^S$. We must show that (8.8) has an optimal solution that is integral. Choose an optimal solution $(\bar{y}^1, \bar{y}^2)$ and define c^1, c^2 as above. Now replace $\bar{y}^i$ by the optimal solution of the single-matroid dual problem for M_i, c^i delivered by the dual Greedy Algorithm. Hence, we may assume that there is a sequence $T_1^1 \subset T_2^1 \subset \ldots \subset T_{m_1}^1$ of subsets of S such that $\bar{y}_A^1 > 0$ only if $A = T_j^1$ for some j, and there is a sequence $T_1^2 \subset T_2^2 \subset \ldots \subset T_{m_2}^2$ of subsets of S such that $\bar{y}_A^2 > 0$ only if $A = T_j^2$ for some j. Therefore, there is an optimal solution to (8.8) that is an optimal solution to the problem obtained from (8.8) by deleting all columns except those corresponding to the T_j^i. The constraint matrix of that problem is totally unimodular. (See Exercise 6.28.) Therefore, since c is integral, this problem has an optimal solution that is integral, by Theorem 6.25, and hence so does (8.8). ∎

Algorithms for Weighted Matroid Intersection

There is a natural way to try to extend the matroid intersection algorithm of the last section to solve (8.6). It is analogous to a version of the primal-dual algorithm for finding a feasible flow of minimum cost. Recall that, if a feasible flow x is of minimum cost for its value, and x' is obtained from x by augmenting on an x-augmenting path of minimum cost, then x' is minimum-cost of its value. Given a common independent set J of M_1 and M_2, we define $G = G(M_1, M_2, c, J)$ to be $G = G(M_1, M_2, J)$ with the following assignment of a cost p_{vw} to each arc vw:

- $p_{ef} = c_f - c_e$ for each M_1-arc ef with $e \notin J$ and $f \in J$;
- $p_{es} = -c_e$ for each M_1-arc es with $e \notin J$;
- $p_{vw} = 0$ for each M_2-arc vw.

(As an example, a weighted matroid intersection problem is created from the problem in Figure 8.3 by defining weights. The problem and the auxiliary digraph corresponding to the common independent set $J = \{a, d, e\}$ are shown in Figure 8.6.) The motivation for the definition is that, if we augment J on

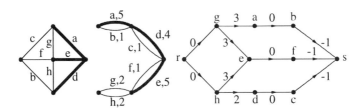

Figure 8.6. Auxiliary digraph for a weighted problem

an (r, s)-dipath P with node-sequence

$$r, e_1, f_1, \ldots, e_m, f_m, e_{m+1}, s$$

to obtain a new common independent set J', then $c(J) - c(J')$ is the cost (with respect to p) of P. So to maximize the increase in the weight of the new independent set, we should use an (r, s)-dipath of least cost. (Of course, this "increase" may be negative. In the example, the least-cost (r, s)-dipath has node-sequence r, h, d, c, s, which suggests augmenting to obtain the set

$J' = \{a, c, e, h\}$.) Notice that in the cardinality case, all such dipaths will have the same cost, namely -1, and we know that not every such dipath will give a J' that is common independent. (We need to choose a chordless one!) So it should not be surprising that we need another condition.

Theorem 8.24 *Let M_1, M_2 be matroids on S, let $c \in \mathbf{R}^S$, let $k \in \mathbf{Z}$, let J be a maximum weight common independent set of size k, let P be a least-cost (r, s)-dipath in $G(M_1, M_2, c, J)$ having as few arcs as possible, and let J' be the set obtained from J by using P. Then J' is a maximum-weight common independent set of size $k + 1$.*

Given this result, it is easy to find a common independent set of maximum weight. Obviously, $\emptyset$ is a maximum-weight common independent set of cardinality 0. So if we begin with $\emptyset$ and repeatedly find dipaths and adjust J as in the theorem, we will obtain an optimal common independent set of size k for every possible k. From these we simply pick the best one. (We will see later that there is another way to handle this last part.) We remind the reader that it is easy to modify a shortest path algorithm to find a least-cost path that has as few arcs as possible. See Chapter 2.

Thus Theorem 8.24 leads to the following weighted matroid intersection algorithm. It is not only easy to state, but also is essentially as efficient as the cardinality algorithm. This form of the algorithm was proposed by Lawler [1975].

Weighted Matroid Intersection Algorithm

Set $k = 0$;
Set $J_k = \emptyset$;
Loop
    Construct $G = G(M_1, M_2, c, J_k)$;
    If there is an (r, s)-dipath in G
        Find a least-cost (r, s)-dipath P having as few arcs
        as possible;
        Augment J_k using P to obtain J_{k+1};
        Replace k by $k + 1$;
    Else
        Choose $J = J_p$, where $c(J_p) \geq c(J_i)$, $1 \leq i \leq k$, and stop.

The algorithm has an attractive simplicity, but there are three complaints that one might make. One is that we have no proof of Theorem 8.24 as yet, and thus no proof of the correctness of the algorithm. (Notice that it is not even clear that the sets J_i are common independent, since it is not clear that the dipaths on which we augment will be chordless. In fact, they will not be chordless, in general.) A second complaint is that the algorithm does not construct an optimal solution of the dual problem (8.8). And a third is that the least-cost dipath problem that we have to solve at each iteration will in general have negative costs. (Admittedly, this is not a very serious complaint.) We will give a more sophisticated version of the same algorithm, which addresses all of these points. It uses a different assignment of costs to the arcs of G, defined by a weight-splitting, but with the property that the same dipaths are least-cost. The weight-splitting will be used to verify that the current J is optimal of its cardinality, and hence will be used to prove Theorem 8.24.

Let (c^1, c^2) be a weight-splitting, let J be a common independent set, and let c_0^i denote $\max\{c_e^i : e \notin J, J \cup \{e\} \in \mathcal{I}_i\}$. (If these are not well-defined, then J is of maximum cardinality, and we do not need to look for a best set of larger size.) We define $G = G(M_1, M_2, J, c^1, c^2)$ to be $G(M_1, M_2, J)$ with arc costs p_{uv} defined by:

- $p_{es} = c_0^1 - c_e^1$ for each M_1-arc es with $e \notin J$;
- $p_{re} = c_0^2 - c_e^2$ for each M_2-arc re with $e \notin J$;
- $p_{ef} = -c_e^1 + c_f^1$ for each M_1-arc ef with $e \notin J$ and $f \in J$;
- $p_{fe} = -c_e^2 + c_f^2$ for each M_2-arc fe with $e \notin J$ and $f \in J$.

(For the problem of Figure 8.6, the common independent set J indicated there is maximum-weight of cardinality 3, and a weight-splitting that verifies this is $c^1 = (3, -1, 0, 2, 3, 1, 2, 2)$, where components are in alphabetical order of the elements, and $c^2 = c - c^1$. The corresponding arc-costs are indicated on the auxiliary digraph of Figure 8.7.) If we augment on an (r, s)-dipath P to

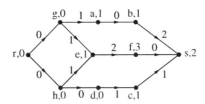

Figure 8.7. Second auxiliary digraph

obtain J' from J, then the cost of P is $c_0^1 + c_0^2 + c(J) - c(J')$. Therefore, a dipath will have least cost with respect to the new costs if and only if it has least cost with respect to the old ones. (We can see that, in the example, the least-cost (r, s)-dipath is the same as before, although its cost is now 2.)

Let us see how the properties of a good weight-splitting are reflected in the arc-costs of G. We say that a weight-splitting (c^1, c^2) *certifies* a common independent set J if, for $i = 1$ and 2, J has maximum c^i-weight over all independent sets of M_i having cardinality $|J|$. Of course, if a weight-splitting certifies a common independent set J, then J is a maximum-weight common independent set of its size.

Proposition 8.25 (c^1, c^2) *certifies J if and only if*

(a) *The arc costs of G are all nonnegative;*

(b) *For $i = 1$ and 2 and $f \in J$, $c_0^i \le c_f^i$.*

Proof: Let $|J| = k$; (c^1, c^2) certifies J if and only if, for $i = 1$ and 2, J is a c^i-optimal basis of the matroid M_i' obtained from M_i by truncating it to rank k. This is true if and only if, by Corollary 8.7,

$$e \notin J, \ f \in J, \ (J \cup \{e\}) \setminus \{f\} \in \mathcal{I}_i \text{ imply } c_e^i \le c_f^i.$$

Consider this condition for a pair e, f such that $J \cup \{e\} \notin \mathcal{I}_i$. Then, if $i = 1$, ef is an M_1-arc a of G, and if $i = 2$, fe is an M_2-arc a of G; in either case the condition is exactly that the cost of a be nonnegative. Now consider the condition for a pair e, f such that $J \cup \{e\} \in \mathcal{I}_i$. Then es is an M_1-arc of G or re is an M_2-arc of G, and the condition is equivalent to (b). ∎

We are going to show how to update the weight-splitting as well as J so that the weight-splitting continues to certify the common independent set. The idea for this is the same as the one used in the least-cost augmenting path algorithm for the minimum-cost flow problem, where the dual solution was modified after solving a shortest-path problem. We remark that Frank [1981] used weight-splitting in a version of the weighted matroid intersection algorithm, so this algorithm can be viewed as a refinement of that algorithm, as well.

Weighted Matroid Intersection Algorithm

Set $k = 0$;
Set $J_k = \emptyset$;
Let $c^1 = c$, $c^2 = 0$;
While J_k is neither an M_1-basis nor an M_2-basis
    Construct $G = G(M_1, M_2, J_k, c^1, c^2)$;
    Find a least-cost dipath from r to v in G of cost d_v for
    each v;
    For all $v \in S$, let $\sigma_v = \min\{d_v, d_s\}$ and replace c_v^1 by
    $c_v^1 - \sigma_v$, c_v^2 by $c_v^2 + \sigma_v$;
    Construct $G = G(M_1, M_2, J_k, c^1, c^2)$;
    If there is an (r, s)-dipath in G
        Find a least-cost (r, s)-dipath P having as few arcs as
        possible;
        Augment J_k on P to obtain J_{k+1};
        Replace k by $k + 1$;
    Else
        Choose $J = J_p$, where $c(J_p) \geq c(J_i)$, $1 \leq i \leq k$, and stop.

(In the digraph of Figure 8.7, we have indicated also the numbers d_v. The resulting new weight-splitting is therefore $c^1 = (2, -2, -1, 2, 2, -1, 2, 2)$, and $c^2 = c - c^1$. The reader can verify that this weight-splitting certifies that $\{a, e, d\}$ is optimal of size 3 *and* that $\{a, c, e, h\}$ is optimal of size 4.) We prove the correctness of this algorithm, and hence the earlier one, through a sequence of results.

Lemma 8.26 *After a change in the weight-splitting, the weight-splitting still certifies J_k.*

Proof: We consider the first time when this fails to hold. We use c^1, c^2 to denote the old weight-splitting, p to denote the old arc-costs, and p' to denote the new arc-costs. We prove the conditions of Proposition 8.25. Suppose that ef is an M_1-arc corresponding to $e \notin J_k$. Then

$$p'_{ef} = -(c_e^1 - \sigma_e) + (c_f^1 - \sigma_f) = p_{ef} + \sigma_e - \sigma_f.$$

If $\sigma_e = d_s$, then
$$p'_{ef} \geq p_{ef} + d_s - d_s = p_{ef} \geq 0.$$

If $\sigma_e = d_e$, then
$$p'_{ef} \geq d_e - d_f \geq 0.$$

(The latter inequality follows from the definition of the d_v.) The proof for M_2-arcs fe is similar, and arcs incident with r and s have nonnegative cost by definition. Therefore (a) is satisfied.

Now suppose that condition (b) of Proposition 8.25 is violated, so there exist $f \in J_k$ and $e \notin J_k$ such that $J_k \cup \{e\} \in \mathcal{I}_1$ and $c_e^1 - \sigma_e > c_f^1 - \sigma_f$. (The other case is similar.) Since $c_e^1 \leq c_f^1$, we must have $\sigma_f > \sigma_e$, so $\sigma_e = d_e$. Therefore,

$$c_e^1 - d_e > c_f^1 - \sigma_f \geq c_f^1 - d_s \geq c_f^1 - (d_e + c_0^1 - c_e^1).$$

It follows that $c_f^1 < c_0^1$, a contradiction. ∎

Lemma 8.27 *Every arc of P has cost zero.*

Proof: Let P' be a least-cost (r, s)-dipath with respect to the arc-costs p determined by the weight-splitting c^1, c^2 at the beginning of the iteration. For an arc uv of P', we have $d_u + p_{uv} = d_v$. It follows that $d_v \leq d_s$ for every node v of P'. Now consider an M_1-arc ef of P' for which $e \notin J_k$ and $f \in J_k$. Then the new cost p'_{ef} is $p_{ef} + d_e - d_f = 0$. Now for an M_1-arc es, consider

$$c_e^1 - \sigma_e \leq c_e^1 - d_e \leq c_e^1 - d_s + p_{es} = c_0^1 - d_s.$$

If e is the second-last node of P', then $d_s \geq d_e$, so $\sigma_e = d_e$ and $d_e + p_{es} = d_s$. Therefore both of the two inequalities above hold with equality for this e. It follows that, among all elements e such that es is an arc, $c_e^1 - \sigma_e$ is maximized by the second-last node of P'. This means that this element will determine the new value of c_0^1, so for this element, $p'_{es} = 0$.

We now know that every least-cost (r, s)-dipath with respect to p has cost zero with respect to p'. It follows that P must have cost zero with respect to p', and so, since $p' \geq 0$, that all of its arcs have cost zero. ∎

The next lemma is harder. We use results from the cardinality case in a slightly tricky way.

Lemma 8.28 *If $J_k \in \mathcal{I}_1 \cap \mathcal{I}_2$, then $J_{k+1} \in \mathcal{I}_1 \cap \mathcal{I}_2$.*

Proof: Notice that this is not at all obvious, since P may have chords. However, it cannot have zero-cost chords. Let $\pi_1 > \pi_2 > \cdots > \pi_\ell$ be the distinct values of c_e^1 that occur in J_k, let T_i denote $\{e \in S : c_e^1 \geq \pi_i\}$ for $1 \leq i \leq \ell - 1$, and let T_ℓ denote $\text{span}_1(J_k)$. It follows from Corollary 8.7 applied to $M_1 \backslash \overline{T_\ell}$ that J_k meets T_i in an M_1-basis of T_i for each i. Thus J_k is independent in the matroid $N = M_1'$ constructed from $M = M_1$ in Proposition 8.12. Similarly, we apply the same construction to M_2, obtaining M_2', and we form the auxiliary digraph G' with respect to $M_1', M_2', J_k, c^1, c^2$.

An M_1'-arc of the form ef with $e \notin J_k$ and $f \in J_k$ appears in G' if and only if ef is an M_1-arc of zero cost in G, and similarly for M_2'-arcs. Therefore, in G', P has no chords, and so J_{k+1} is common independent with respect to M_1', M_2', and so is common independent with respect to M_1, M_2. ∎

Lemma 8.29 *After an augmentation the weight-splitting certifies J_{k+1}.*

Proof: We show that J_{k+1} is c^1-optimal of cardinality $k+1$ in $\mathcal{I}_1$. (The other case is similar.) Let e_{m+1} be the second-last node of P. Then $c_{m+1}^1 = c_0^1$, because by Lemma 8.27 the corresponding arc has cost zero, so $J_k \cup \{e_{m+1}\}$ is c^1-optimal of cardinality $k+1$ in $\mathcal{I}_1$. But because each arc of P has cost zero by Lemma 8.27, each element of J_k on P has the same c^1-weight as the nonelement of J_k that immediately precedes it. It follows that J_{k+1} has the same c^1-weight as $J_k \cup \{e_{m+1}\}$, so J_{k+1} is c^1-optimal of cardinality $k+1$ in $\mathcal{I}_1$, as required. ∎

The proof of Theorem 8.24 is now complete. Therefore, we know that the two versions of the weighted matroid intersection algorithm are correct. More precisely, they generate, for each possible cardinality k, a common independent set J_k having maximum c-weight over all common independent sets of cardinality k. Moreover, the second version of the algorithm generates, for each k, a weight-splitting that validates this conclusion. (It also needs to solve only nonnegative-cost shortest path problems.)

Of course, J, the set that maximizes $c(J_k)$ over all k, solves the maximum-weight common independent set problem. We want to show how the algorithm can be used to find a corresponding weight-splitting, as in Corollary 8.22. First observe that $c_0^2 = 0$ throughout the execution of the algorithm. (Even more, $c_e^2 = 0$ for all e such that re is an arc of G.) Suppose that we run the algorithm until a change of the weight-splitting first makes $c_0^1 \leq 0$; more exactly, we modify the definition of σ_v to be $\min\{d_v, d_s, c_0^1\}$.

We are going to show that, at this point, J_k, c^i have the properties (a),(b),(c) of Corollary 8.6. The nonnegativity of the arc-costs guarantees (c), and $c_0^1 = c_0^2 = 0$ implies (b). If (a) is violated, then $c_f^i < 0$ for some i and some $f \in J_k$. It is easy to see that an augmentation will not cause this to happen, so consider the change in the weight-splitting that did. Then clearly $i = 1$, and $c_f^1 < 0 = c_0^1$, which implies that replacing f by some e with es an arc would give an M_1-independent set with higher c^1-weight, a contradiction to the c^1-optimality of J_k, and we are done. (In the example of Figure 8.7, since $c_0^1 = 1$, we get a different weight-splitting, namely, $c^1 = (2, -2, -1, 2, 2, 0, 2, 2)$ and $c^2 = c - c^1$. The reader can verify that this certifies J as a maximum weight common independent set.)

Corollary 8.30 *Let $m = \min(k : c_0^1 \leq 0$ in iteration $k)$. Them J_m is a c-optimal common independent set.* ∎

Exercises

8.52. Given a digraph G with arc weights, and a node r of G, we want to find a minimum weight directed spanning tree with root r. Show how this problem could be solved using matroid intersection.

8.53. Show how to solve the problem of finding a least-cost (r, s)-dipath in a digraph $G = (V, E)$ (or finding a dicircuit of negative cost) by solving a minimum weight common basis problem for two matroids. (Hint: Add a loop at every node other than r and s and give it cost zero. Define matroids on the arc-set so that the common bases consist of a simple (r, s)-dipath together with node-disjoint dicircuits.)

8.54. Prove that the characteristic vector of any common independent set is a vertex of the set of feasible solutions of (8.7).

8.55. Prove that, if P_1 and P_2 are polyhedra, then every common vertex of P_1 and P_2 is a vertex of the polyhedron $P_1 \cap P_2$.

8.56. Prove that if P_1, P_2, P_3 are the polyhedra associated with three matroids on S, then $P_1 \cap P_2 \cap P_3$ may have a nonintegral vertex. (Hint: There is an example with $|S| = 3$.)

8.57. Give a system of linear inequalities whose solution set is the convex hull of branchings of a digraph.

8.58. Interpret Corollary 8.22 for the special case of maximum-weight bipartite matching.

8.59. Prove Theorem 8.19 from Corollary 8.22.

8.60. Let M_1, M_2 be matroids on S, let $c \in \mathbf{R}^S$, and let q_i denote the weight of a maximum-weight common independent set for $i = 0, 1, \ldots, k$, where k is the maximum size of a common independent set. Prove that there exists j with $0 \leq j \leq k$ and $q_0 \leq q_1 \leq \ldots \leq q_j \geq q_{j+1} \geq \ldots \geq q_k$.

8.61. Show that in the case where each $c_e = 1$, Corollary 8.22 is a consequence of the Matroid Intersection Theorem.

CHAPTER 9

$\mathcal{NP}$ and $\mathcal{NP}$-Completeness

9.1 INTRODUCTION

Probably most readers have some intuitive idea about what a problem is and what an algorithm is, and what is meant by the running time of an algorithm. Although for the greater part of this book this intuition has been sufficient to understand the substance of the matter, in some cases it is important to formalize this intuition. This is particularly the case when we deal with concepts like $\mathcal{NP}$ and $\mathcal{NP}$-*complete*.

The class of problems solvable in polynomial time is usually denoted by $\mathcal{P}$. The class $\mathcal{NP}$, which will be described more precisely below, is a class of problems that might be larger (and many people believe *is* larger). It includes most combinatorial optimization problems, including all problems that are in $\mathcal{P}$. That is: $\mathcal{P} \subseteq \mathcal{NP}$. In particular, $\mathcal{NP}$ does *not* mean "not polynomial time." The letters $\mathcal{NP}$ stand for "nondeterministic polynomial time." The class $\mathcal{NP}$ consists, roughly speaking, of all those questions with the property that for any input that has a positive answer, there is a "certificate" from which the correctness of this answer can be derived in polynomial time.

For instance, the question:

Given an undirected graph G, does G have a Hamiltonian circuit? (9.1)

belongs to $\mathcal{NP}$. If the answer is "yes," we can convince anyone that this answer is correct by just giving a Hamiltonian circuit in G as a certificate. With this certificate, the answer "yes" can be checked in polynomial time. Here it is not required that we are able to *find* the certificate in polynomial time. The only requirement is that there exists a certificate with the required properties.

309

Checking the certificate in polynomial time means checking it in time bounded by a polynomial in the original input. In particular, it implies that the certificate itself has size bounded by a polynomial in the original input.

To elucidate the meaning of $\mathcal{NP}$, it is not known if for any graph G for which question (9.1) has a *negative* answer, there is a certificate from which the correctness of this answer can be derived in polynomial time. So there is an easy way of convincing "your boss" that a certain graph is Hamiltonian (that is, it contains a Hamiltonian circuit), but there is no such way known for convincing this person that a certain graph is non-Hamiltonian.

Within the class $\mathcal{NP}$ there are the "$\mathcal{NP}$-complete" problems. These are by definition the hardest problems in the class $\mathcal{NP}$: A problem Π in $\mathcal{NP}$ is $\mathcal{NP}$-*complete* if every problem in $\mathcal{NP}$ can be "reduced" to Π, in polynomial time. It implies that if one $\mathcal{NP}$-complete problem can be proved to be solvable in polynomial time, then *each* problem in $\mathcal{NP}$ can be solved in polynomial time. In other words, then $\mathcal{P} = \mathcal{NP}$ would follow.

Surprisingly, there are a great many prominent combinatorial optimization problems that are $\mathcal{NP}$-complete, like the traveling salesman problem. This pioneering eye-opener was given by Cook and Karp at the beginning of the 1970s.

Since that time one generally distinguishes between the polynomially solvable problems and the $\mathcal{NP}$-complete problems, although there is no proof that these two concepts really are distinct. For almost every combinatorial optimization problem one has been able to prove either that it is solvable in polynomial time, or that it is $\mathcal{NP}$-complete. But theoretically it is still a possibility that these two concepts are just the same! Thus it is unknown which of the two diagrams in Figure 9.1 applies.

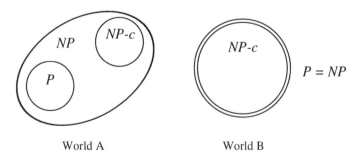

World A   World B

Figure 9.1. Complexity world

Below we make some of the notions more precise. We will not elaborate all technical details fully, but hope that the reader will be able to see the details without too much effort. For precise discussions we refer to the books by Aho, Hopcroft and Ullman [1974] and Garey and Johnson [1979].

9.2 WORDS

If we use the computer to solve a certain graph problem, we usually do not put a picture of the graph into the computer. Rather we put some appropriate encoding of the problem into the computer, by describing it by a sequence of symbols taken from some fixed finite "alphabet" Σ. We can take for Σ for instance the ASCII set of symbols or the set $\{0, 1\}$. It is convenient to have symbols like $(\ ,\ )$, $\{\ ,\ \}$ and the comma in Σ, and moreover some symbol like _ meaning: "blank." Let us fix one alphabet Σ.

We call any ordered finite sequence of elements from Σ a *word*. The set of all words is denoted by Σ^*.

It is not difficult to encode objects like rational numbers, vectors, matrices, graphs, and so on, as words. For instance, the graph given in Figure 9.2 can be encoded by the word:

$$(\{a, b, c, d, e\}, \{\{a, b\}, \{a, c\}, \{b, c\}, \{c, d\}, \{d, e\}, \{e, a\}\}).$$

A function f defined on a finite set X can be encoded by giving the set of

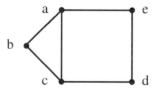

Figure 9.2. Graph

pairs $(x, f(x))$ with $x \in X$. For instance, the following describes a function defined on the edges of the graph above:

$$\{(\{a, b\}, 32), (\{a, c\}, -17), (\{b, c\}, 5/7),$$
$$(\{c, d\}, 6), (\{d, e\}, -1), (\{e, a\}, -9)\}.$$

A pair of a graph and a function can be described by (w, v), where w is the encoding of the graph and v is the encoding of the function.

The *size* of a word w is the number of symbols used in w, counting multiplicities, and is denoted by $\text{size}(w)$. (So the word $abaa32bc$ has size 8.) The size is important when we make estimates on the running time of algorithms.

Note that in encoding numbers (integers or rational numbers), the size depends on the number of symbols necessary to encode these numbers. Thus if we encounter a problem on a graph with numbers defined on the edges, then the size of the input is the total number of symbols necessary to represent this structure. It might be much larger than just the number of nodes and edges of the graph.

Although there are several ways of choosing an alphabet and encoding objects by words over this alphabet, any way chosen is quite arbitrary. We will be dealing with solvability in polynomial time in this chapter, and for that purpose most encodings are equivalent. We will sometimes exploit this flexibility.

9.3   PROBLEMS

What is a problem? Informally, it is a question or a task, for instance, "Does this given graph have a perfect matching?" or "Find a shortest traveling salesman tour in this graph." In fact there are two types of problems: problems that can be answered by "yes" or "no" and those that ask you to find a certain object. We here restrict ourselves to the first type of problems. From a complexity point of view this is not that much of a restriction. For instance, the problem of finding a shortest traveling salesman tour in a graph is really not harder than the question: Given a graph, a length function on the edges, and a rational number r, does there exist a traveling salesman tour of length at most r? If we can answer this second question in polynomial time, we can find the length of a shortest tour in polynomial time, for instance, by binary search.

So we study problems of the form: Given a certain object (or sequence of objects), does it have a certain property? For instance, given an undirected graph G, does it have a perfect matching?

Since we encode objects by words, a problem is nothing but: Given a word w, does it have a certain property? Thus the problem is fully described by describing the "certain property." This, in turn, is fully described by just the set of all words that have the property. Therefore we have the following mathematical definition: a *problem* is any subset Π of Σ^*.

If we consider any problem $\Pi \subseteq \Sigma^*$, the corresponding "informal" problem is:

$$\text{Given word } w, \text{ does } w \text{ belong to } \Pi?$$

In this context, the word w is called an *instance* or the *input*.

9.4   ALGORITHMS AND RUNNING TIME

An algorithm is a list of instructions to solve a problem. The classical mathematical formalization of an algorithm is the *Turing machine*. In this section we will describe a slightly different concept of an algorithm that is useful for our purposes (explaining $\mathcal{NP}$-completeness). In Section 9.9 below we will show that it is equivalent to the notion of a Turing machine.

A basic step in an algorithm is: Replace subword u by u'. It means that if word w is equal to tuv, where t and v are words, we replace w by the word

$tu'v$. Now by definition, an *algorithm* is a finite list of instructions of this type. It thus is fully described by a set

$$\{(u_1, u'_1), \ldots, (u_n, u'_n)\}, \tag{9.2}$$

where $u_1, u'_1, \ldots, u_n, u'_n$ are words. If a word w contains one of $u_1, \ldots, u_n$ as a subword we choose the smallest j such that w contains u_j as a subword. We write $w = tu_jv$, in such a way that the size of t is as small as possible. (So we choose the first occurrence of u_j in w.) Then the *successor* of w is the word tu'_jv. We say that the algorithm *stops* at word w if w has no subword equal to one of $u_1, \ldots, u_n$. So for any word w there is a unique successor of w, or the algorithm stops at w.

A (finite or infinite) sequence of words $w_0, w_1, w_2, \ldots$ is called *allowed* if w_{i+1} is the successor of w_i $(i = 0, 1, 2, \ldots)$, and, when the sequence is finite, the algorithm stops at the last word of the sequence. So for each word w there is a unique allowed sequence starting with w. We say that algorithm A *accepts* w if the allowed sequence starting with w is finite.

For reasons of consistency it is important to have the "empty space" at both sides of a word as part of the word. Thus instead of starting with a word w, we start with $\_w\_$, where $\_$ is a symbol indicating space. We moreover allow that the words in (9.2) contain more symbols than those in the alphabet Σ, and we specify these as the alphabet over which the word w might be chosen.

We say that algorithm A *solves* problem $\Pi \subseteq \Sigma^*$ if Π is the set of words in Σ^* accepted by A. Moreover, A *solves Π in polynomial time* if A solves Π and there exists a polynomial $p(x)$ such that for any word $w \in \Sigma^*$: If A accepts w, then the allowed sequence starting with w contains at most $p(\text{size}(w))$ words.

This definition enables us indeed to decide in polynomial time if a given word w belongs to Π. We just take $w_0 := w$, and next, for $i = 0, 1, 2, \ldots$, we choose "the first" subword u_j in w_i and replace it by u'_j (for some $j \in \{1, \ldots, n\}$) thus obtaining w_{i+1}. If within $p(\text{size}(w))$ iterations we stop, we know that w belongs to Π, and otherwise we know that w does not belong to Π.

Then the symbol $\mathcal{P}$ denotes the set of all problems that can be solved by a polynomial-time algorithm.

Exercises

9.1. Give an algorithm that accepts any word in $\{0, 1\}^*$ if and only if it contains a 1.

9.2. Give an algorithm that accepts any word in $\{0, 1\}^*$ if and only if it does not contain a 1.

9.3. Give an algorithm that accepts any word in $\{0, 1\}^*$ if and only if it is a palindrome. (A *palindrome* is a word that is the same when we read it backwards.)

9.5   THE CLASS $\mathcal{NP}$

We mentioned above that $\mathcal{NP}$ denotes the class of problems for which a positive answer has a "certificate" from which the correctness of the positive answer can be derived in polynomial time. We will now make this more precise.

The class $\mathcal{NP}$ consists of those problems $\Pi \subseteq \Sigma^*$ for which there exist a problem $\Pi' \in \mathcal{P}$ and a polynomial $p(x)$ such that for any $w \in \Sigma^*$:

> $w \in \Pi$ if and only if there exists a word v such that $(w, v) \in \Pi'$ and such that $\text{size}(v) \leq p(\text{size}(w))$.

So the word v acts as a certificate showing that w belongs to Π. With the polynomial-time algorithm solving Π', the certificate proves in polynomial time that w belongs to Π.

As an example, the problems

$$\begin{aligned} \Pi_1 &:= \{G \; : \; G \text{ is an undirected graph having a perfect matching}\} \\ \Pi_2 &:= \{G \; : \; G \text{ is an undirected Hamiltonian graph}\} \end{aligned}$$

(encoding G as above) belong to $\mathcal{NP}$, since the problems

$$\begin{aligned} \Pi_1' &:= \{(G, M) \; : \; G \text{ is an undirected graph and } M \text{ is a perfect} \\ & \qquad \text{matching in } G\} \\ \Pi_2' &:= \{(G, H) \; : \; G \text{ is an undirected graph and } H \text{ is a Hamilto-} \\ & \qquad \text{nian circuit in } G\} \end{aligned}$$

belong to $\mathcal{P}$.

Similarly, the problem

$$\begin{aligned} \Pi_3 &:= \{(G, l, r) \; : \; G \text{ is an undirected graph, } l \text{ is a "length"} \\ & \qquad \text{function on the edges of } G, \text{ and } r \text{ is a rational number} \\ & \qquad \text{such that } G \text{ has a Hamiltonian circuit of length at most } r\} \end{aligned}$$

("the traveling salesman problem") belongs to $\mathcal{NP}$, since the problem

$$\begin{aligned} \Pi_3' &:= \{(G, l, r, H) \; : \; G \text{ is an undirected graph, } l \text{ is a "length"} \\ & \qquad \text{function on the edges of } G, r \text{ is a rational number, and } H \\ & \qquad \text{is a Hamiltonian circuit in } G \text{ of length at most } r\} \end{aligned}$$

belongs to $\mathcal{P}$.

Clearly, $\mathcal{P} \subseteq \mathcal{NP}$, since if Π belongs to $\mathcal{P}$, then we can just take the empty string as a certificate for any word w to show that it belongs to Π. That is, we can take $\Pi' := \{(w,) \; : \; w \in \Pi\}$. Since $\Pi \in \mathcal{P}$, also $\Pi' \in \mathcal{P}$.

The class $\mathcal{NP}$ is apparently much larger than the class $\mathcal{P}$, and there might not be much reason to believe that the two classes are the same. But, as yet, nobody has been able to show that they really are different! This is an intriguing mathematical question, and answering it might also have practical significance. If $\mathcal{P}=\mathcal{NP}$ can be shown, the proof might contain a revolutionary new algorithm or, alternatively, it might imply that the concept of "polynomial-time" is completely useless. If $\mathcal{P} \neq \mathcal{NP}$ can be shown, the proof might give us more insight into the reasons why certain problems are more difficult than others, and might guide us to detect and attack the kernel of the difficulties.

By definition, a problem $\Pi \subseteq \Sigma^*$ belongs to the class co-$\mathcal{NP}$ if the "complementary" problem $\overline{\Pi} := \Sigma^* \setminus \Pi$ belongs to $\mathcal{NP}$.

For instance, the perfect matching problem Π_1 defined above belongs to co-$\mathcal{NP}$, since the problem

$$\Pi_1'' := \{(G, W) : G \text{ is an undirected graph and } W \text{ is a subset of the node set of } G \text{ such that the graph } G \setminus W \text{ has more than } |W| \text{ odd components}\}$$

belongs to $\mathcal{P}$. This follows from Tutte's Matching Theorem 5.3: If an undirected graph G has no perfect matching, then there is a subset W of the node set of G with the properties described in Π_1''. (Here, strictly speaking, the complementary problem $\overline{\Pi}_1$ of Π_1 consists of all words w that either do not represent a graph, or represent a graph having no perfect matching. We assume, however, that there is an easy way of deciding whether a given word represents a graph. Therefore, we might assume that the complementary problem is just $\{G : G \text{ is an undirected graph having no perfect matching}\}$.)

It is not known if the problems Π_2 and Π_3 belong to co-$\mathcal{NP}$.

Since for any problem Π in $\mathcal{P}$ the complementary problem $\overline{\Pi}$ also belongs to $\mathcal{P}$, we know that $\mathcal{P} \subseteq$ co-$\mathcal{NP}$. So $\mathcal{P} \subseteq \mathcal{NP} \cap$ co-$\mathcal{NP}$. The problems in $\mathcal{NP} \cap$ co-$\mathcal{NP}$ are those for which there exist certificates both in case the answer is positive and in case the answer is negative. As we saw above, the perfect matching problem Π_1 is such a problem. Tutte's Matching Theorem gives us the certificates. Therefore, Tutte's Matching Theorem is called a *good characterization*.

In fact, there are very few problems known that are proved to belong to $\mathcal{NP} \cap$ co-$\mathcal{NP}$, but that are not known to belong to $\mathcal{P}$. Most problems having a good characterization have been proved to be solvable in polynomial time. A notable exception for which this is not yet proved is *primality testing*: testing if a given natural number is a prime number. (It is quite easy to see that this problem belongs to co-$\mathcal{NP}$ since a number can be shown to be non-prime by giving one nontrivial divisor. The fact that the problem belongs to $\mathcal{NP}$ can be shown with the help of a theorem of Fermat.)

9.6 $\mathcal{NP}$-COMPLETENESS

The $\mathcal{NP}$-complete problems are defined to be the hardest problems in $\mathcal{NP}$. We first define the concept of a polynomial-time reduction. Let Π and Π' be two problems, and let A be an algorithm. We say that A is a *polynomial-time reduction* of Π' to Π if A is a polynomial-time algorithm ("solving" Σ^*), so that for any allowed sequence starting with w and ending with v, $w \in \Pi'$ if and only if $v \in \Pi$. A problem Π is called $\mathcal{NP}$-*complete*, if for each problem Π' in $\mathcal{NP}$ there exists a polynomial-time reduction of Π' to Π.

It is not very difficult to see that if Π belongs to $\mathcal{P}$ and there exists a polynomial-time reduction of Π' to Π, then Π' belongs to $\mathcal{P}$. It implies that if one $\mathcal{NP}$-complete problem can be solved in polynomial time, then each problem in $\mathcal{NP}$ can be solved in polynomial time. Moreover, if Π belongs to $\mathcal{NP}$, Π' is $\mathcal{NP}$-complete, and there exists a polynomial-time reduction of Π' to Π, then also Π is $\mathcal{NP}$-complete (Exercise 9.4).

Exercises

9.4. Show that if Π belongs to $\mathcal{NP}$, Π' is $\mathcal{NP}$-complete, and there exists a polynomial-time reduction of Π' to Π, then also Π is $\mathcal{NP}$-complete.

9.7 $\mathcal{NP}$-COMPLETENESS OF THE SATISFIABILITY PROBLEM

We now show that there exist $\mathcal{NP}$-complete problems. In fact we show that the so-called *satisfiability problem*, denoted by SAT, is $\mathcal{NP}$-complete.

To define SAT, we need the notion of a *boolean expression*. Examples are:

$$((x_2 \wedge x_3) \vee \neg(x_3 \vee x_5) \wedge x_2), ((\neg x_{47} \wedge x_2) \wedge x_{47}), \neg(x_7 \wedge \neg x_7).$$

Boolean expressions can be defined inductively. First, for each natural number n, the "word" x_n is a boolean expression (using some appropriate encoding of natural numbers and of subscripts). Next, if v and w are boolean expressions, then $(v \wedge w)$, $(v \vee w)$ and $\neg v$ also are boolean expressions. These rules give us all boolean expressions. (If necessary, we may use subscripts other than the natural numbers.)

Now SAT is a subcollection of all boolean expressions; it consists of those boolean expressions that are satisfiable. A boolean expression

$$f(x_1, x_2, x_3, \ldots)$$

is called *satisfiable* if there exist $\alpha_1, \alpha_2, \alpha_3, \ldots \in \{0, 1\}$ such that

$$f(\alpha_1, \alpha_2, \alpha_3, \ldots) = 1,$$

using the well-known identities

$$0 \wedge 0 = 0 \wedge 1 = 1 \wedge 0 = 0, 1 \wedge 1 = 1, \tag{9.3}$$
$$0 \vee 0 = 0, 0 \vee 1 = 1 \vee 0 = 1 \vee 1 = 1,$$
$$\neg 0 = 1, \neg 1 = 0, (0) = 0, (1) = 1.$$

The satisfiability problem SAT trivially belongs to $\mathcal{NP}$: We can take as certificate for a certain $f(x_1, x_2, x_3, \ldots)$ to belong to SAT, the equations $x_i = \alpha_i$ that give f the value 1. (We give equations only for those x_i that occur in f.)

To show that SAT is $\mathcal{NP}$-complete, it is convenient to assume that $\Sigma = \{0, 1\}$. This is not that much of a restriction: We can fix some order of the symbols in Σ, and encode the first symbol by 10, the second one by 100, the third one by 1000, and so on. There is an easy (certainly polynomial-time) way of obtaining one encoding from the other.

The following result is basic for the further proofs.

Theorem 9.1 *Let* $\Pi \subseteq \{0, 1\}^*$ *be in* $\mathcal{P}$. *Then there exist a polynomial* $p(x)$ *and an algorithm that finds for each natural number* n *in time* $p(n)$ *a boolean expression* $f(x_1, x_2, x_3, \ldots)$ *with the property:*

> *any word* $\alpha_1 \alpha_2 \ldots \alpha_n$ *in* $\{0, 1\}^*$ *belongs to* Π, *if and only if the boolean expression* $f(\alpha_1, \ldots, \alpha_n, x_{n+1}, x_{n+2}, \ldots)$ *is satisfiable.*

Proof: Since Π belongs to $\mathcal{P}$, there exists a polynomial-time algorithm A solving Π. So there exists a polynomial $p(x)$ such that a word w belongs to Π if and only if the allowed sequence for w contains at most $p(\text{size}(w))$ words. It implies that there exists a polynomial $q(x)$ such that any word in the allowed sequence for w has size less than $q(\text{size}(w))$.

We describe the algorithm whose existence is claimed by the theorem. Choose a natural number n. Introduce variables $x_{i,j}$ and $y_{i,j}$ for $i = 0, \ldots, p(n)$, $j = 1, \ldots, q(n)$. There exists (see Exercise 9.5) a boolean expression f in these variables with the following properties. Any assignment $x_{i,j} := \alpha_{i,j} \in \{0, 1\}$ and $y_{i,j} := \beta_{i,j} \in \{0, 1\}$ makes f equal to 1, if and only if the allowed sequence starting with the word $w_0 := \alpha_{0,1} \alpha_{0,2} \ldots \alpha_{0,n}$ is a finite sequence $w_0, \ldots, w_k$, so that

(i) $\alpha_{i,j}$ is equal to the jth symbol in the word w_i, for each $i \leq k$ and each $j \leq \text{size}(w_i)$;

(ii) $\beta_{i,j} = 1$ if and only if $i > k$ or $j \leq \text{size}(w_i)$.

The important point is that f can be found in time bounded by a polynomial in n. To see this, we can encode the fact that word w_{i+1} should follow from word w_i by a boolean expression in the "variables" $x_{i,j}$ and $x_{i+1,j}$, representing the different positions in w_i and w_{i+1}. (The extra variables $y_{i,j}$ and $y_{i+1,j}$ are introduced to indicate the sizes of w_i and w_{i+1}.) Moreover, the fact that the algorithm stops at a word w also can be encoded by a boolean expression. Taking the "conjunction" of all these boolean expressions will give us the boolean expression f. ∎

As a direct consequence we have the following.

Corollary 9.2 *Theorem 9.1 also holds if we replace $\mathcal{P}$ by $\mathcal{NP}$ in the first sentence.*

Proof: Let $\Pi \subseteq \{0,1\}^*$ belong to $\mathcal{NP}$. Then, by definition of $\mathcal{NP}$, there exists a problem Π' in $\mathcal{P}$ and a polynomial $r(x)$ such that any word w belongs to Π if and only if (w,v) belongs to Π' for some word v with size$(v) \leq r(\text{size}(w))$. By properly reencoding, we may assume that for each $n \in N$, any word $w \in \{0,1\}^*$ belongs to Π if and only if wv belongs to Π' for some word v of size $r(\text{size}(w))$. Applying Theorem 9.1 to Π' gives the corollary. ∎

Now, Cook's Theorem follows directly.

Corollary 9.3 *(Cook's Theorem) The satisfiability problem SAT is $\mathcal{NP}$-complete.*

Proof: Let Π belong to $\mathcal{NP}$. We describe a polynomial-time reduction to SAT. Let $w = \alpha_1 \ldots \alpha_n \in \{0,1\}^*$. By Corollary 9.2 we can find in time bounded by a polynomial in n a boolean expression f such that w belongs to Π if and only if $f(\alpha_1, \ldots, \alpha_n, x_{n+1}, \ldots)$ is satisfiable. This is the required reduction to SAT. ∎

This remarkable result shows that a polynomial-time algorithm for the innocent-looking satisfiability problem would immediately mean that we could solve in polynomial time any problem that has a short certificate of a "yes" answer.

Exercises

9.5. Let $n \geq 1$ be a natural number and let W be a collection of words in $\{0,1\}^*$ all of length n. Prove that there exists a boolean expression $f(x_1, \ldots, x_n)$ in the variables $x_1, \ldots, x_n$ such that for each word $w = \alpha_1 \ldots \alpha_n$ in the symbols 0 and 1 one has: $w \in W$ if and only if

$$f(\alpha_1, \ldots, \alpha_n) = 1.$$

9.8 $\mathcal{NP}$-COMPLETENESS OF SOME OTHER PROBLEMS

At the current time, most problems in combinatorial optimization for which we do not know a polynomial-time algorithm have been shown to be $\mathcal{NP}$complete. (Remember, this does not imply that there does not exist a polynomial-time algorithm.) Cook's theorem forms here a basis from which the $\mathcal{NP}$-completeness of other problems is derived by sequences of reductions. Fundamental work on this was done by Karp [1972], who found the corollaries below.

First we show that the *3-satisfiability problem* 3-SAT is $\mathcal{NP}$-complete. Let B_1 be the set of all words $x_1, \neg x_1, x_2, \neg x_2, \ldots$. Let B_2 be the set of all words

$(w_1 \vee \ldots \vee w_k)$, where $w_1, \ldots, w_k$ are words in B_1, and $1 \leq k \leq 3$. Let B_3 be the set of all words $w_1 \wedge \ldots \wedge w_k$, where $w_1, \ldots, w_k$ are words in B_2. We say that a word $f(x_1, x_2, \ldots) \in B_3$ is *satisfiable* if there exists an assignment $x_i := \alpha_i \in \{0, 1\}$ $(i = 1, 2, \ldots)$ such that $f(\alpha_1, \alpha_2, \ldots) = 1$ (using the identities (9.3)).

Now the 3-satisfiability problem 3-SAT is: Given a word $f \in B_3$, decide if it is satisfiable.

Corollary 9.4 *The 3-satisfiability problem* 3-SAT *is $\mathcal{NP}$-complete.*

Proof: We give a polynomial-time reduction of SAT to 3-SAT. Let

$$f(x_1, x_2, \ldots)$$

be a boolean expression. Introduce a variable y_g for each subword g of f that is a boolean expression.

Now f is satisfiable if and only if the following system is satisfiable:

$$y_g = y_{g'} \vee y_{g''} \text{ (if } g = (g' \vee g''))$$
$$y_g = y_{g'} \wedge y_{g''} \text{ (if } g = (g' \wedge g''))$$
$$y_g = \neg y_{g'} \text{ (if } g = \neg g')$$
$$y_f = 1.$$

Now $y_g = y_{g'} \vee y_{g''}$ can be equivalently expressed by: $y_g \vee \neg y_{g'} = 1, y_g \vee \neg y_{g''} = 1, \neg y_g \vee y_{g'} \vee y_{g''} = 1$. Similarly, $y_g = y_{g'} \wedge y_{g''}$ can be equivalently expressed by: $\neg y_g \vee y_{g'} = 1, \neg y_g \vee y_{g''} = 1, y_g \vee \neg y_{g'} \vee \neg y_{g''} = 1$. The expression $y_g = \neg y_{g'}$ is equivalent to: $y_g \vee y_{g'} = 1, \neg y_g \vee \neg y_{g'} = 1$.

By renaming variables, we thus obtain words $w_1, \ldots, w_k$ in B_2, so that f is satisfiable if and only if the word $w_1 \wedge \ldots \wedge w_k$ is satisfiable. $\blacksquare$

We next derive that the *exact cover problem* EXACTCOVER is $\mathcal{NP}$-complete. This is the problem: Given a collection $\mathcal{C}$ of subsets of a finite set X, is there a subcollection of $\mathcal{C}$ that forms a partition of X?

Corollary 9.5 EXACTCOVER *is $\mathcal{NP}$-complete.*

Proof: We give a polynomial-time reduction of 3-SAT to EXACTCOVER. Let $f = w_1 \wedge \ldots \wedge w_k$ be a word in B_3, where $w_1, \ldots, w_k$ are words in B_2. Let $x_1, \ldots, x_m$ be the variables occurring in f. Make a graph G with nodes $w_1, \ldots, w_k$ and $x_1, \neg x_1, \ldots, x_m, \neg x_m$, and edges $\{w_i, x_j\}$ if x_j occurs in w_i, $\{w_i, \neg x_j\}$ if $\neg x_j$ occurs in w_i, and $\{x_j, \neg x_j\}$ for $j = 1, \ldots, m$. Let X be the set of nodes $w_1, \ldots, w_k$ and all edges of G.

Let $\mathcal{C}'$ be the collection of all sets $\{w_i\} \cup E'$, where E' is a nonempty subset of the edge set incident with w_i. Let $\mathcal{C}''$ be the collection of all sets $\delta(\alpha)$, where $\alpha \in \{x_1, \neg x_1, \ldots, x_m, \neg x_m\}$ (with $\delta(\alpha)$ being the collection of edges incident with α).

Now f is satisfiable, if and only if the collection $\mathcal{C}' \cup \mathcal{C}''$ contains a subcollection that partitions X. Thus we have a reduction of 3-SAT to EXACT-COVER. ∎

We derive the $\mathcal{NP}$-completeness of the *directed Hamiltonian circuit problem* DIRECTED HAMILTONIAN CIRCUIT: Given a directed graph, does it have a directed Hamiltonian circuit?

Corollary 9.6 DIRECTED HAMILTONIAN CIRCUIT *is $\mathcal{NP}$-complete.*

Proof: We give a polynomial-time reduction of EXACTCOVER to DIRECTED HAMILTONIAN CIRCUIT. Let $\mathcal{C} = \{C_1, \ldots, C_m\}$ be a collection of subsets of the set $X = \{x_1, \ldots, x_k\}$. Introduce "nodes"

$$r_0, r_1, \ldots, r_m, s_0, s_1, \ldots, s_k.$$

For each $i = 1, \ldots, m$ we do the following. Let $C_i = \{x_{j_1}, \ldots, x_{j_t}\}$. We construct a directed graph on the nodes $r_{i-1}, r_i, s_{j_1-1}, s_{j_1}, \ldots, s_{j_t-1}, s_{j_t}$, and $3t$ new nodes, as indicated in Figure 9.3. (If $s_{j_h-1} = s_{j_h-1}$ the nodes will

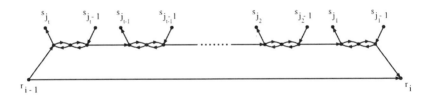

Figure 9.3. Graph for Directed Hamilton Tour

coincide in their picture.) Moreover, we make arcs from r_m to s_0 and from s_k to r_0.

Let D be the directed graph arising. Then it is not difficult to check that there exists subcollection $\mathcal{C}'$ of $\mathcal{C}$ that partitions X, if and only if D has a directed Hamiltonian circuit C. (We take: $(r_{i-1}, r_i) \in C \Longleftrightarrow C_i \in \mathcal{C}'$.) ∎

From this we derive the $\mathcal{NP}$-completeness of the *undirected Hamiltonian circuit problem* UNDIRECTED HAMILTONIAN CIRCUIT: Given an undirected graph, does it have a Hamiltonian circuit?

Corollary 9.7 UNDIRECTED HAMILTONIAN CIRCUIT *is $\mathcal{NP}$-complete.*

Proof: We give a polynomial-time reduction of DIRECTED HAMILTO-NIAN CIRCUIT to UNDIRECTED HAMILTONIAN CIRCUIT. Let D be a directed graph. Replace each node v by three nodes v', v'', v''', and make edges $\{v', v''\}$ and $\{v'', v'''\}$. Moreover, for each arc (v_1, v_2) of D, make an edge $\{v_1', v_2'''\}$. This makes the undirected graph G. One easily checks that D has a directed Hamiltonian circuit, if and only if G has an (undirected) Hamiltonian circuit. ∎

This trivially implies the $\mathcal{NP}$-completeness of the *traveling salesman problem* TSP: Given a complete graph $G = (V, E)$, a "length" function l on E, and a rational r, does there exist a Hamiltonian circuit of length at most r?

Corollary 9.8 TSP *is $\mathcal{NP}$-complete.*

Proof: We give a polynomial-time reduction of UNDIRECTED HAMIL-TONIAN CIRCUIT to TSP. Let G be an undirected graph. Let G' be the complete graph on V. Let $l(e) := 0$ for each edge e of G, and let $l(e) := 1$ for each edge of G' that is not an edge of G. Then G has a Hamiltonian circuit, if and only if G' has a Hamiltonian circuit of length at most 0. ∎

9.9 TURING MACHINES

In Section 9.4 we gave a definition of "algorithm." How adequate is this definition? Can any computer program be modeled after that definition?

To study this question, we need to know what we understand by a "computer." Turing gave in 1936 the following computer model, now called a *Turing machine* or a *one-tape Turing machine*.

A Turing machine consists of a "processor," which can be in a finite number of "states," and of a "tape," of infinite length (in two ways). Moreover, there is a "read-write head," which can read symbols on the tape (one at a time). Depending on the state of the processor and the symbol read, the processor passes to another (or the same) state, the symbol on the tape is changed (or not), and the tape is moved one position "to the right" or "to the left."

The whole system can be described by just giving the dependence mentioned in the previous sentence. So, mathematically, a Turing machine is just a function

$$T : M \times \Sigma \longrightarrow M \times \Sigma \times \{+1, -1\}.$$

Here M and Σ are finite sets: M is interpreted as the set of states of the processor, while Σ is the set of symbols that can be written on the tape. The function T describes an "iteration": $T(m, \sigma) = (m', \sigma', +1)$ should mean that if the processor is in state m and the symbol read on the tape is σ, then the next state will be m', the symbol σ is changed to the symbol σ', and the tape is moved one position to the right. $T(m, \sigma) = (m', \sigma', -1)$ has a similar meaning; now the tape is moved one position to the left.

Thus if the processor is in state m and has the word $w'\alpha'\sigma\alpha''w''$ on the tape, where the symbol indicated by σ is read, and if $T(m,\sigma) = (m',\sigma',+1)$, then next the processor will be in state m' and has the word $w'\alpha'\sigma'\alpha''w''$ on the tape, where the symbol indicated by α'' is read; similarly if $T(m,\sigma) = (m',\sigma',-1)$.

We assume that M contains a certain "start state" 0 and a certain "halting state" ∞. Moreover, Σ is assumed to contain a symbol $_-$ meaning "blank." (This is necessary to identify the beginning and the end of a word on the tape.)

We say that the Turing machine T *accepts* a word $w \in (\Sigma \setminus \{_-\})^*$ if, when starting in state 0 and with word w on the tape (all other symbols being blank), so that the read-write head is reading the first symbol of w, then after a finite number of iterations, the processor is in the halting state ∞. (If w is the empty word, the symbol read initially is the blank symbol $_-$.)

Let Π be the set of words accepted by T. So Π is a problem. We say that T *solves* Π. Moreover, we say that T *solves* Π *in polynomial time* if there exists a polynomial $p(x)$ such that if T accepts a word w, it accepts w in at most $p(\text{size}(w))$ iterations.

It is not difficult to see that the concept of algorithm defined in Section 9.4 above is at least as powerful as that of a Turing machine. We can encode any state of the computer model (processor+tape+read-write head) by a word (w', m, w''). Here m is the state of the processor and $w'w''$ is the word on the tape, while the first symbol of w'' is read. We define an algorithm A by:

> replace subword m,σ by σ',m', whenever $T(m,\sigma) = (m',\sigma',+1)$
> and $m \neq \infty$;
> replace subword α,m,σ by $m',\alpha\sigma'$, whenever $T(m,\sigma) = (m',\sigma',-1)$
> and $m \neq \infty$.

To be precise, we should assume here that the symbols indicating the states in M do not belong to Σ. Moreover, we assume that the symbols (and) are not in Σ. Furthermore, to give the algorithm a start, it contains the tasks of replacing subword $_-\alpha$ by the word $(,0,\alpha$, and subword α_- by $\alpha)$ (for any α in $\Sigma \setminus \{_-\}$). Then, when starting with a word w, the first two iterations transform it to the word $(,0,w)$. After that, the rules stated above simulate the Turing machine iterations. The iterations stop as soon as we arrive at state ∞.

So T accepts a word w if and only if A accepts w—in (about) the same number of iterations. That is, T solves a problem Π (in polynomial time), if and only if A solves Π (in polynomial time).

This shows that the concept of "algorithm" defined in Section 9.4 is at least as powerful as that of a Turing machine. Conversely, it is not hard (although technically somewhat complicated) to simulate an algorithm by a Turing machine. But how powerful is a Turing machine?

One could think of several objections to the Turing machine. It uses only one tape, which should serve as an input tape, and as a memory, and as an output tape. We have only limited access to the information on the tape (we can shift only one position at a time). Moreover, the computer program seems to be implemented in the "hardware" of the computer model; the Turing machine solves only one problem.

To counter these objections, several other computer models have been proposed that model a computer more realistically: multitape Turing machines, random access machines (RAMs), the universal Turing machine. However, from a polynomial-time algorithmic point of view, these models all turn out to be equivalent. Any problem that can be solved in polynomial time by any of these computer models, can also be solved in polynomial time by some one-tape Turing machine, and hence by an algorithm in the sense of Section 9.4. We refer to Aho, Hopcroft and Ullman [1974] for an extensive discussion.

APPENDIX A

Linear Programming

Linear programming concerns the problem of finding a vector x that maximizes a given linear function $c^T x$, where x ranges over all vectors satisfying a given system $Ax \leq b$ of linear inequalities. So among all vectors x satisfying $Ax \leq b$, we want to find one with the largest value $c^T x$.

This is a problem that occurs in a variety of practical situations, and for which several fast solution methods have been developed. The most prominent of these is the *simplex method*, designed by G.B. Dantzig in the late 1940s. In geometrical terms, it consists of making a "trip" along the vertices and edges of the polyhedron $\{x : Ax \leq b\}$, until an "optimal" vertex is attained.

The simplex method turns out to work very well in practice. However, no one has been able to prove that the number of vertices traversed can be bounded by a polynomial in the problem dimensions (the dimensions of the matrix A).

It was an unanswered question whether or not the linear-programming problem can be solved in polynomial time, until L.G. Khachiyan answered this in the affirmative in 1979 using a variant of the "ellipsoid method" for nonlinear programming. This method, however, turned out to be slow in practice. In 1984, N. Karmarkar published his revolutionary "interior-point method," which not only was proved to be a polynomial-time algorithm, but also works very fast in practice, in some cases faster than the simplex method.

In this book, knowledge of the polynomial-time methods for linear programming may be helpful, but is not a strict prerequisite. However, the duality theory of linear programming and the related theory of linear inequalities are used very often. Therefore, in this appendix, we give an overview of these theoretical aspects. We also give a brief discussion of the simplex method. For a detailed treatment of these topics see Chvátal [1983] and Schrijver [1986].

325

Farkas' Lemma

Let A be an $m \times n$ matrix, and let $b \in \mathbf{R}^m$. With Gaussian elimination one can prove that $Ax = b$ has a solution x, if and only if there is no solution y for the system of linear equations

$$y^T A = 0, y^T b = -1.$$

An analogous characterization for the existence of a solution to $Ax \leq b$ can be derived from an elimination procedure for systems of linear inequalities known as *Fourier-Motzkin elimination*.

To start this procedure, note that since we may multiply inequalities in $Ax \leq b$ by positive scalars without altering the set of solutions, we may assume (after reordering the inequalities) that the system has the form

$$
\begin{array}{rcll}
x_1 + (a_i')^T x' & \leq & b_i, & \text{for all } i \in \{1, \ldots, m'\} \\
-x_1 + (a_i')^T x' & \leq & b_i, & \text{for all } i \in \{m'+1, \ldots, m''\} \\
(a_i')^T x' & \leq & b_i, & \text{for all } i \in \{m''+1, \ldots, m\}
\end{array}
$$

where $x' = (x_2, \ldots, x_n)$ and $(a_i')^T$ is the ith row of A with the first entry deleted. This system has a solution if and only if

$$
\begin{array}{rcll}
(a_j')^T x' - b_j & \leq & b_i - (a_i')^T x', & \text{for all } i \in \{1, \ldots, m'\} \text{ and} \\
& & & j \in \{m'+1, \ldots, m''\} \\
(a_i')^T x' & \leq & b_i, & \text{for all } i \in \{m''+1, \ldots, m\}
\end{array}
$$

has a solution (since we may choose an appropriate value for x_1). Writing this reduced system in the usual form we have that $Ax \leq b$ has a solution if and only if

$$
\begin{array}{rcll}
(a_i' + a_j')^T x' & \leq & b_i + b_j, & \text{for all } i \in \{1, \ldots, m'\} \text{ and} \\
& & & j \in \{m'+1, \ldots, m''\} \\
(a_i')^T x' & \leq & b_i, & \text{for all } i \in \{m''+1, \ldots, m\}
\end{array}
$$

has a solution. So we have eliminated variable x_1 (at the expense of having many more inequalities). We use this in an inductive proof of the following result.

Theorem A.1 *(Farkas' Lemma for Inequalities) The system $Ax \leq b$ has a solution x, if and only if there is no vector y satisfying $y \geq 0, y^T A = 0$ and $y^T b < 0$.*

Proof: Suppose $Ax \leq b$ has a solution $\tilde{x}$, and suppose there exists a vector $\tilde{y} \geq 0$ satisfying $\tilde{y}^T A = 0$ and $\tilde{y}^T b < 0$. Then we obtain the contradiction

$$0 > \tilde{y}^T b \geq \tilde{y}^T (A\tilde{x}) = (\tilde{y}^T A)\tilde{x} = 0.$$

Now suppose that $Ax \le b$ has no solution. If A has only one column, then the result is easy. Otherwise, apply Fourier-Motzkin elimination to obtain a system $A'x' \le b'$ with one less variable. Since $A'x' \le b'$ also has no solution, we can assume by induction that there exists a vector $y' \ge 0$ satisfying $y'^T A' = 0$ and $y'^T b' < 0$. Now since each inequality in $A'x' \le b'$ is the sum of positive multiples of inequalities in $Ax \le b$, we can use y' to construct a vector y satisfying the conditions in the theorem. ∎

There exist several variants that can be easily derived from Theorem A.1. Results of this form are often called "Farkas' Lemma," since Gy. Farkas first proved such a statement in 1894. The variant that Farkas actually proved was the following.

Corollary A.2 *(Farkas' Lemma) The system $Ax = b$ has a nonnegative solution, if and only if there is no vector y satisfying $y^T A \ge 0$ and $y^T b < 0$.*

Proof: Define

$$A' = \begin{pmatrix} A \\ -A \\ -I \end{pmatrix}, b' = \begin{pmatrix} b \\ -b \\ 0 \end{pmatrix},$$

where I denotes the identity matrix. Then $Ax = b$ has a nonnegative solution x, if and only if the system $A'x' \le b'$ has a solution x'. Applying Farkas' Lemma for Inequalities to $A'x' \le b'$ gives the corollary. ∎

A consequence of Farkas' Lemma is the following result.

Corollary A.3 *Suppose the system $Ax \le b$ has at least one solution. Then every solution x of $Ax \le b$ satisfies $c^T x \le \delta$, if and only if there exists a vector $y \ge 0$ such that $y^T A = c^T$ and $y^T b \le \delta$.*

Proof: If such a vector y exists, then for every vector x one has

$$Ax \le b \implies y^T Ax \le y^T b \implies c^T x \le y^T b \implies c^T x \le \delta.$$

On the other hand, suppose such a vector y does not exist. This means that the following system of linear equations in the variables y and λ has no solution $(y^T \ \lambda) \ge (0 \ 0)$:

$$(y^T \ \lambda) \begin{pmatrix} A & b \\ 0 & 1 \end{pmatrix} = (c^T \ \delta).$$

According to Farkas' Lemma this implies that there exists a vector $\begin{pmatrix} z \\ \mu \end{pmatrix}$ so that

$$\begin{pmatrix} A & b \\ 0 & 1 \end{pmatrix} \begin{pmatrix} z \\ \mu \end{pmatrix} \ge \begin{pmatrix} 0 \\ 0 \end{pmatrix} \text{ and } (c^T \ \delta) \begin{pmatrix} z \\ \mu \end{pmatrix} < 0.$$

Suppose that $\mu = 0$. Then $Az \geq 0$ and $c^T z < 0$. However, by assumption, $Ax \leq b$ has a solution x^0. Then, for large enough $\tau \in \mathbf{R}$ we have

$$A(x^0 - \tau z) \leq b \text{ and } c^T(x^0 - \tau z) > \delta.$$

This contradicts the fact that $Ax \leq b$ implies $c^T x \leq \delta$.

Now suppose that $\mu > 0$. Then for $x = -\mu^{-1}z$ one has:

$$Ax \leq b \text{ and } c^T x > \delta.$$

Again this contradicts the fact that $Ax \leq b$ implies $c^T x \leq \delta$.   ∎

Duality Theorem

One of the standard forms of a linear-programming (LP) problem is

$$\text{Maximize } c^T x \tag{A.1}$$
$$\text{subject to}$$
$$Ax \leq b.$$

So linear programming can be considered as maximizing a "linear function" $c^T x$ over a polyhedron $P = \{x : Ax \leq b\}$. Geometrically, this can be seen as shifting a hyperplane to its "highest" level, under the condition that it intersects P.

Problem (A.1) corresponds to determining

$$\max\{c^T x : Ax \leq b\}. \tag{A.2}$$

Clearly, any *minimization* problem can be transformed to form (A.2), since

$$\min\{c^T x : Ax \leq b\} = -\max\{-c^T x : Ax \leq b\}.$$

We say that x is a *feasible solution* of (A.2) if x satisfies $Ax \leq b$. If x moreover attains the maximum, x is called an *optimum solution*.

The value of the maximum (A.2) is related to the minimum value of another, so-called *dual* LP problem:

$$\min\{y^T b : y \geq 0, y^T A = c^T\}.$$

An easy result is the following Weak Duality Theorem.

Theorem A.4 *(Weak Duality Theorem) Let A be an $m \times n$ matrix, $b \in \mathbf{R}^m$, $c \in \mathbf{R}^n$. Suppose that $\tilde{x}$ is a feasible solution to $Ax \leq b$ and $\tilde{y}$ is a feasible solution to $y \geq 0, y^T A = c^T$. Then*

$$c^T \tilde{x} \leq \tilde{y}^T b.$$

Proof: This follows from the computation $c^T \tilde{x} = (\tilde{y}^T A)\tilde{x} = \tilde{y}^T(A\tilde{x}) \leq \tilde{y}^T b.$ ∎

The stronger Duality Theorem of Linear Programming, due to von Neumann in 1947, can be derived from Farkas' Lemma for Inequalities.

Theorem A.5 *(Duality Theorem) Let A be an $m \times n$ matrix, $b \in \mathbf{R}^m$, $c \in \mathbf{R}^n$. Then*

$$\max\{c^T x : Ax \le b\} = \min\{y^T b : y \ge 0, y^T A = c^T\}, \qquad (A.3)$$

provided that both sets are nonempty.

Proof: First note that the Weak Duality Theorem implies that

$$\sup\{c^T x : Ax \le b\} \le \inf\{y^T b : y \ge 0, y^T A = c^T\}. \qquad (A.4)$$

Define $\delta = \sup\{c^T x : Ax \le b\}$. Then by (A.4) we have $\delta \le \inf\{y^T b : y \ge 0, y^T A = c^T\}$.

Now, by the definition of δ, we know that

$$\text{if } Ax \le b \text{ then } c^T x \le \delta.$$

So, by Corollary A.3, there exists a vector y such that

$$y \ge 0, y^T A = c^T, y^T b \le \delta.$$

This implies that the infimum in (A.4) is attained, and is equal to δ. So the minimum in (A.3) exists, and is equal to δ.

We finally should prove that $\max\{c^T x : Ax \le b\}$ exists, and equals δ. Thus, we should prove that there exists a vector x satisfying

$$Ax \le b, c^T x \ge \delta,$$

that is,

$$Ax \le b, -c^T x \le -\delta.$$

Suppose such a vector x does not exist. Then there exists, by Farkas' Lemma for Inequalities, a vector $z \ge 0$ and a real $\lambda \ge 0$ such that:

$$z^T A - \lambda c^T = 0, z^T b - \lambda \delta < 0.$$

Since by assumption, $Ax \le b$ has a solution x_0, we know $\lambda > 0$ (since if $\lambda = 0$ then $0 = z^T A x_0 \le z^T b < 0$). Let $y = z/\lambda$. Then $y \ge 0, y^T A = c^T$ and $y^T b < \delta$. So the infimum in (A.4) is strictly less than δ, contradicting (A.4). ∎

It is easy to derive variants of the Duality Theorem for other forms of LP problems (see also the Exercises). In this way, we obtain the notion of a "dual LP" for any form of LP problem.

Corollary A.6 *Let A be an $m \times n$ matrix, $b \in \mathbf{R}^m, c \in \mathbf{R}^n$. Then*

$$\max\{c^T x : x \geq 0, Ax = b\} = \min\{y^T b : y^T A \geq c^T\},$$

provided that both sets are nonempty.

Proof: Define

$$\tilde{A} = \begin{pmatrix} A \\ -A \\ -I \end{pmatrix}, \tilde{b} = \begin{pmatrix} b \\ -b \\ 0 \end{pmatrix}.$$

Then

$$
\begin{aligned}
\max\{c^T x \quad &: \quad x \geq 0, Ax = b\} \\
&= \quad \max\{c^T x : \tilde{A} x \leq \tilde{b}\} \\
&= \quad \min\{z^T \tilde{b} : z \geq 0, z^T \tilde{A} = c^T\} \\
&= \quad \min\{u^T b - v^T b + w^T 0 : u, v, w \geq 0, u^T A - v^T A - w^T = c^T\} \\
&= \quad \min\{y^T b : y^T A \geq c^T\}.
\end{aligned}
$$

The last equality follows by taking $y = u - v$. ∎

We finally note the idea of *complementary slackness*. Let x^* and y^* be feasible solutions to $\max\{c^T x : Ax \leq b\}$ and $\min\{y^T b : y \geq 0, y^T A = c^T\}$, respectively. Denote the rows of A by $a_1, \ldots, a_m$.

Suppose that x^* and y^* attain the maximum and minimum, respectively, and that for a certain $i \in \{1, \ldots, m\}$, $a_i x^* < b_i$ holds. Then we must have $y_i^* = 0$, since if $y_i^* > 0$ we have the contradiction

$$y^{*T} b - c^T x^* = y^{*T}(b - Ax^*) \geq (y_i^*)^T(b_i - a_i x^*) > 0.$$

Conversely, if

$$\text{for each } i \in \{1, \ldots, m\}, \text{ either } y_i^* = 0 \text{ or } a_i x^* = b_i, \qquad (A.5)$$

then

$$y^{*T} b - c^T x^* = y^{*T}(b - Ax^*) = \sum_{i=1}^{m} y_i^*(b_i - a_i x^*);$$

together with the Weak Duality Theorem, this implies that x^* and y^* are optimal solutions to the two LP problems.

The conditions stated in (A.5) are known as the *complementary slackness conditions* for the pair of LP problems.

Theorem A.7 *(Complementary Slackness Theorem) Let x^* be a feasible solution of $\max\{c^T x : Ax \leq b\}$, and let y^* be a feasible solution of $\min\{y^T b : y \geq 0, y^T A = c^T\}$. Then x^* and y^* are optimum solutions for the maximum and minimum, respectively, if and only if the complementary slackness conditions (A.5) hold.* ∎

Simplex Algorithm

The simplex algorithm is currently the most widely used method to solve linear-programming problems in practice. We give an outline of the algorithm and refer the reader to the nice description in Chvátal [1983] for more details.

We will work with LP problems of the form

$$\text{Maximize } c^T x \tag{A.6}$$
$$\text{subject to}$$
$$Ax = b$$
$$x \geq 0$$

where A is an m by n matrix and $rank(A) = m$. (The simplex algorithm can also be described for other forms of LP problems.)

Suppose that the columns of A are indexed by the set T. For any set $B \subseteq T$, let A_B denote the matrix consisting of the columns in B, let c_B denote the corresponding components of the vector c, and let x_B denote the corresponding variables. If $|B| = m$ and the columns of A_B are linearly independent, then B is called a *basis* of A.

Given a basis B, we define a solution x to the system $Ax = b$ by setting $x_i = 0$ for all $i \in T \setminus B$ and finding the unique solution to the system $A_B x_B = b$. We call this the *basic solution* corresponding to the basis B. We also define a solution y to the dual linear-programming problem

$$\text{Minimize } y^T b \tag{A.7}$$
$$\text{subject to}$$
$$y^T A \geq c.$$

by finding the unique solution to the linear system $y^T A_B = c_B$.

The primal-dual pair of solutions corresponding to B has a nice property. Namely, if any variable x_i is positive, then the corresponding dual constraint $y^T a_i \geq c_i$ holds as an equation. Thus, if x and y are feasible solutions to (A.6) and (A.7) respectively, then by complementary slackness, we know that they are optimal solutions.

A basis B is called *feasible* if x_B is nonnegative, that is, when the corresponding basic solution is feasible for (A.6). The simplex algorithm generates a sequence of feasible bases. At each step, the dual solution y is checked for feasibility. If y is feasible, we can stop since we know that x and y are optimal.

So suppose we have a basis B and basic solution x that is feasible, but not optimal, for (A.6). Choose some index $i \in T$ such that $y^T a_i < c_i$. We will modify B by adding i and removing some other index j. To this end, we find the unique solution to the system $A_B z = a_i$. So z describes the entering column i in terms of the current basis.

For any nonnegative $\epsilon \in R$, we can obtain a new solution to $Ax = b$ by

$$\text{replacing } x_B \text{ by } x_B - \epsilon z \text{ and letting } x_i = \epsilon. \qquad \text{(A.8)}$$

The objective value of the new solution is

$$
\begin{aligned}
c_B^T(x_B - \epsilon z) + c_i \epsilon &= c_B^T x_B + \epsilon(c_i - c_B^T z) \qquad \text{(A.9)} \\
&= c_B^T x_B + \epsilon(c_i - y^T A_B z) = c_B^T x_B + \epsilon(c_i - y^T a_i).
\end{aligned}
$$

Now since $c_i - y^T a_i > 0$, any positive value of ϵ will improve the solution. Since we want our solution to stay feasible, we choose the largest ϵ such that $x_B - \epsilon z \geq 0$.

Notice that if there is no such largest ϵ, then we can make the objective value arbitrarily large by letting ϵ take on arbitrarily large values, that is, the LP problem is *unbounded*. So in this case we can stop the algorithm. Otherwise, there must be some index $j \in B$ such that $z_j > 0$ and the jth component of $x_B - \epsilon z$ is 0. We choose one such index j to leave the basis.

The process of selecting j is often called the *ratio test*, since j is any index in B such that $z_j > 0$ and j minimizes the ratio x_k / z_k over all $k \in B$ having $z_k > 0$.

Let $B' = (B \cup \{i\}) \setminus \{j\}$. Notice that B' is a basis (since z_j is nonzero) and that the basic solution corresponding to B' is just the solution we described in (A.8). We replace B by B' and x_B by the solution given in (A.8) and repeat the procedure.

The algorithm can be summarized as follows:

Simplex Algorithm with a Starting Basis

Loop
    Find the unique solution to $y^T A_B = c_B$;
    If $y^T a_i \geq c_i$ for all $i \in T \setminus B$
        Stop, the current solution is optimal.
    Else
        Choose i such that $y^T a_i < c_i$;
        Find the unique solution to $A_B z = a_i$;
        Find the largest ϵ such that $x_B - \epsilon z \geq 0$;
        If ϵ does not exist
            Stop, the LP problem is unbounded.
        Else
            Choose $j \in B$ such that $z_j > 0$ and the
            jth component of $x_B - \epsilon z$ is 0;
            Replace B by $(B \cup \{i\}) \setminus \{j\}$ and
            x_B by $x_B - \epsilon z$ and $x_i = \epsilon$;

From the above discussion, it is clear that if the simplex algorithm terminates then it either finds an optimal solution or shows that the LP problem is unbounded. It is not clear however, that the algorithm will eventually terminate. By equation (A.9) we see that the objective value will never decrease during the course of the algorithm. If we knew that the objective function actually increased, then we could conclude that the method will indeed terminate since there are only finitely many choices for B. (The increasing objective means that we will never repeat a basis B.)

Unfortunately, the ratio test may give us $\epsilon = 0$. So the objective value may stay the same from one iteration to the next. Indeed, it is possible to construct examples where the simplex algorithm will run through a *cycle* of basic solutions, all giving the same objective value and ending up with the same basis that started the cycle.

A way to handle the problem of cycling is to make use of the flexibility we have in choosing the entering index i and leaving index j. A recipe for selecting i and j is known as a "pivot rule." There are a number of such rules that will prevent cycling. (See Chvátal [1983].)

Another issue that needs to be handled is the fact that we need an initial feasible basis to begin the procedure. Indeed, there may not exist any feasible solutions to the LP problem. Moreover, even if there are solutions, if A is not of full row rank then there are no basic solutions.

These problems can be dealt with by multiplying some of the equations $Ax = b$ by -1 if necessary so that we may assume that $b \geq 0$, then considering the auxiliary problem

$$\text{Maximize } \sum(-x'_i \; : \; i = 1, \ldots n) \qquad\qquad \text{(A.10)}$$

$$\text{subject to}$$

$$(\; A \quad I \;) \begin{pmatrix} x \\ x' \end{pmatrix} = b$$

$$x \geq 0, x' \geq 0.$$

The original LP problem (A.6) has a feasible solution if and only if the optimal value of (A.10) is 0. So we can test whether or not (A.6) is feasible by solving (A.10) with the simplex algorithm, starting with the basis corresponding to the new variables x'.

Suppose that the algorithm terminates with an optimal basis B that gives a solution with objective value 0. Let y be the corresponding dual solution. By applying complementary slackness to (A.10), we can conclude that for any index $i \in T$ such that $y_i^T a_i > 0$ the corresponding variable x_i must be 0 in any feasible solution to (A.6). So these variables can be removed from the problem. Moreover, if we extend (A.6) by adding the variables x'_k such that $k \in B \setminus T$ (setting $c_k = 0$), then complementary slackness applied to (A.10) implies that these new variables will take on the value 0 in every feasible solution to the extended LP. Therefore, we can solve the extended problem, using B as our initial feasible basis.

Exercises

A.1. Prove that there exists a vector $x \geq 0$ such that $Ax \leq b$, if and only if for each $y \geq 0$ satisfying $y^T A \geq 0$ one has $y^T b \geq 0$.

A.2. (Stiemke's Theorem) Prove that there exists a vector $x > 0$ such that $Ax = 0$, if and only if for each y satisfying $y^T A \geq 0$ one has $y^T A = 0$.

A.3. (Gordan's Theorem) Prove that there exists a vector $x \neq 0$ satisfying $x \geq 0$ and $Ax = 0$, if and only if there is no vector y satisfying $y^T A > 0$.

A.4. Prove that there exists a vector x satisfying $Ax < b$, if and only if $y = 0$ is the only solution for $y \geq 0, y^T A = 0, y^T b \leq 0$.

A.5. (Motzkin's Theorem) Prove that there exists a vector x satisfying $Ax < b$ and $A'x \leq b'$, if and only if for all vectors $y, y' \geq 0$ one has:

(i) if $y^T A + y'^T A' = 0$ then $y^T b + y'^T b' \geq 0$, and

(ii) if $y^T A + y'^T A' = 0$ and $y \neq 0$ then $y^T b + y'^T b' > 0$.

A.6. Prove that

$$\max\{c^T x : x \geq 0, Ax \leq b\} = \min\{y^T b : y \geq 0, y^T A \geq c^T\}$$

assuming that both sets are nonempty.

A.7. Prove that

$$\max\{c^T x : Ax \geq b\} = \min\{y^T b : y \leq 0, y^T A = c^T\}$$

assuming that both sets are nonempty.

A.8. Let $A, B, C, D, E, F, G, H, K$ be matrices, a, b, c column vectors, and d, e, f row vectors (of appropriate dimensions) and prove that

$$
\begin{aligned}
\max\{dx + ey + fz : \quad & x \geq 0, z \leq 0, \\
& Ax + By + Cz \leq a, \\
& Dx + Ey + Fz = b, \\
& Gx + Hy + Kz \geq c\} \\
= \quad \min\{ua + vb + wc : \quad & u \geq 0, w \leq 0, \\
& uA + vD + wG \geq d, \\
& uB + vE + wH = e, \\
& uC + vF + wK \leq f\},
\end{aligned}
$$

assuming that both sets are nonempty.

A.9. (i) Prove that for any matrix A and vectors b and c (of appropriate dimensions) one has

$$\sup\{c^T x : Ax \leq b\} = \inf\{y^T b : y \geq 0, y^T A = c^T\},$$

provided that *at least one* of the sets is nonempty.

(ii) Give an example of a matrix A and vectors b and c for which both $\{x : Ax \leq b\}$ and $\{y : y \geq 0, y^T A = c^T\}$ are empty.

A.10. Prove that if $\max\{c^T x : Ax \leq b\}$ has an optimal solution, then $\min\{y^T b : y \geq 0, y^T A = c^T\}$ also has an optimal solution.

A.11. State and prove a complementary slackness theorem for the LP problems given in Exercise A.6.

A.12. Describe the simplex algorithm for linear-programming problems of the form $\max\{w^T x : Ax = b, 0 \leq x \leq u\}$, where $u \in \mathbf{R}^n$ is a vector of upper bounds on the variables.

Bibliography

[1974] A.V. Aho, J.E. Hopcroft, and J.D. Ullman, *The Design and Analysis of Computer Algorithms*, Addison-Wesley, Reading, 1974.

[1987] R.P. Anstee, "A polynomial algorithm for b-matchings: An alternative approach," *Information Processing Letters* 24 (1987) 153–157.

[1995] D. Applegate, R. Bixby, V. Chvátal, and W. Cook, "Finding cuts in the TSP," *DIMACS Technical Report* 95-05, 1995.

[1993] D. Applegate and W. Cook, "Solving large-scale matching problems," in: *Algorithms for Network Flows and Matching* (D.S. Johnson and C.C. McGeoch, eds.), American Mathematical Society, 1993, pp. 557–576.

[1983] J. Aráoz, W.H. Cunningham, J. Edmonds, and J. Green-Krótki, "Reductions to 1-matching polyhedra," *Networks* 13 (1983) 455–483.

[1983] M.O. Ball and U. Derigs, "An analysis of alternative strategies for implementing matching algorithms," *Networks* 13 (1983) 517–549.

[1989] F. Barahona and W.H. Cunningham, "On dual integrality in matching problems," *Operations Research Letters* 8 (1989) 245–249

[1958] R.E. Bellman, "On a routing problem," *Quarterly of Applied Mathematics* 16 (1958) 87–90.

[1957] C. Berge, "Two theorems in graph theory," *Proceedings of the National Academy of Sciences* (U.S.A.) 43 (1957) 842–844.

[1958] C. Berge, "Sur le couplage maximum d'un graphe," *Comptes Rendus de l'Académie des Sciences Paris, series 1, Mathématique* 247 (1958), 258–259.

[1995] D. Bertsimas, C. Teo, and R. Vohra, "Nonlinear relaxations and improved randomized approximation algorithms for multicut problems," in: *Proceedings of the 4th IPCO Conferences* (E. Balas and J. Clausen, eds.), *Lecture Notes in Computer Science* 920, Springer, 1995, pp. 29–39.

[1946] G. Birkhoff, "Tres observaciones sobre el algebra lineal," *Revista Facultad de Ciencias Exactas, Puras y Aplicadas Universidad Nacional de Tucuman, Serie A (Matematicas y Fisica Teorica)* 5 (1946) 147–151.

[1961] R.G. Busacker and P.J. Gowen, "A procedure for determining a family of minimal cost network flow patterns," *Technical Paper* 15, Operations Research Office, 1961.

[1983] R.W. Chen, Y. Kajitani, and S.P. Chan, "A graph theoretic via minimization algorithm for two-layer printed circuit boards," *IEEE Transactions on Circuits and Systems* 30 (1983) 284–299.

[1997] C.S. Chekuri, A.V. Goldberg, D.R. Karger, M.S. Levine, and C. Stein, "Experimental study of minimum cut algorithms," in: *Proceedings of the 8th Annual ACM-SIAM Symposium on Discrete Algorithms*, 1997, pp. 324–333.

[1989] J. Cheriyan and S.N. Maheshwari, "Analysis of preflow push algorithms for maximum network flow," *SIAM Journal on Computing* 18 (1989) 1057–1086.

[1976] N. Christofides, "Worst-case analysis of a new heuristic for the travelling salesman problem," Report 388, Graduate School of Industrial Administration, Carnegie Mellon University, Pittsburgh, PA, 1976.

[1973] V. Chvátal, "Edmonds polytopes and a hierarchy of combinatorial problems," *Discrete Mathematics* 4 (1973) 305–337.

[1973a] V. Chvátal, "Edmonds polytopes and weakly hamiltonian graphs," *Mathematical Programming* 5 (1973) 29–40.

[1983] V. Chvátal, *Linear Programming*, Freeman, New York, 1983.

[1985] V. Chvátal, "Cutting planes in combinatorics," *European Journal of Combinatorics* 6 (1985) 217–226.

[1989] V. Chvátal, W. Cook and M. Hartmann, "On cutting-plane proofs in combinatorial optimization," *Linear Algebra and its Applications* 114/115 (1989) 455–499.

[1997] W. Cook and A. Rohe, "Computing minimum-weight perfect matchings," Report Number 97863, Research Institute for Discrete Mathematics, Universität Bonn, 1997.

[1976] W.H. Cunningham, "A network simplex method," *Mathematical Programming* 11 (1976) 105–116.

[1979] W.H. Cunningham, "Theoretical properties of the network simplex method," *Mathematics of Operations Research* 4 (1979) 196–208.

[1978] W.H. Cunningham and A.B. Marsh III, "A primal algorithm for optimum matching," *Mathematical Programming Study* 8 (1978) 50–72.

[1954] G. Dantzig, R. Fulkerson, and S. Johnson, "Solution of a large-scale traveling-salesman problem," *Operations Research* 2 (1954) 393–410.

[1991] U. Derigs and A. Metz, "Solving (large scale) matching problems combinatorially," *Mathematical Programming* 50 (1991) 113–122.

[1959] E. Dijkstra, "A note on two problems in connexion with graphs," *Numerische Mathematik* 1 (1959) 269–271.

[1970]   E.A. Dinits, "Algorithm for solution of a problem of maximum flow in a network with power estimation," *Soviet Mathematics Doklady* 11 (1970) 1277–1280.

[1987]   H. Edelsbrunner, *Algorithms in Combinatorial Geometry*, Springer-Verlag, Berlin, 1987.

[1965]   J. Edmonds, "Paths, trees, and flowers," *Canadian Journal of Mathematics* 17 (1965) 449–467.

[1965a] J. Edmonds, "Maximum matching and a polyhedron with 0,1-vertices," *Journal of Research of the National Bureau of Standards* (B) 69 (1965) 125–130.

[1970]   J. Edmonds, "Matroids, submodular functions, and certain polyhedra," in: *Combinatorial Structures and Their Applications* (R.K. Guy, H. Hanani, N. Sauer, and J. Schönheim, eds.), Gordon and Breach, New York, 1970, pp. 69–87.

[1971]   J. Edmonds, "Matroids and the greedy algorithm," *Mathematical Programming* 1 (1971) 127–136.

[1965]   J. Edmonds and D.R.Fulkerson, "Transversal and matroid partition," *Journal of Research of the National Bureau of Standards* B 69 (1965) 147–153.

[1977]   J. Edmonds and R. Giles, "A min-max relation for submodular functions on graphs," in: *Studies in Integer Programming* (P.L. Hammer, *et al.* eds.), *Annals of Discrete Mathematics* 1 (1977) 185–204.

[1970]   J. Edmonds and E.L. Johnson, "Matching: A well-solved class of integer linear programs," in: *Combinatorial Structures and their Applications* (R.K. Guy, H. Hanani, N. Sauer, and J. Schönheim, eds.), Gordon and Breach, New York, 1970, pp. 89–92.

[1973]   J. Edmonds and E.L. Johnson, "Matching, Euler tours, and the Chinese postman," *Mathematical Programming* 5 (1973) 88–124.

[1969]   J. Edmonds, E.L. Johnson, and S.C. Lockhart, "Blossom I, a code for matching," unpublished report, IBM T.J. Watson Research Center, Yorktown Heights, New York.

[1972]   J. Edmonds and R.M. Karp, "Theoretical improvements in algorithmic efficiency for network flow problems," *Journal of the Association for Computing Machinery* 19 (1972) 248–264.

[1956]   L.R. Ford, Jr., "Network flow theory," Paper P-923, RAND Corporation, Santa Monica, California, 1956.

[1956]   L.R. Ford, Jr. and D.R. Fulkerson, "Maximal flow through a network," *Canadian Journal of Mathematics* 8 (1956) 399–404.

[1957]   L.R. Ford, Jr. and D.R. Fulkerson, "A primal-dual algorithm for the capacitated Hitchcock problem," *Naval Research Logistics Quarterly* 4 (1957) 47–54.

[1958] L.R. Ford, Jr. and D.R. Fulkerson, "Suggested computation for maximal multi-commodity network flows," *Management Science* 5 (1958) 97–101.

[1962] L.R. Ford, Jr. and D.R. Fulkerson, *Flows in Networks*, Princeton University Press, Princeton, New Jersey, 1962.

[1981] A. Frank, "A weighted matroid intersection algorithm," *Journal of Algorithms* 2 (1981) 328–336.

[1986] S. Fujishige, "A capacity-rounding algorithm for the minimum-cost circulation problem: A dual framework for the Tardos algorithm," *Mathematical Programming* 35 (1986) 298–308.

[1990] H. Gabow, "Data structures for weighted matching and nearest common ancestors," in: *Proceedings of the 1st Annual ACM-SIAM Symposium on Discrete Algorithms*, ACM, New York, 1990, pp. 434–443.

[1957] D. Gale, "A theorem on flows in networks," *Pacific Journal of Mathematics* 7 (1957) 1073–1082.

[1986] G. Gallo and S. Pallottino, "Shortest path methods: A unifying approach," *Mathematical Programming Study* 26 (1986) 38–64.

[1979] M.R. Garey and D.S. Johnson, *Computers and Intractability: A Guide to the Theory of NP-completeness*, Freeman, San Francisco, 1979.

[1995] A.M.H.Gerards, "Matching," in: *Network Models* (M.O. Ball, T.L. Magnanti, C.L. Monma, and G.L. Nemhauser, eds.), North Holland, Amsterdam, 1995.

[1986] A.M.H. Gerards and A. Schrijver, "Matrices with the Edmonds-Johnson property," *Combinatorica* 6 (1986) 403–417.

[1979] F.R. Giles and W.R. Pulleyblank, "Total dual integrality and integer polyhedra," *Linear Algebra and its Applications* 25 (1979) 191–196.

[1995] M.X. Goemans and D.P. Williamson, "A general approximation technique for constrained forest problems," *SIAM Journal on Computing* 24 (1995) 296–317.

[1985] A.V. Goldberg, "A new max-flow algorithm," *Technical Report* MIT/-LCS/TM 291, Laboratory for Computer Science, M.I.T. (1985).

[1988] A.V. Goldberg and R.E. Tarjan, "A new approach to the maximum flow problem," *Journal of the Association for Computing Machinery* 35 (1988) 921–940.

[1989] A.V. Goldberg and R.E. Tarjan, "Finding minimum-cost circulations by canceling negative cycles," *Journal of the Association for Computing Machinery* 33 (1989) 873–886.

[1960] R.E. Gomory, "Solving linear programming problems in integers," in: *Combinatorial Analysis* (R. Bellman and M. Hall, eds.), *Proceedings of Symposia in Applied Mathematics* X, American Mathematical Society, Providence, 1960, pp. 211–215.

[1961] R.E. Gomory and T.C. Hu, "Multi-terminal network flows," *SIAM Journal on Applied Mathematics* 9 (1961) 551–556.

[1988] M. Grötschel and O. Holland, "Solution of large-scale symmetric travelling salesman problems," *Mathematical Programming* 51 (1991) 141–202.

[1988] M. Grötschel, L. Lovász, and A. Schrijver, *Geometric Algorithms and Combinatorial Optimization*, Springer-Verlag, Berlin, 1988.

[1979] M. Grötschel and M.W. Padberg, "On the symmetric travelling salesman problem II: Lifting theorems and facets," *Mathematical Programming* 16 (1979) 282–302.

[1985] M. Grötschel and M.W. Padberg, "Polyhedral theory," in: *The Traveling Salesman Problem* (E.L. Lawler, J.K. Lenstra, A.H.G. Rinnooy Kan, and D. Shmoys, eds.), Wiley, Chichester, 1985, pp. 251–306.

[1986] M. Grötschel and W.R. Pulleyblank, "Clique tree inequalities and the symmetric travelling salesman problem," *Mathematics of Operations Research* 11 (1986) 537–569.

[1990] D. Gusfield, "Very simple methods for all pairs network flow analysis," *SIAM Journal on Computing* 19 (1990) 143–155.

[1975] F.O. Hadlock, "Finding a maximum cut in a planar graph in polynomial time," *SIAM Journal on Computing* 4 (1975) 221–225.

[1992] X. Hao and J.B. Orlin, "A faster algorithm for finding the minimum cut in a graph," *Proceedings of the 3rd SIAM-ACM Symposium on Discrete Algorithms*, 1992, pp. 165–174.

[1970] M. Held and R.M. Karp, "The traveling-salesman problem and minimum spanning trees," *Operations Research* 18 (1970) 1138–1162.

[1971] M. Held and R.M. Karp, "The traveling-salesman problem and minimum spanning trees: Part II," *Mathematical Programming* 1 (1971) 6–25.

[1974] M. Held, P. Wolfe, and H.P. Crowder, "Validation of subgradient optimization," *Mathematical Programming* 6 (1974) 62–88.

[1960] A.J. Hoffman, "Some recent applications of the theory of linear inequalities to extremal combinatorial analysis," *Proceedings of Symposia on Applied Mathematics* 10 (1960) 113–127.

[1974] A.J. Hoffman, "A generalization of max flow-min cut," *Mathematical Programming* 6 (1974) 352–359.

[1956] A.J. Hoffman and J.B. Kruskal, "Integral boundary points of convex polyhedra," in: *Linear Inequalities and Related Systems* (H.W. Kuhn and A.W. Tucker, eds.), Princeton University Press, Princeton, 1956, pp. 223–246.

[1987] O. Holland, *Schnittebenenverfahren für Travelling-Salesman- und verwandte Probleme*, Ph.D. Thesis, Universität Bonn, Germany, 1987.

[1973] J.E. Hopcroft and R.M. Karp, "An $n^{5/2}$ algorithm for maximum matching in bipartite graphs," *SIAM Journal on Computing* 2 (1973) 225–231.

[1930] V. Jarníik, "O jistém problému minimálníim," *Práce Moravské Přírodovědecké Společnosti* 6 (1930) 57–63.

[1997] D.S. Johnson, J.L. Bentley, L.A. McGeoch, and E.E. Rothberg, in preparation.

[1997] D.S. Johnson and L.A. McGeoch, "The traveling salesman problem: A case study in local optimization," in: *Local Search in Combinatorial Optimization* (E.H.L. Aarts and J.K. Lenstra, eds.), Wiley, New York, 1997, pp. 215–310.

[1993] M. Jünger and W. Pulleyblank, "Geometric duality and combinatorial optimization," in: *Jahrbuck Überblicke Mathematik* (S.D. Chatterji, B. Fuchssteiner, U. Kluish, and R. Liedl, eds.), Vieweg, Brunschweig/Wiesbaden, 1993, pp. 1–24.

[1995] M. Jünger, G. Reinelt, and G. Rinaldi, "The traveling salesman problem," in: *Handbook on Operations Research and Management Sciences: Networks* (M. Ball, T. Magnanti, C.L. Monma, and G. Nemhauser, eds.), North-Holland, 1995, pp. 225–330.

[1994] M. Jünger, G. Reinelt, and S. Thienel, "Provably good solutions for the traveling salesman problem," *Zeitschrift für Operations Research* 40 (1994) 183–217.

[1942] L.V. Kantorovich, "On the translocation of masses", CR de l'Academie des Sciences de l'URSS, 1942.

[1993] D.R. Karger, "Global min-cuts in $\mathcal{RNC}$ and other ramifications of a simple mincut algorithm," in: *Proceedings of the 4th Annual ACM-SIAM Symposium on Discrete Algorithms*, ACM-SIAM, 1993, pp. 84–93.

[1993] D.R. Karger and C. Stein, "An $\tilde{O}(n^2)$ algorithm for minimum cuts," in: *Proceedings of the 25th ACM Symposium on the Theory of Computing*, ACM Press, 1993. pp. 757–765.

[1972] R.M. Karp, "Reducibility among combinatorial problems," in: *Complexity of Computer Computations* (R.E. Miller and J.W. Thatcher, eds.), Plenum Press, New York, 1972, pp. 85–103.

[1931] D. König, "Graphok és matrixok," *Matematikai és Fizikai Lapok* 38 (1931) 116–119.

[1990] B. Korte, L. Lovász, H.J. Prömel, and A. Schrijver, eds., *Paths, Flows, and VLSI-Layout*, Springer, Berlin, 1990.

[1956] A. Kotzig, "Súvislost' a Pravidelná Súvislost' Konečných Grafor," Bratislava: *Vysoká Škola Ekonomická* (1956).

[1956] J.B. Kruskal, "On the shortest spanning subtree of a graph and the traveling salesman problem," *Proceedings of the American Mathematical Society* 7 (1956) 48–50.

[1955] H.W. Kuhn, "The Hungarian method for the assignment problem," *Naval Research Logistics Quarterly* 2 (1955) 83–97.

[1975] E.L. Lawler, "Matroid intersection algorithms," *Mathematical Programming* 9 (1975) 31–56.

[1985] E.L. Lawler, J.K. Lenstra, A.H.G. Rinnooy Kan, and D. Shmoys, *The Traveling Salesman Problem*, Wiley, Chichester, 1985.

[1973] S. Lin and B.W. Kernighan, "An effective heuristic algorithm for the traveling salesman problem," *Operations Research* 21 (1973) 498–516.

[1979] L. Lovász, "Graph theory and integer programming," in: *Discrete Optimization I* (P.L. Hammer, E.L. Johnson, and B.H. Korte, eds), *Annals of Discrete Mathematics* 4 (1979) 146–158.

[1986] L. Lovász and M. Plummer, *Matching Theory*, North Holland, Amsterdam, 1986.

[1993] K.-T. Mak and A.J. Morton, "A modified Lin-Kernighan traveling-salesman heuristic," *Operations Research Letters* 13 (1993) 127–132.

[1979] A.B. Marsh III, *Matching Algorithms*, Ph.D. Thesis, Johns Hopkins University, Baltimore.

[1996] O. Martin and S.W. Otto, "Combining simulated annealing with local search heuristics," *Annals of Operations Research* 63 (1996) 57–75.

[1992] O. Martin, S.W. Otto, and E.W. Felten, "Large-step Markov chains for the TSP incorporating local search heuristics," *Operations Research Letters* 11 (1992) 219–224.

[1980] S. Micali and V.V. Vazirani, "An $O(\sqrt{|V|}|E|)$ algorithm for finding maximum matching in general graphs," *Proceedings of the 21st Annual Symposium on Foundations of Computer Science*, IEEE, 1980, pp. 17–27.

[1995] D.L. Miller and J.F. Pekney, "A staged primal-dual algorithm for perfect b-matching with edge capacities," *ORSA Journal on Computing* 7 (1995) 298–320.

[1957] J. Munkres, "Algorithms for the assignment and transportation problems," *SIAM Journal on Applied Mathematics* 5 (1957) 32–38.

[1990] D. Naddef, "Handles and teeth in the symmetric traveling salesman polytope," in: *Polyhedral Combinatorics* (W. Cook and P.D. Seymour, eds.), American Mathematical Society, 1990, pp. 61–74.

[1992] H. Nagamochi and T. Ibaraki, "Computing edge connectivity in multigraphs and capacitated graphs," *SIAM Journal on Discrete Mathematics* 5 (1992) 54–66.

[1966] C.St.J.A. Nash-Williams, "An application of matroids to graph theory," in: *Theory of Graphs* (P. Rosenstiehl, ed.) Dunod, Paris, 1966, pp. 263–265.

[1988] G.L. Nemhauser and L.A. Wolsey, *Integer and Combinatorial Optimization*, Wiley, New York, 1988.

[1985] J.B. Orlin, "On the simplex algorithm for networks and generalized networks," *Mathematical Programming Study* 24 (1985) 166–178.

[1988] J.B. Orlin, "A faster strongly polynomial minimum cost flow algorithm," *Proceedings of the 20th ACM Symposium on Theory of Computing*, ACM Press, 1988, pp. 377–387.

[1992] J.G. Oxley, *Matroid Theory*, Oxford University Press, Oxford, 1992.

[1982] M.W. Padberg and M.R. Rao, "Odd minimum cut-sets and b-matchings," *Mathematics of Operations Research* 7 (1982) 67–80.

[1990] M. Padberg and G. Rinaldi, "An efficient algorithm for the minimum capacity cut problem," *Mathematical Programming* 47 (1990) 19–36.

[1991] M. Padberg and G. Rinaldi, "A branch-and-cut algorithm for the resolution of large-scale symmetric traveling salesman problems," *SIAM Review* 33 (1991) 60–100.

[1977] C. Papadimitriou and K. Steiglitz, "On the complexity of local search for the traveling salesman problem," *SIAM Journal on Computing* 6 (1977) 76–83.

[1976] J.-C. Picard, "Maximal closure of a graph and applications to combinatorial problems," *Management Science* 22 (1976) 1268–1272.

[1984] R.Y. Pinter, "Optimal layer assignment for interconnect," *Journal VLSI Comput. Syst.* 1 (1984) 123–137.

[1900] H. Poincaré, "Second complément à l'analysis situs," *Proceedings of the London Mathematical Society* 32 (1900) 277–308.

[1967] B.T. Polyak, "A general method of solving extremum problems" (in Russian), Doklady Akademmi Nauk SSSR 174 (1) (1967) 33–36.

[1957] R.C. Prim, "Shortest connection networks and some generalizations," *Bell System Technical Journal* 36 (1957) 1389–1401.

[1973] W.R. Pulleyblank, *Faces of Matching Polyhedra*, Ph.D. Thesis, Department of Combinatorics and Optimization, University of Waterloo, Waterloo, Ontario, 1973.

[1974] W.R. Pulleyblank and J. Edmonds, "Facets of 1-matching polyhedra," in: *Hypergraph Seminar* (C. Berge and D. Ray-Chaudhuri, eds.), Springer, Berlin, 1974, pp. 214–242.

[1942] R. Rado, "A theorem on independence relations," *Quarterly Journal of Mathematics Oxford* 13 (1942) 83–89.

[1957]   R. Rado, "A note on independence functions," *Proceedings of the London Mathematical Society* 7 (1957) 300–320.

[1991]   G. Reinelt, "TSPLIB—A traveling salesman problem library," *ORSA Journal on Computing* 3 (1991) 376–384.

[1994]   G. Reinelt, *The Traveling Salesman: Computational Solutions for TSP Applications*, Springer-Verlag, Berlin, 1994.

[1970]   J. Rhys, "A selection problem of shared fixed costs and network flows," *Management Science* 17 (1970) 200–207.

[1977]   D.J. Rosenkrantz, R.E. Stearns, and P.M. Lewis II, "An analysis of several heuristics for the traveling salesman problem," *SIAM Journal on Computing* 6 (1977) 563–581.

[1980]   A. Schrijver, "On cutting planes," in: *Combinatorics 79, Part II* (M. Deza and I.G. Rosenberg, eds.), *Annals of Discrete Mathematics* 9 (1980) 291–296.

[1983]   A. Schrijver, "Short proofs on the matching polytope," *Journal of Combinatorial Theory (Series B)* 34 (1983) 104–108.

[1983a] A. Schrijver, "Min-max results in combinatorial optimization," in: *Mathematical Programming, the State of the Art: Bonn 1982* (A. Bachem, M. Grötschel, and B. Korte, eds.), Springer-Verlag, Berlin, 1983, pp. 439–500.

[1984]   A. Schrijver, "Total dual integrality from directed graphs, crossing families, and sub- and supermodular functions," in: *Progress in Combinatorial Optimization* (W.R. Pulleyblank, ed.), Academic Press, Toronto, 1984, pp. 315–361.

[1986]   A. Schrijver, *Theory of Linear and Integer Programming*, Wiley, Chichester, 1986.

[1980]   P.D. Seymour, "Decomposition of regular matroids," *Journal of Combinatorial Theory (Series B)* 28 (1980) 305–359.

[1981]   P.D. Seymour, "On odd cuts and plane multicommodity flows," *Proceedings of the London Mathematical Society* (1981) 178–192.

[1977]   T.H.C. Smith and G.L. Thompson, "A LIFO implicit enumeration search algorithm for the symmetric traveling salesman problem using Held and Karp's 1-tree relaxation," in: *Studies in Integer Programming* (P.L. Hammer , *et al.* eds.), *Annals of Discrete Mathematics* 1 (1977) 479–493.

[1994]   M. Stoer and F. Wagner, "A simple min cut algorithm," to appear.

[1987]   R. Tamassia, "On embedding a graph in the grid with the minimum number of bends," *SIAM Journal on Computing* 16 (1987) 421–444.

[1985]   E. Tardos, "A strongly polynomial minimum cost circulation algorithm," *Combinatorica* 5 (1985) 247–255.

[1983] R.E. Tarjan, *Data Structures and Network Algorithms*, SIAM, Philadelphia, 1983.

[1994] L. Tunçel, "On the complexity of preflow-push algorithms for the maximum flow problem," *Algorithmica* 11 (1994) 353–359.

[1947] W.T. Tutte, "The factorization of linear graphs," *Journal of the London Mathematical Society* 22 (1947) 107–111.

[1954] W.T. Tutte, "A short proof of the factor theorem for finite graphs," *Canadian Journal of Mathematics* 6 (1954) 347–352.

[1965] W.T. Tutte, "Lectures on matroids," *Journal of Research of the National Bureau of Standards* (B) 69 (1965) 1–47.

[1968] A.F. Veinott and G.B. Dantzig, "Integral extreme points," *SIAM Review* 10 (1968) 371–372.

[1976] D.J.A. Welsh, *Matroid Theory*, Academic Press, London, 1976.

[1996] D.B. West, *Introduction to Graph Theory*, Prentice Hall, Upper Saddle River, 1996

[1973] N. Zadeh, "A bad network problem for the simplex method and other minimum cost flow algorithms," *Mathematical Programming* 5 (1973) 255–266.

Index

WILEY-INTERSCIENCE
SERIES IN DISCRETE MATHEMATICS AND OPTIMIZATION

ADVISORY EDITORS

RONALD L. GRAHAM
AT & T Bell Laboratories, Murray Hill, New Jersey, U.S.A.

JAN KAREL LENSTRA
Department of Mathematics and Computer Science,
Eindhoven University of Technology, Eindhoven, The Netherlands

ROBERT E. TARJAN
Princeton University, New Jersey, and
NEC Research Institute, Princeton, New Jersey, U.S.A.

AARTS AND KORST • Simulated Annealing and Boltzmann Machines: A Stochastic Approach to Combinatorial Optimization and Neural Computing

AARTS AND LENSTRA • Local Search in Combinatorial Optimization

ALON, SPENCER, AND ERDÖS • The Probabilistic Method

ANDERSON AND NASH • Linear Programming in Infinite-Dimensional Spaces: Theory and Application

ASENCOTT • Simulated Annealing: Parallelization Techniques

BARTHÉLEMY AND GUÉNOCHE • Trees and Proximity Representations

BAZARRA, JARVIS, AND SHERALI • Linear Programming and Network Flows

CHONG AND ZAK • An Introduction to Optimization

COFFMAN AND LUEKER • Probabilistic Analysis of Packing and Partitioning Algorithms

COOK, CUNNINGHAM, PULLEYBLANK, AND SCHRIJVER • Combinatorial Optimization

DASKIN • Network and Discrete Location: Modes, Algorithms and Applications

DINITZ AND STINSON • Contemporary Design Theory: A Collection of Surveys

ERICKSON • Introduction to Combinatorics

GLOVER, KLINGHAM, AND PHILLIPS • Network Models in Optimization and Their Practical Problems

GOLSHTEIN AND TRETYAKOV • Modified Lagrangians and Monotone Maps in Optimization

GONDRAN AND MINOUX • Graphs and Algorithms *(Translated by S. Vajdā)*

GRAHAM, ROTHSCHILD, AND SPENCER • Ramsey Theory, Second Edition

GROSS AND TUCKER • Topological Graph Theory

HALL • Combinatorial Theory, Second Edition

JENSEN AND TOFT • Graph Coloring Problems

LAWLER, LENSTRA, RINNOOY KAN, AND SHMOYS, Editors • The Traveling Salesman Problem: A Guided Tour of Combinatorial Optimization

LEVITIN • Perturbation Theory in Mathematical Programming Applications

MAHMOUD • Evolution of Random Search Trees

MARTELLO AND TOTH • Knapsack Problems: Algorithms and Computer Implementations

McALOON AND TRETKOFF • Optimization and Computational Logic

MINC • Nonnegative Matrices

MINOUX • Mathematical Programming: Theory and Algorithms *(Translated by S. Vajdā)*

MIRCHANDANI AND FRANCIS, Editors • Discrete Location Theory

NEMHAUSER AND WOLSEY • Integer and Combinatorial Optimization

NEMIROVSKY AND YUDIN • Problem Complexity and Method Efficiency in Optimization *(Translated by E. R. Dawson)*

PACH AND AGARWAL • Combinatorial Geometry

PLESS • Introduction to the Theory of Error-Correcting Codes, Second Edition

ROOS AND VIAL • Ph. Theory and Algorithms for Linear Optimization: An Interior Point Approach

SCHEINERMAN AND ULLMAN • Fractional Graph Theory: A Rational Approach to the Theory of Graphs

SCHRIJVER • Theory of Linear and Integer Programming

TOMESCU • Problems in Combinatorics and Graph Theory *(Translated by R. A. Melter)*

TUCKER • Applied Combinatorics, Second Edition

YE • Interior Point Algorithms: Theory and Analysis